Post-Silicon Validation and Debug

Prabhat Mishra · Farimah Farahmandi
Editors

Post-Silicon Validation and Debug

Springer

Editors
Prabhat Mishra
Department of Computer and Information Science and Engineering
University of Florida
Gainesville, FL, USA

Farimah Farahmandi
Department of Computer and Information Science and Engineering
University of Florida
Gainesville, FL, USA

ISBN 978-3-030-07454-8 ISBN 978-3-319-98116-1 (eBook)
https://doi.org/10.1007/978-3-319-98116-1

This Springer imprint is published by the registered company Springer Nature Switzerland AG
The registered company address is: Gewerbestrasse 11, 6330 Cham, Switzerland

Preface

In our day-to-day activities, we interact with a wide variety of computing systems. When we use desktops or laptops, we are aware of the fact that these systems are computing something for us. However, in many systems, the computing is embedded in them, such as cyber-physical systems and Internet-of-Things (IoT) devices. When we drive a car or fly in an airplane, many computing devices seamlessly work together to ensure a pleasant and safe journey. Similarly, when we perform any financial transaction or share personal details using a smartphone, embedded computing devices try to ensure the security and privacy of these transactions. Can we assume that these computing devices are correct by construction and therefore we can safely rely on them? A short answer is that no one can prove the absolute infallibility of today's computing systems. This book provides a clear insight into the fundamental challenges associated with validation and debug of computing systems. This book also provides effective solutions to address these challenges.

Most of these computing systems consist of software (application programs), firmware, and hardware. The brain behind these computing systems is called System-on-Chip (SoC). A typical SoC includes one or more processor cores, coprocessors, caches, memory, controllers, converters, peripherals, input/output devices, sensors, and so on. To understand why SoC validation is so challenging, let us consider one of the simplest components in a SoC—an adder. An adder adds two input values and produces the result. Typically, the input values are 64-bit integers. Therefore, to verify this adder, we have to simulate several trillions ($2^{64} \times 2^{64}$) of test vectors. Clearly, it is infeasible to apply trillions of test vectors to verify an adder. If we cannot completely verify a simple adder, what is the hope that we can verify complex SoCs. During the design stage (before fabrication), pre-silicon validation techniques try to identify and fix functional errors as well as nonfunctional requirements.

In spite of extensive efforts, it is not always possible to detect all the errors during pre-silicon validation. Post-silicon validation is used to detect design flaws including the escaped functional errors as well as nonfunctional requirements (such as security vulnerabilities). Post-silicon validation is widely acknowledged as a major bottleneck for complex SoC designs. Various industrial studies indicate that the post-silicon validation effort consumes more than 50% of an SoC's overall design effort. To understand why post-silicon validation is challenging, consider a real-world post-silicon bug. Consider a scenario when a secure firmware is getting executed in a processor pipeline and an asynchronous reset disables the locking mechanism, which is supposed to restrict all unauthorized accesses during firmware execution. Once lock is disabled, a user application can access the unencrypted firmware from the instruction cache. After painful debugging for several days, the problem turned out to be that the instruction memory was not cleared after the asynchronous reset. It is important to highlighting three important facts at this point.

- *Escape from Pre-silicon*: In this case, it was infeasible for pre-silicon validation to cover all possible reset sequences in an automotive SoC consisting of approximately 200 Intellectual Property (IP) blocks, 20 reset domains, and a total of few thousand reset signals.
- *Long Debug Time*: There are many reasons for long debug in this case. First, the error was reported in a very different context. Next, setting up the system to reproduce the failure is a tedious process involving many trials and errors. Finally, it was time-consuming to localize the error due to the very limited observability of the internal signals. To understand this better, think of the complexity of debugging software consisting of millions of lines. In case of software debug, you can observe the values of any of the millions of variables during the execution. In case of post-silicon debug, you have limited or no visibility at all. In other words, you have to debug a very complex scenario in a billion-gate SoC with visibility of few hundred signals through in-built trace buffer.
- *Complex Interactions*: The SoC design is very complex with too many components interacting with firmware and software. In this example, the integrity of the circuit is compromised by unauthorized access to the unencrypted firmware. The vulnerability needs to be identified and fixed during post-silicon debug. The debug complexity is expected to get worse as the industry continues to move to even smaller geometries.

This book describes the fundamental challenges associated with post-silicon validation and debug of SoCs. It also covers the state-of-the-art techniques as well as ongoing research efforts to drastically reduce the overall post-silicon validation and debug effort. The first chapter provides an overview of SoC design methodology and highlights the challenges in post-silicon validation and debug. The next five chapters describe efficient techniques to design debug architecture including on-chip instrumentation and signal selection. Chapters 7–10 present effective techniques to generate tests and assertions. The next five chapters provide automated approaches for localizing, detecting, and fixing post-silicon errors. Chapters 16–17 describe post-silicon validation efforts using two case studies, a network-on-chip, and IBM

Power8 processor. The next chapter discusses the inherent conflict between design-for-debug and security vulnerabilities. Finally, Chap. 19 concludes the book with a discussion on future post-silicon debug challenges and opportunities.

Gainesville, Florida, USA

Prabhat Mishra
Farimah Farahmandi

Acknowledgements

This book would not be possible without the contributions of many researchers and experts in the field of post-silicon validation and debug. We would like to gratefully acknowledge the contributions from the following authors who have contributed book chapters:

- Alif Ahmed, University of Florida, USA
- Amir Nahir, Amazon, Israel
- Dr. Azadeh Davoodi, University of Wisconsin, USA
- Dr. Debapriya Chatterjee, IBM, USA
- Dr. Doowon Lee, University of Michigan, USA
- Dr. Kai Cong, Intel, USA
- Dr. Kamran Rahmani, Box, USA
- Dr. Kanad Basu, New York University, USA
- Dr. Pouya Taatizadeh, Synopsys, Canada
- Dr. Sandeep Chandran, Indian Institute of Technology, Delhi, India
- Dr. Xiaobing Shi, McMaster University, Canada
- Dr. Yuanwen Huang, VMware, USA
- Hillel Mendelson, IBM, Israel
- Qinhao Wang, University of Tokyo, Japan
- Yusuke Kimura, University of Tokyo, Japan
- Prof. Fei Xie, Portland State University, USA
- Prof. Georg Weissenbacher, TU Wien, Austria
- Prof. Masahiro Fujita, University of Tokyo, Japan
- Prof. Nicola Nicolici, McMaster University, Canada
- Prof. Preeti Ranjan Panda, Indian Institute of Technology, Delhi, India
- Prof. Sandip Ray, University of Florida, USA
- Prof. Sharad Malik, Princeton University, USA
- Prof. Valeria Bertacco, University of Michigan, USA

- Subodha Charles, University of Florida, USA
- Tom Kolan, IBM, Israel
- Vitali Sokhin, IBM, Israel
- Yangdi Lyu, University of Florida, USA

This work was partially supported by National Science Foundation under grant CCF-1218629. Any opinions, findings, conclusions, or recommendations presented in this book are those of the authors and do not necessarily reflect the views of the National Science Foundation.

Contents

Abbreviations (Acronyms)

AMBA	Advanced Microcontroller Bus Architecture
BDD	Binary Decision Diagram
BMC	Bounded Model Checking
BSR	Boundary Scan Registers
BSTW	Bentley–Sleator–Tarjan–Wei
BTS	Branch Trace Store
CAM	Content Addressable Memory
CDFG	Control and Data Flow Graph
CDV	Coverage Driven Verification
CEGIS	Counterexample Guided Inductive Synthesis
CMP	Chip Multi-Processor
CNF	Conjunctive Normal Form
CR	Cycling Register
CRSG	Constrained-Random Stimuli Generator
CUV	Circuit Under Validation
DCA	Device Coverage Analyzer
DfD	Design-for-Debug
DMA	Direct Memory Access
DRAM	Dynamic Random-access Memory
DUT	Design Under Test
DUV	Design Under Verification
DVFS	Dynamic Voltage and Frequency Scaling
ECC	Error Correction Codes
EoA	Exercisers on Accelerators
ETM	Embedded Trace Macrocell
FIFO	First In First Out
FPGA	Field Programmable Gate Arrays
FPU	Floating Point Unit
GB	Gigabyte
GOOLO	Globally out of order, locally ordered

HDL	Hardware Description Language
HLD	High-level Design
IAV	ID and Address Verification
IC	Integrated Circuit
IFRA	Instruction Footprint Recording and Analysis
ILA	Iterative Logic Array
IoT	Internet of Things
IP	Intellectual Property
IPI	Inter-Processor Interrupt
ISA	Instruction Set Architecture
ISS	Instruction Set Simulator
ITRS	International Technology Roadmap for Semiconductors
JTAG	Joint Test Action Group
KB	Kilobyte
K-map	Karnaugh map
KNL	Knight's Landing
LFSR	Linear Feedback Shift Register
LFU	Least Frequently Used
LLC	Last-level Cache
LRU	Least Recently Used
LSB	Least Significant Byte
LSQ	Load–Store Queue
LSU	Load–Store Unit
LUT	Lookup Table
LZW	Lempel–Ziv–Welch
MC	Memory Controller
MCM	Memory Consistency Model
MCS	Minimal Correction Set
MISR	Multiple-Input Signature Register
MMU	Memory Management Unit
NI	Network Interface
NoC	Network-on-Chip
ODIN	Off die interconnect networks
OS	Operating System
PCI	Peripheral Component Interconnect
PCIe	PCI-express
PRNG	Pseudo-Random Number Generator
PSL	Property Specification Language
QBF	Quantified Boolean Formula
QED	Quick Error Detection
QoS	Quality of Service
QPI	Quick Path Interconnect
RTL	Register-Transfer Level
RTP	Real-time Transport Protocol
SAT	Satisfiability

SC	Sequential Consistency
SEE	Symbolic Execution Environment
SMT	Simultaneous Multithreading
SoC	System-on-Chip
SRR	State Restoration Ratio
SSL	Storage Specification Language
SVA	System Verilog Assertions
TLB	Translation Lookaside Buffer
TLM	Transaction-level Modeling
TLP	Transaction Layer Packet
TSO	Total Store Order
UHT	Update History Table
VC	Virtual Channel
VD	Virtual device
VHDL	VHSIC Hardware Description Language
VHSIC	Very High Speed Integrated Circuit
VLIW	Very Large Instruction Word
VM	Virtual Machine
VPO	Virtual Power On

Part I
Introduction

Chapter 1
Post-Silicon SoC Validation Challenges

Farimah Farahmandi and Prabhat Mishra

1.1 Introduction

People have embraced a wide variety of mobile devices in their daily lives in addition to their traditional desktop computers and laptops. Considering the fabric of Internet of Things (IoT) [30], where the number of connected devices exceeds the human population, we should all agree to the fact that computing devices pervade every aspect of our lives. In IoT devices, integrated electronics, sensors, sophisticated software and firmware, and learning algorithms are employed to make physical objects smart and adjustable to their environment. These highly complex and smart IoT devices are embedded everywhere—starting from household items (e.g., refrigerators, slow cookers, ceiling fans), wearable devices (e.g., fitness trackers, smart glasses, air bugs), and medical devices (e.g., insulin pump, asthma monitoring, ventilator) to cars. These IoT devices are connected to each other as well as cloud in order to provide a real-time aid on a daily basis. Given the diverse and critical applications of these computing devices, it is crucial to verify the correctness, security, and reliability of these devices.

Modern computing devices are designed using System-on-Chip (SoC) technology. In other words, SoC is the backbone for most of the IoT devices. An SoC architecture typically consists of several predesigned Intellectual Property (IP) blocks, where each IP implements a specific functionality of the overall design. Figure 1.1 shows a typical SoC with its associated IPs. These IPs communicate with each other through Network-on-Chip (NoC) or standard communication fabrics. The IP-based design approach is popular today since it enables a low-cost design while meeting

---

F. Farahmandi (✉) · P. Mishra
Department of Computer and Information Science and Engineering, University of Florida, Gainesville, FL, USA
e-mail: ffarahmandi@ufl.edu

P. Mishra and F. Farahmandi (eds.), *Post-Silicon Validation and Debug*,
https://doi.org/10.1007/978-3-319-98116-1_1

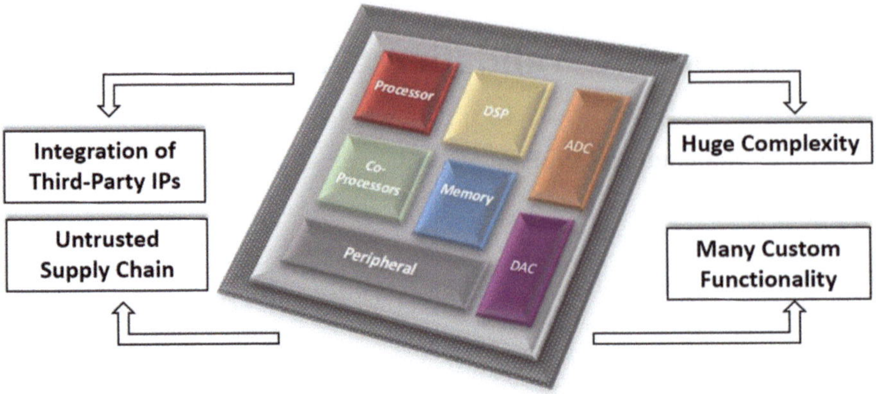

Fig. 1.1 An SoC design integrates a wide variety of IPs in a chip. It can include one or more processor cores, on-chip memory, digital signal processor (DSP), analog-to-digital (ADC), and digital-to-analog converters (DAC), controllers, input/output peripherals, and communication fabric. Huge complexity, many custom designs, distributed supply chain, and integration of untrusted third-party IPs make post-silicon validation challenging

stringent time-to-market requirements. Validation of SoC designs includes assuring their functional correctness, meeting power and performance constraints, checking for security properties and robustness against electrical noises as well as physical and thermal stresses. In other words, validation efforts need to ensure the correct and safe behavior of the design while keeping area, power and timing overheads under control [1].

Validation is widely acknowledged as a major bottleneck in the SoC Design—many studies suggest that about 70% of the overall time, efforts, and resources are spent during SoC validation and verification [23]. The integrity of hardware designs is ensured using pre- and post-silicon validation as well as in-field debugging efforts. Pre-silicon validation refers to attempts for verifying the correctness and sufficiency of the design *model* before sending the design for fabrication. On the other hand, post-silicon validation refers to the validation efforts of manufactured chips in the actual application environment to ensure correct functionality under operating conditions before sending the design for mass production [24]. Post-silicon validation is responsible for detecting design flaws including the escaped functional errors, various forms of in-field security vulnerabilities as well as electrical bugs by using different tests and design-for-debug infrastructure.

Due to limited observability and controllability of the actual silicon as well as complex components in a SoC design, post-silicon validation is extremely challenging. Moreover, post-silicon validation is usually performed under tight schedule regarding time-to-market constraints. Post-silicon validation efforts can be classified into three major steps: (i) preparing for post-silicon validation and debug, (ii) detecting a problem by applying test programs, (iii) localizing and root-causing the exact source of the problem, and fixing the problem. In this chapter, we review the spectrum of the design validation from pre-silicon to post-silicon and in-field debug. Then, we briefly discuss different steps in post-silicon validation as well as their associated challenges.

1.2 Validation Activities

The design validation efforts can be broadly divided into three categories: pre-silicon validation, post-silicon validation, and in-field debugging. Figure 1.2 shows these categories. Validation activities start from pre-silicon, and as we go toward post-silicon and in-field debug, we can observe several key differences. First, the error scenarios can become both realistic and complicated. In other words, some of these errors could not be modeled or detected in previous validation stages. Next, the observability and controllability of the design become drastically reduced. For example, a designer is able to observe all the signals in a pre-silicon validation framework of register-transfer level (RTL) or gate-level models, however, only few hundred (out of millions) can be observed (traced) in a post-silicon environment involving a fabricated SoC. Therefore, root-causing the source of the error becomes more challenging. All of these factors lead to a significant increase in the cost of finding a bug in the later stages of the validation. As a result, it is crucial to find and fix bugs as

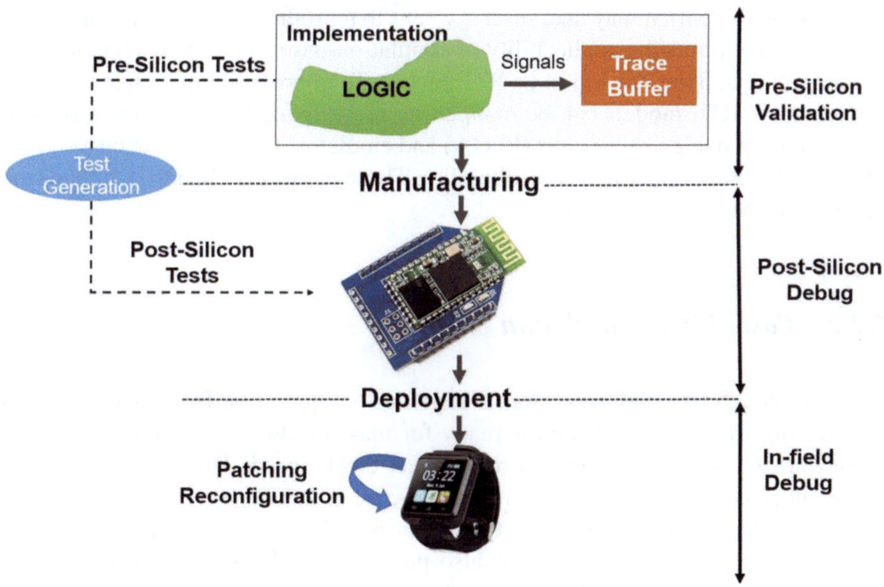

Fig. 1.2 Three important stages of SoC validation: pre-silicon validation, post-silicon validation, and on-field debug

early as possible. In this section, we provide a high-level overview of the validation spectrum.

1.2.1 Pre-silicon Validation

Pre-silicon validation refers to the overall validation and verification efforts prior to sending the design for fabrication. Pre-silicon validation activities involve functional validation, assertion coverage, and code reviewing (code coverage). The validation goals are achieved using design simulation with different types of stimulus such as random, constraint-random, and directed test as well as static analysis of the design using formal and semiformal approaches.

A test plan is prepared in order to perform validation. Test plan contains test-bench architectures, functional requirements, use-case scenarios, corner cases, stimuli types, abstraction models, verification methodologies, and coverage closures. Different design models are considered to validate the design in various stages of its life cycle. Architectural models, as well as high-level software models, are generated for inter-component communication validation (coarse-grained validation), while Register-Transfer Level (RTL) and gate-level models are used to verify the implementation of IPs (components) using simulation and formal methods. Note that simulation of RTL models are significantly slower than executing on an actual chip. For example, if we have traces of few seconds execution from a chip (like booting an operating system), it may take several weeks to reproduce it using RTL simulator. This drawback limits the applicability of simulation-based validation approaches to test executing software on top of an RTL model. To improve simulation (execution) performance, RTL models can be mapped on reconfigurable architectures such as Field Programmable Gate Arrays (FPGA) and emulators [12, 13] at the cost significant loss in observability and controllability. These models are hundred to thousand times faster than RTL simulators.

1.2.2 Post-silicon Validation

Post-silicon validation efforts refer to activities to check first few silicon samples and making sure that the design is ready for mass production. Post-silicon debug framework is used to test the design in target clock speed. Therefore, it is possible to check complex hardware/software use-case scenarios such as booting whole operating system, monitoring security options as well as power management across all existing IPs in few seconds. It is also possible to validate nonfunctional characteristics of the design such as peak power, temperature tolerance, and electrical noise margin. However, unlike RTL simulators where the value of all of the internal signals can be observed quickly, it is difficult to watch and control the states of the design during run-time. In FPGAs and emulators, the observable architecture can be

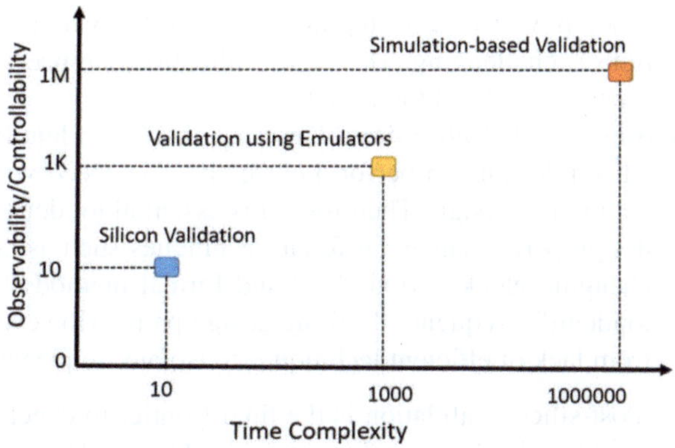

Fig. 1.3 Comparison of simulation-based, emulation, and silicon validation approaches regarding execution time and observability/controllability capabilities

configured such that the value of hundreds or thousands of internal signals can be visible during run-time. However, only few hundred of internal signals (out of millions or even billions of signal in a typical SoC) can be observed in silicon. Figure 1.3 compares simulation-based, emulation, and silicon validation approaches regarding time complexity and observability/controllability capabilities.

Post-silicon validation involves a diverse range of activities from checking the functional requirements to nonfunctional design constraints such as timing and energy behaviors. While the validation engineer has to focus on a wide a variety of post-silicon activities, we briefly outline five of them.

- **Power-on-Debug**: The first activity in post-silicon validation is to test the design when the power is on. Powering the device is a challenging task since any problem in powering the system cannot be root-caused easily. The reason is that most of the options in the design do not work without power, therefore, diagnosing is difficult. As a result, power-on-debug is usually done using high controllable and configurable custom boards. In the first step, bare minimum options are considered, and gradually complex features and Design-for-Debug (DfD) options are added until the whole design can successfully access the power.
- **Logic Validation**: After power-on-debug, the next step is to ensure that the hardware works as intended. This step involves testing specific behaviors and features of the design as well as corner cases using random, constrained-random and directed tests. The tests are required to not only check different options of IPs but also check features that involve working multiple IPs and their communications together.
- **Hardware/software Co-validation**: In this step, the compatibility of the silicon with the operating system, application software, various network protocols, communication infrastructures and peripherals is checked. This is a complicated step since there can be dozen operating system versions, hundred peripherals, and many applications that should be validated.
- **Electrical Validation**: This step is responsible for ensuring electrical characteristics such as clock, analog/mixed-signal, and power delivery under worst-case operating conditions. Similar to hardware/software co-validation phase, the parameter

space is vast in this step and covering the whole spectrum of operating conditions is challenging. Therefore, validation engineers try to identify most essential scenarios and test them first.
- **Speed-path Validation**: This step involves identifying the speed at which the silicon design can perform correctly. The speed will be imposed by the slowest path in the design. Therefore, it is essential to identify such paths to optimize the design performance. Different techniques such as laser-assisted techniques [31], changing clock periods [35] and formal methods [15, 26] have been proposed to identify frequency-limiting design paths. However, modern designs still suffer from lack of efficient techniques to isolate frequency limiting paths.

Post-silicon validation is the final frontier to check the correct behavior and the integrity of the design before mass production. However, the start date of mass production is imposed by several factors mostly by marketing reasons such as launching time of competitor products, holiday times, back to school time frame, etc. Missing such deadlines may cause millions to billions of dollars in revenue loss or may cause missing the whole market in the worst case. Therefore, high-quality post-silicon validation efforts should be performed in a very limited timeframe. Otherwise, the reputation of the company will be at risk. Based on the type of post-silicon bugs and the difficulty to fix them, important decisions are taken to either send the design for mass production or abandon the line of output. Therefore, post-silicon validation should be able to aggregate such essential data to enable the decision on mass production.

1.2.3 In-Field Debug

In-field debugging refers to activities to fix and mitigate the errors observed during the execution of the silicon after deployment. The Note that in-field failures can be catastrophic as they may be exploited as security vulnerabilities causing damages to the company reputation. As a result, it is crucial to detect and fix in-field bugs. The capability of the in-field debug is dependent on the design-for-debug (DfD) architectures that are primarily employed for facilitating post-silicon debug.

The mitigation techniques are classified into two groups: (i) patching, (ii) reconfiguring the design. In-field debug efforts and activities depend on DfD infrastructures and configurability options. DfD infrastructures are extra hardware components that are designed to facilitate silicon debug. DfD helps to observe the effect of the error as well as root-causing the source of the bug. To fix the error, the design must have features for reconfigurability in order to fix the functionality through software or firmware updates. On the other hand, designing efficient DfD and reconfigurability options is extremely challenging and requires a highly creative process in order to have a flexible, debug-friendly, trustworthy, and secure silicon designs. There are certain challenges associated with in-field debug. As we mentioned earlier, limited observability and controllability is the main reason that makes in-field debug a complex task. Moreover, new techniques should be employed to fix the bugs after silicon deployment due to the limited time frame.

In order to decrease the debug complexity, plan for post-silicon readiness becomes a necessary task. In the next section, we discuss briefly DfD architecture to mitigate the complexity of validation and debugging efforts.

1.3 Planning for Post-silicon Readiness

Creating test plans is the first step of post-silicon validation. Test plans include the test architecture, debug software, functionality requirements, corner cases, coverage targets, and coverage closures. Post-silicon test plans mostly target system-level use cases of the design which cannot be tested in pre-silicon validation. Test plans are created concurrently with design planning. Initially, test plans do not consider implementation details, and they rely on high-level architectural specifications. Gradually, test plans get mature alongside with design implementation and more design features are added to the test plan.

Designing debug software is another crucial component of post-silicon validation readiness. It includes any infrastructure that is needed to run post-silicon tests, trace failures, and triage. The software consists of the following essential components.

- **Instrumented System Software**: Specialized operating systems are implemented to perform post-silicon debug. The target of such operating systems is to decrease the complexity of modern operating systems (e.g., MacOS, Windows, Android, Linux) to test underlying hardware issues. The specialized operating system contains some hooks and instrumentations to improve debug, observability and controllability of the system.
- **Configuration Software**: Customized software tools are designed for controlling and configuring the internal states of the silicon. The software is used to configure registers and triggering trace buffers and coverage monitors to facilitate observing particular scenarios during debugging time.
- **Access Software**: A software is needed to transport the debug data from silicon. Debug data can be transferred off-chip either using the available pins, or through available ports from the platform (e.g., USB and PCIe). However, the ports may be unavailable due to power management features. The access software ensures that debug data can be transported while facilitating the power-down functionality of the hardware to be exercised during silicon validation.
- **Analysis Software**: Software tools should be designed to perform different analysis on the transported data. These tools include encrypting the trace data [4], aggregating the trace data to high-level data structures, and visualizing hardware/software coordination.

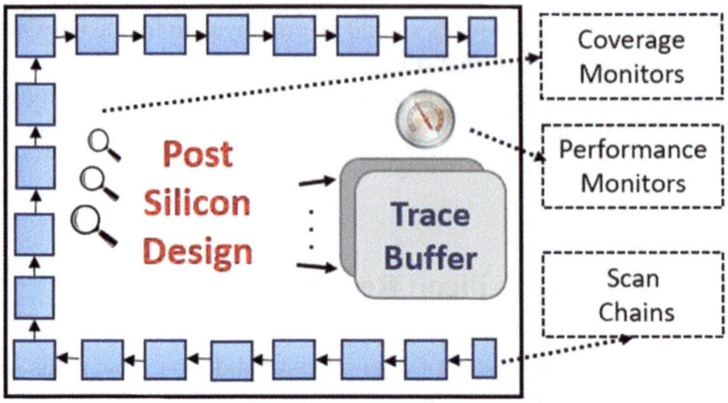

Fig. 1.4 Debug infrastructure for post-silicon validation

1.4 Post-silicon Debug Infrastructure

These days, most of the designs come with post-silicon validation and debug infrastructures. Design-for-Debug (DfD) are extra components that are embedded to facilitate validation efforts of the silicon. As shown in Fig. 1.4, they either monitor some particular functionality during the run-time, measure design performance (e.g., number of cache misses and number of branch misprediction), enhance the observability of the internal states and signals in the design, or improve the design controllability to test different components. ARM Coresight architecture [36] and Intel Platform Analysis Tools [11] are two examples of standard post-silicon observability architectures. They contain a set of hardware and software IPs which provide a way to trigger, collect, synchronize, timestamp, transport, and analyze observability data. While such standardization is helpful, it should be noted that the current state of such tools is rather primitive and a lot of manual effort is required for achieving the validation objective. In this section, we focus on planning for post-silicon readiness to design efficient trace buffers to address observability limitation in post-silicon debug.

There is significant research on improving silicon observability through trace buffers. Trace buffers are extra memory units that store the value of some selected signals during run-time. The stored values can be used offline in order to restore the values of other (untraced) internal signals. Note that since the speed of accessing the input/output ports (for example using JTAG) is significantly slower than the speed of the execution, it is not possible to dump the value of trace signals at run-time. Trace buffer values plus restored values can be beneficial in post-silicon debug as they enhance the overall design observability. Trace buffers have two physical characteristics, *width* and *depth*. Trace buffer width defines the number of selected signals that the trace buffer can sample at a time. On the other hand, trace buffer depth defines the number of clock cycles that the trace buffer can store the value of the selected signals. The depth and width of the trace buffers are limited due to area and power constraints. As a result, it is typical for a design with millions of signals, only few hundred of signals are traced in few thousands clock cycles. Therefore, the main question is that how to choose a small set of signals to maximize the post-silicon observability.

Trace signals can be selected based on different metrics. Different coverage goals impose selection of different set of signals. Trace signals can be selected based on restoration ratio [21] such that they aim to increase the observability of all internal signals in the design. Restoration ratio has been introduced to measure the ratio of the restored design states using the restored values to the number of states that are tracked [16]. On the other hand, signals can be selected based on their merit on design functional coverage [8, 22]. Error detection capability is also a helpful metric for trace signal selection [17, 34]. In this metric, signals are selected such that they improve error detection latency.

The efficiency of trace-based validation and debug is dependent on the quality of the selected signals. Traditional approaches select trace signals manually based on the opinion of the designers and their experiences. They select signals to increase the observability in the scenarios that are more vulnerable to errors. However, manual selection of trace signals cannot guarantee their quality as an error may happen in unexpected scenarios. Therefore, automated signal selection algorithms are introduced. These algorithms can be broadly classified into three groups as follows.

- Metric-Based Signal Selection: Metric-based signal selection algorithms select trace signals based on the structure of the design to increase restoration ratio [2, 16].
- Simulation-based Signal Selection: Simulation-based approaches measure the capability of selected signals using the information gathered by simulation [5]. Simulation-based approaches are more precise in comparison with metric-based approaches due to the similarity of the simulated design with the actual behavior of the design. However, they are extremely slow, and can be unsuitable for large and complex designs.
- Hybrid Signal Selection: To address the limitations of metric-based and simulation-based signal selection algorithms, hybrid approaches are defined [14, 19, 32]. These approaches select an initial set of candidates for trace signals using the structure of the design. The final set of trace signals is selected by applying simulation-based algorithms on the initial set of candidates. The quality of the selected signals using these approaches is higher than metric-based techniques. However, hybrid methods sacrifice the quality of selected signals in comparison with simulated-based techniques to reduce the required time for signal selection. Machine learning based techniques are can improve both restoration ratio and signal selection time [27–29].

We refer the readers to the second part of the book for a detailed discussion on each of these approaches and associated challenges.

1.5 Generation of Tests

The quality of post-silicon and debug is dependent on the set of test vectors. The test should be effective to examine different use-case scenarios and expose hidden errors and vulnerabilities of the design. Billions of random and constrained-random tests are used to exercise the system for unexpected scenarios. Directed tests are carefully designed to check particular behaviors of the design. Directed tests are very promising in reducing the overall validation effort since a drastically small number of directed tests are required compared to random tests to obtain the same coverage goal. Directed test generation is mostly performed by human intervention. Handwritten tests entail laborious and time-consuming effort of verification engineers who have deep knowledge of the design under verification. Due to the manual development, it is both error-prone and infeasible to generate all directed tests to achieve a coverage goal. Automatic directed test generation based on a comprehensive coverage metric is the alternative to address this problem. Therefore, it is very important to generate efficient tests to not only activate the bugs but also propagate the effect to the observable points [6].

Post-silicon tests can be generated by making use of pre-silicon stimuli to reduce test generation efforts. Pre-silicon tests are designed based on the test plan and specification templates to exercise different features of the design. These templates should be mapped to silicon scenarios according to the processor architecture and actual memory addresses. Moreover, the pre-silicon tests usually do not consider the propagation of the effect of bug to the observable points since the observability is not an issue during pre-silicon validation [7]. Additionally, tests should be designed in such a way that they reduce the latency of observing the effect of a bug. For instance, consider a memory write to some address that places an incorrect value. The effect of this bug may not be observed unless the value written is read (it may take hundreds or even thousands of clock cycles) and the wrong value of the previous write is propagated to an observable point. To address the latency, quick error detection (QED) technique has been proposed [10, 20]. The idea is to transform a pre-silicon test into another one with lower latency between bug excitation and failure in silicon. For instance, for the memory read example above, a QED test would transform the original test by introducing a memory read immediately after each memory write; thus, an error introduced by the write would be excited immediately by the corresponding read.

Pre-silicon tests should be incorporated with observability features to be applied on silicon. Verification engineers usually favor tests that include exciting verification events (e.g., register dependency, memory collisions). However, the distribution of the generated tests should not be uniform to ensure that corner cases have been considered [25].

1.6 Post-silicon Debug

After the effect of the bug is observed using post-silicon tests, the next step is to localize the source of error, root-cause bugs, and fix them. When the effect of a bug is manifested in the observable points, validation engineers try to use DfD and trace information to localize the source of errors. Different techniques can be used to effectively debug the buggy silicon. Trace buffer information as well as coverage monitors and scan chains data are analyzed using satisfiability solvers [37] and virtual prototypes [3, 18] to localize the fault. The path from observing a failure to root-causing the error consists of the following steps.

- Repeating the failure: After a failure is observed, some sanity checks (e.g., checking the set up of the SoC and power connectivity) are done. If the problem is not resolved using sanity checks, we need to reproduce the failure to discover the recipe for it. Repeating a failure is not trivial and it involves executing the test several times using different hardware, software conditions until sighting the failure.
- Failure deposition: after defining the exact case of failure (when the source of failure with the conditions that caused it are known), the debugging team is assigned to create a plan for addressing the problem. The plan typically involves creating workarounds using features from architecture, design, and implementation to mitigate the problem.
- Bug resolution: Once the plan is developed for a failure, a team is assigned to follow the plan promptly to address it. In this step, the failure is called a bug. It is required to determine whether the bug comes from a design issue or it comes from silicon and manufacturing issue. Moreover, it is necessary to group the bug fix since an error might have been manifested in different forms of failure. Fixing the bug may resolve several failures. Therefore, it is very important not to waste resources to debug a failure twice. Finally, it is very important to validate the bug fix to ensure that it did not introduce any new error.

In the final step, high-level methods such as using programmable circuits (e.g., lookup tables) are deployed to patch post-silicon bugs in post-silicon debug [9, 33]. Templates are automatically generated using high-level description to patch the design. After patching, the functionality of the debugged implementation may be different from the specification. Therefore, the design should be analyzed to ensure that the correct design behavior is maintained under operating use-case scenarios.

In post-silicon, there is no time for sequentially finding and fixing bugs. When a bug is found, two things should happen in parallel. One group should work to fix the bug. The other group should find a creative way to workaround the previous bug and continue the debugging in order to find new bugs. Finding efficient work-arounds is challenging. Furthermore, debugging should account for the effect of physical characteristics such as temperature and electrical noise. For instance, errors may be masked in the presence of glitches, thermal effects, and voltage scaling. Therefore, various tuning should be done to make the error reproducible. Considering the vast parameter space, error reproducibility is challenging. Last but not least, sophisticated security features to protect SoC assets (e.g., encryption keys) and power management mechanisms make the debugging difficult. Security mechanisms try to decrease the observability to safeguard the design assets against adversaries. Similarly, power management features also disable specific components to save energy. Therefore, they reduce design observability and the complexity of debugging increases drastically in the presence of these options.

1.7 Summary

This chapter provided a thorough discussion on different steps of post-silicon validation and debug in the modern era of SoC design. We highlighted the importance of post-silicon validation as well as various critical steps of it. We also outlined existing challenges in each stage and provide novel and practical solutions to address them. We briefly described post-silicon readiness, debug infrastructure, test generation approaches, debug methodologies, CAD flows, etc. We believe this overview on post-silicon validation challenges will motivate the readers to explore further in this domain. Moreover, this introductory material will provide the necessary context for the readers to understand the remaining chapters in this book.

References

1. A. Adir, A. Nahir, A. Ziv, C. Meissner, J. Schumann, Reaching coverage closure in post-silicon validation, in *Haifa Verification Conference*. (Springer, Berlin, 2010), pp. 60–75
2. K. Basu, P. Mishra, Efficient trace signal selection for post silicon validation and debug, in *2011 24th International Conference on VLSI Design (VLSI Design)*. (IEEE, 2011), pp. 352–357
3. P. Behnam, B. Alizadeh, S. Taheri, M. Fujita., Formally analyzing fault tolerance in datapath designs using equivalence checking. In *2016 21st Asia and South Pacific Design Automation Conference (ASP-DAC)*. (IEEE, 2016), pp. 133–138
4. S. Chandran, P.R. Panda, S.R. Sarangi, A. Bhattacharyya, D. Chauhan, S. Kumar, Managing trace summaries to minimize stalls during postsilicon validation. IEEE Trans. Very Large Scale Integr. (VLSI) Syst. **25**(6), 1881–1894 (2017)
5. D. Chatterjee, C. McCarter, V. Bertacco, Simulation-based signal selection for state restoration in silicon debug, in *Proceedings of the International Conference on Computer-Aided Design*. (IEEE Press, 2011), pp. 595–601
6. M. Chen, X. Qin, H.-M. Koo, P. Mishra, *System-Level Validation: High-level Modeling and Directed Test Generation Techniques* (Springer Science & Business Media, New York, 2012)
7. F. Farahmandi, P. Mishra, S. Ray, Exploiting transaction level models for observability-aware post-silicon test generation, in *2016 Design, Automation & Test in Europe Conference & Exhibition (DATE)*. (IEEE, 2016), pp. 1477–1480
8. F. Farahmandi, R. Morad, A. Ziv, Z. Nevo, P. Mishra, Cost-effective analysis of post-silicon functional coverage events, in *2017 Design, Automation & Test in Europe Conference & Exhibition (DATE)*. (IEEE, 2017), pp. 392–397

9. M. Fujita, H. Yoshida, Post-silicon patching for verification/debugging with high-level models and programmable logic, in *2012 17th Asia and South Pacific Design Automation Conference (ASP-DAC)*. (IEEE, 2012), pp. 232–237
10. T. Hong, Y. Li, S.-B. Park, D. Mui, D. Lin, Z.A. Kaleq, N. Hakim, H. Naeimi, D.S. Gardner, S. Mitra, Qed: Quick error detection tests for effective post-silicon validation, in *2010 IEEE International Test Conference (ITC)*. (IEEE, 2010), pp. 1–10
11. https://software.intel.com/en-us/intel-platform-analysis-library. Intel Platform Analysis Library
12. https://www.mentor.com/products/fv/emulation-systems/veloce. Veloce2 Emulator
13. http://www.synopsys.com/tools/verification/hardware-verification/emulation/Pages/default.aspx. Zebu
14. E. Hung, S.J. Wilton, Scalable signal selection for post-silicon debug. IEEE Trans. Very Large Scale Integr. (VLSI) Syst. **21**(6), 1103–1115 (2013)
15. D. Kaiss, J. Kalechstain. Post-silicon timing diagnosis made simple using formal technology, in *2014 Formal Methods in Computer-Aided Design (FMCAD)*. (IEEE, 2014), pp. 131–138
16. H.F. Ko, N. Nicolici, Automated trace signals identification and state restoration for improving observability in post-silicon validation, in *Design, Automation and Test in Europe, 2008. DATE'08*. (IEEE, 2008), pp. 1298–1303
17. B. Kumar, A. Jindal, V. Singh, M. Fujita, A methodology for trace signal selection to improve error detection in post-silicon validation, in *2017 30th International Conference on VLSI Design and 2017 16th International Conference on Embedded Systems (VLSID)*. (IEEE, 2017), pp. 147–152
18. L. Lei, F. Xie, K. Cong, Post-silicon conformance checking with virtual prototypes, in *Proceedings of the 50th Annual Design Automation Conference*. (ACM, New York, 2013), p. 29
19. M. Li, A. Davoodi, Multi-mode trace signal selection for post-silicon debug, in *2014 19th Asia and South Pacific Design Automation Conference (ASP-DAC)*. (IEEE, 2014), pp. 640–645
20. D. Lin, T. Hong, F. Fallah, N. Hakim, S. Mitra, Quick detection of difficult bugs for effective post-silicon validation, in *2012 49th ACM/EDAC/IEEE Design Automation Conference (DAC)*. (IEEE, 2012), pp. 561–566
21. X. Liu, Q. Xu, *Trace-based Post-silicon Validation for VLSI Circuits* (Springer, Berlin, 2016)
22. S. Ma, D. Pal, R. Jiang, S. Ray, S. Vasudevan, Can't see the forest for the trees: State restoration's limitations in post-silicon trace signal selection, in *Proceedings of the IEEE/ACM International Conference on Computer-Aided Design*. (IEEE Press, 2015), pp. 1–8
23. P. Mishra, R. Morad, A. Ziv, S. Ray, Post-silicon validation in the soc era: a tutorial introduction. IEEE Des. Test **34**(3), 68–92 (2017)
24. S. Mitra, S. A. Seshia, N. Nicolici, Post-silicon validation opportunities, challenges and recent advances, in *Proceedings of the 47th Design Automation Conference*. (ACM, New York, 2010), pp. 12–17
25. Y. Naveh, M. Rimon, I. Jaeger, Y. Katz, M. Vinov, E. Marcus, G. Shurek, Constraint-based random stimuli generation for hardware verification. AI Mag. **28**(3), 13 (2007)
26. O. Olivo, S. Ray, J. Bhadra, V. Vedula, A unified formal framework for analyzing functional and speed-path properties, in *2011 12th International Workshop on Microprocessor Test and Verification (MTV)*. (IEEE, 2011), pp. 44–45
27. K. Rahmani, P. Mishra, Feature-based signal selection for post-silicon debug using machine learning. IEEE Trans. Emerg. Top. Comput. (2017)
28. K. Rahmani, P. Mishra, S. Ray, Scalable trace signal selection using machine learning, in *2013 IEEE 31st International Conference on Computer Design (ICCD)*. (IEEE, 2013), pp. 384–389
29. K. Rahmani, S. Ray, P. Mishra, Postsilicon trace signal selection using machine learning techniques. IEEE Trans. Very Large Scale Integr. (VLSI) Syst. **25**(2), 570–580 (2017)
30. S. Ray, Y. Jin, A. Raychowdhury, The changing computing paradigm with internet of things: a tutorial introduction. IEEE Des. Test **33**(2), 76–96 (2016)
31. J.A. Rowlette, T.M. Eiles, Critical timing analysis in microprocessors using near-ir laser assisted device alteration (lada), in *ITC* (2003), pp. 264–273

32. H. Shojaei, A. Davoodi, Trace signal selection to enhance timing and logic visibility in post-silicon validation, in *2010 IEEE/ACM International Conference on Computer-Aided Design (ICCAD)*. (IEEE, 2010), pp. 168–172
33. P. Subramanyan, Y. Vizel, S. Ray, S. Malik, Template-based synthesis of instruction-level abstractions for soc verification, in *Proceedings of the 15th Conference on Formal Methods in Computer-Aided Design*. (FMCAD Inc., Austin, 2015), pp. 160–167
34. P. Taatizadeh, N. Nicolici, Emulation infrastructure for the evaluation of hardware assertions for post-silicon validation. IEEE Trans. Very Large Scale Integr. (VLSI) Syst. **25**(6), 1866–1880 (2017)
35. S. Tam, S. Rusu, U.N. Desai, R. Kim, J. Zhang, I. Young, Clock generation and distribution for the first ia-64 microprocessor. IEEE J. Solid-State Circuits **35**(11), 1545–1552 (2000)
36. www.arm.com. CoreSight On-Chip Trace & Debug Architecture
37. C.S. Zhu, G. Weissenbacher, S. Malik, Post-silicon fault localisation using maximum satisfiability and backbones, in *Proceedings of the International Conference on Formal Methods in Computer-Aided Design*. (FMCAD Inc., 2011), pp. 63–66

Part II
Debug Infrastructure

Chapter 2
SoC Instrumentations: Pre-Silicon Preparation for Post-Silicon Readiness

Sandip Ray

2.1 Introduction

A fundamental requirement of debugging and validation is to observe, comprehend, and analyze the internal behavior of the target system as it executes. In fact, this requirement is so fundamental and inherent, that we rarely give a thought to it during ordinary debug activities. For traditional software program debug, we can realize this requirement by either inserting print statements at various points in the program code, or relying on a debugger that enables evaluation of various internal variables under specific execution conditions. For pre-silicon hardware design (e.g., RTL), this need is effectively addressed by the RTL simulator: we can suspend the execution of the design at any time or under any predefined condition, and inspect the values of various internal design variables.

Unfortunately, this requirement becomes challenging to satisfy for post-silicon validation. The essence of the so-called "limited observability" problem—discussed in virtually any publication on post-silicon validation—is that our ability to observe or control internal design variables as the system executes is severely limited during silicon execution [14]. There is a variety of reasons for this. In particular, what do we really mean by "observing a variable" when the system under debug is a silicon system? We probably mean, "getting the value of the variable out of the silicon system, possibly through a pin or port". This already suggests the problem: there is only a very small number of pins or ports in a silicon chip that we can utilize for this purpose. Another possibility is to use some portion of the system memory to store these values, and transport them off-chip subsequently. This approach enables recording more variables, at the obvious cost that one can only replay the recording *a posteriori* after the execution is complete. Nevertheless, note that there are billions to trillions (depending on how one counts it) of hardware signals, and a post-silicon

S. Ray (✉)
Department of Electrical and Computer Engineering, University of Florida, Gainesville, FL 32611, USA
e-mail: sandip@ece.ufl.edu

P. Mishra and F. Farahmandi (eds.), *Post-Silicon Validation and Debug*,
https://doi.org/10.1007/978-3-319-98116-1_2

execution can go from hours to days at a Gigahertz clock speed. Whatever means one can use to record, store, and transport values of internal signals would be very small compared to what we need to observe in order to debug arbitrary potential design bugs.

In addition to this problem of scale, there is another crucial issue that affects observability in post-silicon debug, viz., the immutability of a silicon implementation. During the debugging of a traditional software program or (pre-silicon) hardware design, we would like to observe different variables (or signals) at successive executions. For instance, suppose during a System-on-Chip (SoC) design validation, one encounters a scenario where the power management unit never turns the system to low-power mode. To debug and root-cause the problem, one will first want to ascertain if the power management unit received such a request. This can be done by potentially observing the input interface of the block. Once (if) it is ascertained that such a request is indeed received, the debugger will subsequently want to observe various components of the internal design logic within the power management unit to determine why the request is not correctly processed. The key message is that debugging is an iterative process and at different iterations, different signals, or variables in the program are of interest. Furthermore, one obviously needs to observe different signals for debugging different failures. During RTL or software, it is easy to observe different signals in different iterations: one simply needs to direct the simulator (or debugger) to display the variables of interest at each execution. Doing so in silicon, however, is a nontrivial task. Any signal that one needs to observe during post-silicon validation must have hardware logic attached to route or transport its value to an observation point such as the memory or output pin.

The above problems are addressed in current practice through an activity known as *post-silicon readiness*. Activities related to post-silicon readiness are done as part of the pre-silicon validation, and developed concurrently with the (functional) design flow. Note that if the readiness activities are in fact imperfect, its effect would be observed much later—during post-silicon, possibly through an inability to debug or validate a specific scenario. However, that would be too late to fix such problems, e.g., if a deficiency in observability is found during post-silicon debug then fixing that would require another silicon spin which would be potentially infeasible. So it is of crucial importance to ensure that the readiness is performed in a disciplined manner and can cover all potential scenarios that may be encountered later in post-silicon. It goes without saying that achieving this is a nontrivial exercise. Indeed, the elaborate and complex nature of readiness is one of the crucial components that markedly distinguishes post-silicon validation from pre-silicon, and is also a major contributor to the post-silicon validation cost.

This chapter is about post-silicon readiness. We provide an overview of the various readiness activities, and how they are carried out at different stages in the production life cycle in current industrial practice. The goal is not to be exhaustive but to give the reader a sense of the scope and complexity of the activities. We then delve in particular into one critical aspect of readiness, the approaches to on-chip instrumentations themselves. The amount and type of instrumentation developed for observability is varied, including tracing, triggers, interrupts, control, off-chip transport, etc. We provide a sampling of these techniques and discuss some of their applications.

2.2 Post-silicon Planning and Development Life Cycle

Post-silicon validation spans the entire SoC design life cycle. A previous paper [12] covers the various post-silicon activities more comprehensively. Figure 2.1 provides a general overview of the scope of this effort.

An important observation from this schedule is that majority of the activities involves readiness. Post-silicon readiness ensures that actual validation phase, when it comes later as the preproduction silicon becomes available, goes smoothly and in a streamlined manner. In the remaining sections, we delve a bit deeper into one specific component, viz., instrumentation. But here, we give a short summary of the slew of other activities to give a sense of their scope.

Test Plans. Developing test plans is one of the most elaborate and critical component of *any* validation activity. It entails defining coverage targets, corner cases to be exercised, the core functionality that one wants to test at different stages of validation. That said, post-silicon test plans are much more elaborate, subtle, and complex than pre-silicon ones. This is because post-silicon tests are significantly deeper and more "probing" than pre-silicon tests, e.g., involving millions to billions of cycles, potentially including behavior of a number of hardware and software blocks. Since

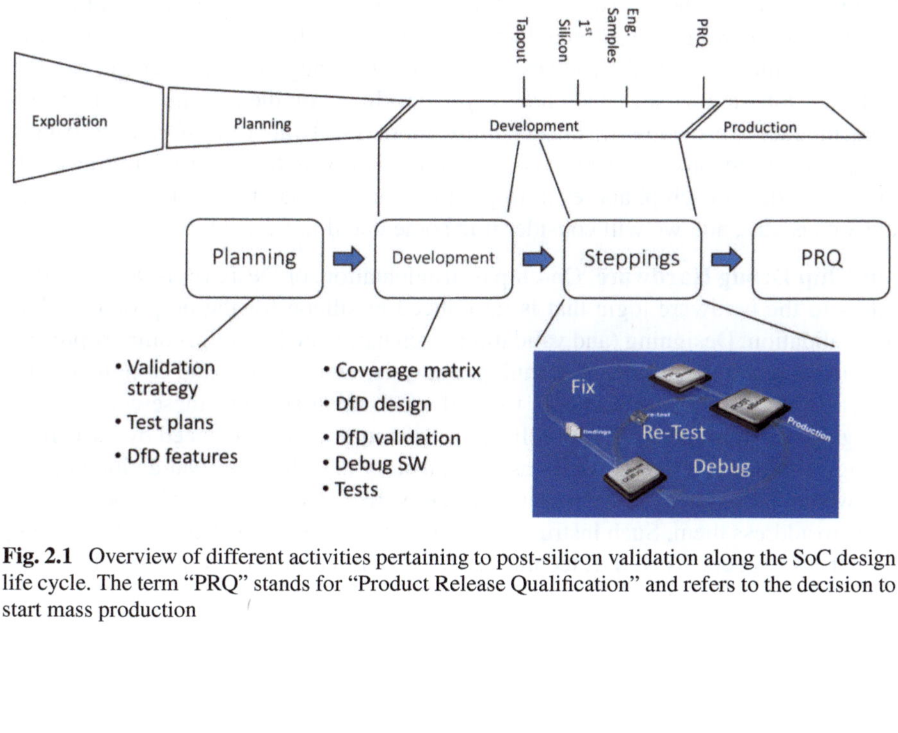

Fig. 2.1 Overview of different activities pertaining to post-silicon validation along the SoC design life cycle. The term "PRQ" stands for "Product Release Qualification" and refers to the decision to start mass production

test plans typically direct how the subsequent post-silicon activities (e.g., the actual tests, necessary test cards or boards, the instrumentation required to observe or control the behavior of the test, etc.) would go, it is critical that test planning starts early. Typically, this starts together with the architecture definition, often even before the microarchitecture is developed. Consequently, initial test planning has to be high-level, exercising system features that are defined at the point of the activity. This high-level plan is then subsequently refined as the design matures. Note that it is critical that the planning activity is in synchrony with the design so that the test plans remain viable and effective as the design matures.

Tests. The central component of silicon debug is the set of tests to run. For the validation to be effective, the tests must expose potential vulnerabilities of the design, and exercise different corner cases and configurations. Post-silicon tests are categorized into either (1) focused tests, i.e., tests created manually for exercising specific design features; and (2) random tests, where the goal is to exercise the system in ways not conceived by humans. Focused tests are written to check for specific register configurations, address decoding, and various power management configurations. Random tests are used to validate the behavior of the system on random instruction sequence, exercise various concurrent interleavings, etc. All such tests, of course, make use of the evolving test plans and must be easily updatable if the test plan changes. Updating test plans may be tricky in cases when they influence focused tests. In practice, the set of targets for focused tests are often left fixed for that reason, and test plans are carefully made to ensure that these targets do not get modified late in the design phase. Furthermore, note that application of certain platform-level tests requires specialized peripherals, boards, and test cards.

Debug Software. The term *debug software* encompasses any software tool or infrastructure that is necessary to enable post-silicon validation. This includes tools to enable running the tests themselves, or to facilitate debug, triage, coverage, etc. For instance, software patches and "hooks" are developed on the operating system running, in order to provide extra observability; there is software to set up various trigger conditions for recording system traces; a slew of software tools is used for transporting debug data off-chip, and even to perform various analytics on it. The scope of the area is vast, and we will consider it in some detail in Sect. 2.6.

On-Chip Debug Hardware. On-chip instrumentation, or Design-for-Debug (DfD), refers to the hardware logic that is introduced in silicon for the purpose of debug and validation. Designing (and validating) such hardware is a long, complex process that itself requires significant planning, analysis, architecture, and design components. In the subsequent sections, we will touch upon some of these components. A significant component of post-silicon instrumentation is influenced by ideas from Design-for-Test (DfT) that itself has a rich history [1]. However, validation induces its own unique challenges and a significant amount of instrumentation has been developed to address them. Such instrumentations include diverse approaches for tracing, numerous triggers, as well as techniques for off-chip transport of debug data.

2.3 Instrumentation Prehistory: Design-for-Test Structures

Augmenting a hardware design with additional logic to enable investigation or diagnosis has a long history in the area of Design-for-Test (DfT). DfT refers to instrumentations that were developed to facilitate manufacturing tests. Discussing the spectrum of DfT features will take us too far afield from the topic of this article. However, some DfT features are also repurposed for post-silicon validation and it is worth a brief mention.

2.3.1 Scan Chains

Scan chain is a structure that is introduced in a hardware design to facilitate observability of all the design flip-flops. The basic idea is to introduce two modes of design execution, the normal *functional mode* and *scan mode*. In the scan mode, all flip-flops are connected in a linear chain forming a (giant) shift register. This is implemented by including the following critical set of signals.

- Signals *scan_in* and *scan_out* define the input and output of the scan chain. They are the inputs and outputs of the first and last flip-flops in the chain, respectively.
- A *scan_enable* is a special signal that is added to a design. When this signal is asserted, the scan mode is activated turning the design into a shift register.
- A *scan_clock* is a clock input used to control all flip-flops during the scan mode.

Given this instrumentation, one can use significant control and observation in the design execution. For example, assuming "full scan", i.e., all flip-flops in the design are part of the scan chain, one can perform the following set of actions.

- Assert scan mode, and shift in the desired input sequence to all flip-flops via scan_in.
- De-assert scan mode, and apply one clock. The effect of that is to compute the next state of all the flip-flops from the state defined by the shifted sequence.
- Reassert scan mode, shift out the data from the flip-flops through scan_out, and diagnose if the circuit has computed the next state correctly.

In the absence of full scan, the process is a bit more involved, e.g., it is necessary to provide a *sequential* pattern and observe the effect after multiple cycles in that case.

Scan provides an interesting observability and controllability feature for the circuit operation during post-silicon validation: it enables the validator to specify/control the *entire system state* at a specific instant, and completely observe the effect of the hardware logic on that state in one cycle. Sometimes, one can use scan simply as an observability feature, i.e., the system is allowed to operate normally until a given instant when scan mode is asserted and the data in the scan chain is transported. However, note that the act of scanning in and out has a disruptive effect on the system execution. Thus, scan is useful for debugging when a "wide" (i.e., full system) view of the design is desired for one specific instant. Such observability is also sometimes referred to as *frozen observability*. In Sect. 2.4, we will consider another form of observability called tracing that can be thought of as the dual of scan in that a narrow portion of the state is observed for a number of cycles.

2.3.2 JTAG Architecture

JTAG, which was named after the *Joint Test Action Group* that created it, is an industry standard architecture for testing (and validating) circuits. It was codified as IEEE Standard 1149.1 and is officially called *Standard Test Access Port and Boundary Scan Architecture* [8]. The standard has been extended by many semiconductor chip manufacturers with specialized variants that provide many vendor-specific features.

The JTAG standard was designed to assist with device, board, and system testing, diagnosis, and fault isolation. JTAG boundary scan technology provides access to many logic signals, including the device pins. The signals are represented in the Boundary Scan Register (BSR) accessible through a specialized port called Test Access Port (TAP). This permits testing as well as controlling the states of the signals for testing and debugging. JTAG is used as the primary means of accessing various subblocks of integrated circuits, making it an essential mechanism for debugging embedded systems and SoC designs. On most systems, JTAG-based debugging is available from the very first instruction after CPU reset; this enables use of JTAG for debugging early boot software. In complex SoC designs, JTAG is also used as a transport mechanisms to access other on-chip debug modules.

A JTAG interface is a special interface added to a chip. Depending on the version of JTAG, two, four, or five pins are added. The four and five pin interfaces are designed so that multiple devices can have JTAG interfaces daisy-chained. Figure 2.2 shows a representative five-pin configuration, and Fig. 2.3 shows how they can be daisy-chained.

JTAG is the primary debug component of processor cores. Many silicon architectures such as PowerPC, MIPS, ARM, and x86 have a significant infrastructure for software debug, instruction tracing, and data tracing on top of the JTAG architecture. Some key examples of JTAG-based debug infrastructure include ARM Coresight™ [6], Intel® Processor Trace [9], etc. JTAG enables processors to be halted, single-stepped, or executed with various breakpoints in code.

Scan Dump. Scan and JTAG are often used in concert for creating scan dump. Scan dump can be used, e.g., in case of a hang during system initialization. To obtain a scan dump, one uses a special JTAG instruction in Master TAP while putting the design in scan mode. The effect is to dump the content of the scan flip-flops via TDO pin.

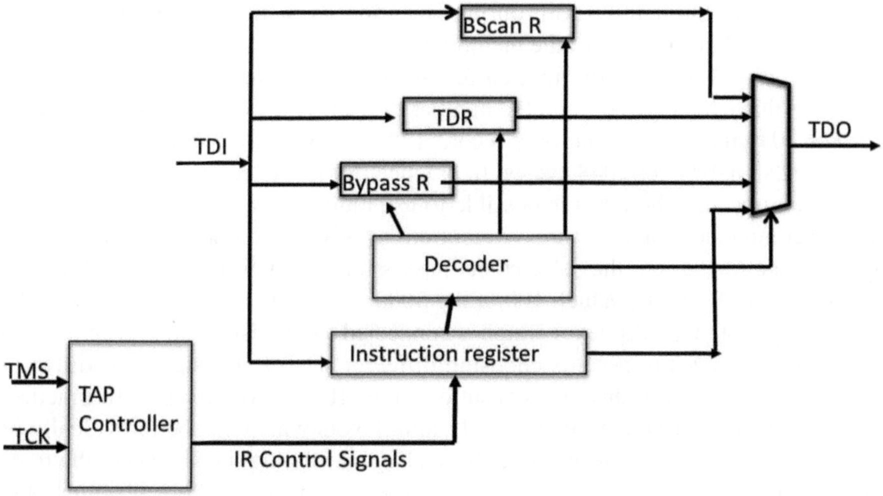

Fig. 2.2 Daisy-Chaining JTAG interfaces from multiple devices. The pins are TDI (Test Data in), TDO (Test Data Out), TCK (Test Clock), TMS (Test Most Select), and TRST (Test Reset). The TRST pin is optional and usually asynchronous

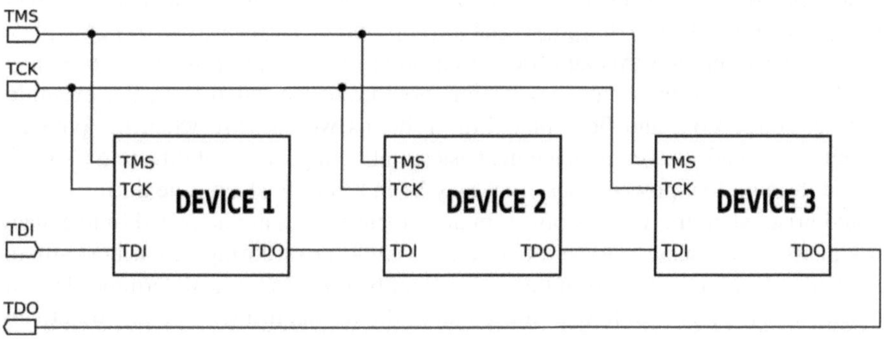

Fig. 2.3 Daisy-Chaining JTAG interfaces from multiple devices

2.4 Tracing

Hardware tracing is another critical infrastructure for post-silicon observability. The basic idea here is to monitor a small set of design signals over a long period of time as the system executes. Contrast this to the observability provided by scan, where the entire design state is observed at one specific point. The observability provided by tracing is often referred to as *streaming observability* to contrast it with the frozen observability provided by scan.

An important requirement in tracing is to determine how to stream or store the traced signals. Since I/O speed (using JTAG, for example) is significantly slower than speed of execution (e.g., MHz vs. GHz), it is not possible in most cases to dump the traced values through I/O ports during execution. Typically, the traced signals are therefore stored in a portion of the memory dedicated for this purpose, referred to as the *trace buffer*. The size of the trace buffer imposes a strong restriction on the number of signals that can be traced (and the number of cycles they can be traced for), e.g., a 128×2048 trace buffer can only record 128 signals (among millions to billions of internal signals) over 2048 cycles. In practice, for any execution about 100 or so signals are traced, although it is possible to use multiplexors to enable observation of a larger number of signals across executions as follows. Suppose we choose 5000 signals to be traced, but the selected set passes through a 5000×100 multiplexor that outputs 100 signals. Which 100 of the 5000 is actually selected is given by the selector input of the multiplexor which is connected to a register that is programmed at run time. This setup permits choosing different sets of 100 signals at different executions, although in any one execution, only 100 are observable. In practice, signals may pass through a hierarchy of multiplexors which can be individually or collectively programmed at run time. Note also that routing the signals to the trace buffer in a modern complex SoC design may face congestion, floor planning, and other layout-related issues.

The above discussion should make it clear that tracing, while critical, is a highly complex and elaborate activity requiring significant upfront planning. In particular, for a signal to be traced during post-silicon validation, it has to be selected in advance during pre-silicon (RTL) design, so that the appropriate hardware can be put in place to route the signal towards an observation point (e.g., output pins or trace buffer). On the other hand, problems in selection would only be found later. For example, routing, congestion, and floor planning problems would arise when the physical layout is created; power management issues affecting observability (e.g., signals routed through low-power blocks that may be in a certain sleep state during debug) would arise when the various power management modes are activated; inadequacy of signals or the need to trace other signals might arise during actual post-silicon validation. Consequently, disciplined investigation is necessary to determine (1) what signals to select, and (2) how to make use of the results of the selection effectively for post-silicon debug.

2.4.1 Trace Signal Selection

The process of selecting appropriate hardware signals for tracing is one of the most well-researched topics in post-silicon readiness. There are numerous treatises and articles on the subject [2, 5, 10, 11, 15, 17], and is covered in other chapters of this book. For this reason, we do not go into details of the various technologies, but point to the general approaches in the context of on-chip instrumentation and discuss some of their limitations.

The primary approach to signal selection entails a metric called *State Restoration Ratio* (SRR). The idea behind SRR is that (1) it is possible to reconstruct other signal values from the signals observed (e.g., one can deduce that the two inputs of an

AND gate are both '1' if the output shows up '1'), and (2) it is desirable to trace those signals which enable maximum reconstruction of design signals. Thus, SRR is defined as the ratio of the set of signals traced and reconstructed to the set of signals traced. Most of the literature on signal selection deals with various selection algorithms and heuristics such that the selected signals have a high value of SRR.

Unfortunately, from the point of view of quality of instrumentations, high SRR may not correspond directly to "debuggability". Consequently, SRR-based methods suffer from a number of inadequacies. First, given a design with large memory arrays, these methods typically cannot distinguish between reconstructing values of individual array elements and reconstructing the control to the array; the latter is often much more critical for debugging. Second, these methods are designed for low-level netlists. Apart from the obvious scalability limitations induced as a result, this can result in spurious conclusions on what signals can be reconstructed. For instance, netlists in an industrial flow have scan chains inserted: consequently, a structural analysis might suggest that (in case of full scan) observing one flip-flop in the scan chain for a sufficient number of cycles would reconstruct the entire system state, ignoring the reality that tracing is not performed in scan mode. Third, signals selected for maximum SRR may not be actually selected due to physical constraints (e.g., routing, congestion, etc.) or security reasons (e.g., signals representing a confidential key); it is unclear for most SRR-based methods how gracefully or sharply the restoration quality would degrade if some of the signals suggested by a signal selection algorithm are not included among the trace signals. Finally—and perhaps most importantly—SRR-based methods do not account for the fact that there are other DfD structures in the system (other than tracing) which also provide observability into some aspects of the design. For example, we discuss below the array freeze and dump technology, which makes it unnecessary for hardware tracing to also reconstruct memory arrays.

In spite of the above deficiencies, SRR continues to be useful as a guiding principle for trace signal selection. There has been recent work on addressing some of its limitations [11]. It is of critical importance to develop selection techniques that directly address debuggability of the design.

2.4.2 Software/Firmware Tracing and Trace Consolidation

Although hardware tracing has received most of the attention in the research community, it is important to realize that the goal of tracing is to enable system-level debug which includes software, hardware, and firmware executions. Consequently, in addition to tracing hardware signals, it is obviously important to also trace software and firmware executions, as well as messages sent by various IPs during system execution. It is important for these traces to be collated and a unified view is presented to the debugger on the state of the hardware and software executions. This is usually addressed through an IP that is responsible for collecting, collating, and consolidating all the traces. In industrial practice, one such IP is the Intel® Trace Hub

(ITH) [9]. Architecturally, ITH is simply an IP that is connected to a communication fabric in an SoC design; however, its purpose is to collect traces from various sources (e.g., messages by different IPs, software and firmware print statements, hardware traces, etc.), timestamp these traces, and provide a consolidated output through one of the system outputs (see Sect. 2.5.4). Similar functionality is provided by the ARM Coresight™ architecture for ARM-based systems.

2.5 Array Access, Triggers, and Patching

In addition to scan and trace, there is a large number of other instrumentations that are added to an SoC design to support post-silicon validation, with estimates running to 20% or more in silicon real estate in some cases. In this section, we provide a flavor of these instrumentations.

2.5.1 Array Freeze and Dump

Most modern SoC designs include large memory arrays. In some cases, it is useful for the debugger to review the content of these arrays during silicon validation. On the other hand, it is inconvenient and difficult to put large arrays into a scan chain, or choose entries from those arrays for signal tracing. The array freeze and dump feature, introduced in Intel for Pentium CPU's [4], is a dedicated observability infrastructure for arrays. The idea is to provide a special JTAG instruction for "freezing" the arrays. On receipt of this instruction, arrays disable their write logic; thus, the state of the array cannot change. The array content can then be dumped out through the read port.

2.5.2 Triggers

The term *trigger* refers to logic that can be programmed at run time to initiate or customize various observability features. Modern SoC designs contain a large number of triggers for various conditions. For instance, there are triggers to control hardware tracing; the debugger can define conditions under which tracing should be initiated, the signals to be traced (if they are multiplexed as discussed in Sect. 2.4), and when the recording ought to stop. Similar mechanisms are often available for most other DfD structures. Generally, triggers are implemented by creating an enable logic for the target observability: the enabling logic is connected to a register that can be programmed to monitor various conditions. The debugger sets up the register appropriately at run time for a specific condition for which it wants to use the target observability *for that execution*.

2.5.3 Microcode Patch

Much of the hardware functionality in modern-day SoC designs are implemented through microcode rather than custom hardware. This provides a means to update functionality at silicon time by changing the microcode flows. Such updates are facilitated by microcode path. Roughly, microcode patch enables the processor to load a software to change microcode. The actual mechanism for the patching varies, depending on the specific IP in which the microcode is executed, e.g., if it is a microcode of the CPU it may be directly changed, but if it is a microcode for a different IP patching may require several message communications between the CPU and the target IP, etc.

Irrespective of the implementation details however, microcode patch is a critical mechanism that is available in most modern SoC designs. In particular, it addresses a key aspect of post-silicon debug, *error sequentiality*. The challenge of error sequentiality is that once a bug is found in hardware during silicon validation, it is often not possible to fix it immediately. Note that each update to the silicon would entail an additional silicon spin. Consequently, the debugger must find some work-around for a discovered bug so that one can continue validation without requiring respin and explore other bugs. Microcode patch provides the key ammunition for the debugger in performing such work-around. Microcode patch is also one of the key methods for fixing an error that is observed in-field; without patching, fixing a silicon implementation might involve costly recalls costing significant damage to profitability of the product and even manufacturer reputation.

2.5.4 Transport

It is obviously inadequate to have logic that "merely" records internal states or state signals; one also needs mechanisms for transporting such data off-chip. Traditional transport mechanisms entailed use of specialized pins or JTAG, e.g., both scan and array freeze and dump make use of JTAG mechanisms. However, the use of JTAG means one has to debug the silicon in isolation. In recent days, there has been a general move towards verifying a form factor factor, e.g., a laptop, tablet, or mobile phone in its entirety. This of course imposes additional restrictions on how data can be transported: one must reuse some of the existing ports in the device rather than depend on a special-purpose port or use an SoC on a custom platform with additional pins. For instance, in recent SoC designs, data transport mechanisms repurpose the universal serial bus (USB) port. Note that this requires a thorough understanding of both the functionality and the validation use cases to ensure that they do not interfere when using the same interface. Finally, there is instrumentation to provide controllability of the execution, e.g., by overriding the system configuration, updating microcode on-the-fly during execution, etc. As we get newer and smaller form factors, e.g., wearables with no USM port, transport mechanisms have to depend on more sophisticated communication, e.g., wireless or bluetooth.

2.5.5 Custom Monitors

In addition to the generic or programmable infrastructures discussed above, most SoC designs have a number of monitors and controllers for custom properties. These are developed ad hoc, based on designers' experience of the complexity of the target property, confidence in their correct implementations, etc. For instance, many modern multiprocessor systems would include monitors to ensure cache coherence, memory ordering, etc.

Custom monitors have also been the topic of a significant amount of research. One of the earliest work is due to Gopalakrishnan and Chou [7]. They use constraint solving and abstract interpretation to compute state estimates for memory protocols. Park and Mitra [13] develop an architecture called IFRA for disciplined on-chip observability of pipelined microprocessors. Boule et al. [3] present architectures for post-silicon assertion checkers. Ray and Hunt [16] present an architecture for on-chip monitor circuits for validation of specific concurrent protocols.

2.6 Integration with Debug Software

The addition of DfD logic is one key aspect of post-silicon readiness. Another key component is debug software. Indeed, DfD hardware and debug software are complementary: the software is required to exploit the effects of the hardware during silicon debug.

What constitutes as debug software? The answer in general is "any software that facilitates silicon debug". This umbrella includes on the one hand tools and scripts that control various DfD mechanisms (e.g., program registers for triggers, etc.), and on the other hand, various analysis software to postprocess, triage, or root-cause from data obtained by various instrumentation. Here, we give a sampling of the various types of software involved.

Instrumented System Software: A key requirement for debug is the need to analyze hardware/software interactions. Thus one needs to know what application was running and in what stage when a certain hardware trace has been collected. Doing this requires that the underlying operating system (or other low-level system software) records various control paths of the program, calls to the system libraries, etc., and transmits them in concert with hardware trace for debugging. To address this need, one typically needs customized operating system to run on the silicon that has less complexity than an off-the-field operating system while including hooks to facilitate debug, observability, and control. Such system software may be written by silicon debug teams from scratch, or by significantly modifying the off-the-shelf implementations.

Triggering Software: We discussed hardware support for triggers in Sect. 2.5.2. Effective utilization of such hardware requires significant software tools. In particular, there are tools for querying or configuring specific hardware registers, setting triggers for various hardware tracing, etc.

Transport Software: Transporting data off-chip may require significant software assistance depending on the mode of transport. For example, transporting through the USB port requires instrumentation of the USB driver to interpret and route the debug data while ensuring the USB functionality is not affected during normal execution.

Analysis Software: The last important class of software tools are those that perform analysis of the post-silicon data. Such tools can range from the raw signal or trace data into high-level data structures (e.g., interpreting signal streams from the communication fabric in the SoC as messages or transactions among IPs) [18], comprehending and visualizing hardware/software coordinations, as well as tools to analyze such traced and observed data for further high-level debug.

2.7 Summary

This chapter has focused on pre-silicon activities that must be performed for post-silicon readiness. We focused primarily on hardware instrumentations, i.e., logic introduced primarily for helping in debug and validation. We hope the treatment in this chapter will help provide an understanding of the current industrial practice (as well as research direction) in this area.

Today's SoC designs include a large amount of such instrumentation—developing, managing, and using these features is a complex and expensive component of validation. The scenario gets particularly murky in current practice as a significant amount of instrumentation carries over from one product to another for legacy reason together with functional components, leading to significant hardware complexity, potential bloat in design, and even bugs introduced by incorrect use of the features themselves. Furthermore, as instrumentations accumulate, the slew of observability features is provided to the debugger as a grab-bag, without clear delineation on what should be used for what scenarios. Clearly, significant research is necessary to address this and make the design of DfD streamlined in the future.

References

1. M. Abramovici, P. Bradley, K. Dwarkanath, P. Levin, G. Memi, D. Miller, A reconfigurable design-for-debug infrastructure for SoCs, in *Proceedings of the 43rd Design Automation Conference (DAC 2006)*. (ACM/IEEE, 2006), pp. 7–12

2. K. Basu, P. Mishra, Efficient trace signal selection for post silicon validation and debug, in *24th International Conference on VLSI Design* (2011), pp. 352 –357
3. M. Boule, J. Chenard, Z. Zilic, Adding debug enhancements to assertion checkers for hardware emulation and silicon debug, in *International Conference on Computer Design*. (IEEE Computer Society, 2006), pp. 294–299
4. A. Carbine, D. Feltham, Pentium Pro processor design for test and debug. IEEE Des. Test Comput. **15**(3), 77–82 (1998)
5. D. Chatterjee, C. McCarter, V. Bertacco, Simulation-based signal selection for state restoration in silicon debug. ICCAD **2011**, 595–601 (2011)
6. CoreSight On-chip Trace and Debug Architecture, www.arm.com
7. G. Gopalakrishnan, C. Chou, The post-silicon verification problem: designing limited observability checkers for shared memory processors, in *5th International Workshop on Designing Correct Circuits (DCC 2004)* ed. by M. Sheeran, T. Melham (2004)
8. IEEE Joint Test Action Group: IEEE Standard Test Access Port and Boundary Scan Architecture. IEEE Std. **1149**(1) (2001)
9. Intel® Platform Analysis Library. https://software.intel.com/en-us/intel-platform-analysis-library
10. H.F. Ko, N. Nicolici, Automated trace signals identification and state restoration for improving observability in post-silicon validation, in *Proceedings of the Design, Automation and Test in Europe DATE '08* (2008), pp. 1298–1303. https://doi.org/10.1109/DATE.2008.4484858
11. S. Ma, D. Pal, R. Jiang, S. Ray, S. Vasudevan, Can't see the forest for trees: state restoration's limitations in post-silicon trace signal selcection. ICCAD **2015**, 1–8 (2015)
12. P. Mishra, R. Morad, A. Ziv, S. Ray, Post-silicon validation in the SoC era: a tutorial introduction. IEEE Des. Test Comput. **34**(3), 68–92 (2017)
13. S. Park, S. Mitra, IFRA: instruction footprint and recording for post-silicon bug localization in processors, in *Proceedings of the 45th Design Automation Conference (DAC 2008)*. (ACM/IEEE, 2008), pp. 373–378
14. P. Patra, On the cusp of a validation wall. IEEE Des. Test Comput. **24**(2), 193–196 (2007)
15. K. Rahmani, P. Mishra, S. Ray, Scalable trace signal selection using machine learning, in *31st International Conference on Computer Design (ICCD 2013)* (2013), pp. 384–389
16. S. Ray, W.A. Hunt Jr., Connecting pre-silicon and post-silicon verification, in *Proceedings of the 9th International Conference on Formal Methods in Computer-Aided Design (FMCAD 2009)* ed. by A. Biere, C. Pixley. (IEEE Computer Society, Austin, TX, 2009), pp. 160–163
17. J. Yang, N. Touba, Automated selection of signals to observe for efficient silicon debug, in *Proceedings of the VLSI Test Symposium* (2009), pp. 79–84
18. H. Zheng, Y. Cao, S. Ray, J. Yang, Protocol-guided analysis of post-silicon traces under limited observability, in *17th International Symposium on Quality Electronic Design (ISQED 2016)* (2016), pp. 301–306

Chapter 3
Structural Signal Selection for Post-Silicon Validation

Kanad Basu

3.1 Introduction

Traditionally, functional verification is performed using either simulation or formal techniques. However, both these approaches have their imperfections. While formal verification is restricted by state space, simulation is extremely slow compared to actual hardware implementation. With more complexity and functionality being introduced in design, both these methods are not sufficient to identify all functional bugs. As a result, more often than not, the first taped silicon is buggy [1]. Post-silicon validation is required to detect these bugs once the chip is fabricated.

Post-silicon validation can be compared with manufacturing testing techniques, which are applied post-manufacturing as well. However, there are some essential distinctions between the two. While manufacturing testing deals with electrical faults like stuck-at-faults or delay faults, post-silicon validation is applied to detect functional faults. Since a functional bug will be present in all the dies, post-silicon debug on a single die is sufficient. However, electrical faults might differ from one chip to the other. Hence, manufacturing testing should be performed on all the dies individually. Finally, post-silicon validation has to be at speed, and hence, start and stop techniques like scan chains, which are applied for manufacturing test, cannot be directly used.

A serious issue with post-silicon validation is limited observability. Since the chip is completely manufactured, it is not possible to observe all the internal signal states. This is analogous to inserting breakpoints in software programs and monitoring internal variable values. Various Design-for-Debug techniques have been suggested over the years to enhance internal observability. Trace buffer based debug is an *ELA* mechanism to observe some of the internal signal states. The trace buffer is a dedicated piece of hardware manufactured within the chip. It stores some of the

K. Basu (✉)
Electrical and Computer Engineering, New York University, New York, NY, USA
e-mail: kb150@nyu.edu

P. Mishra and F. Farahmandi (eds.), *Post-Silicon Validation and Debug*,
https://doi.org/10.1007/978-3-319-98116-1_3

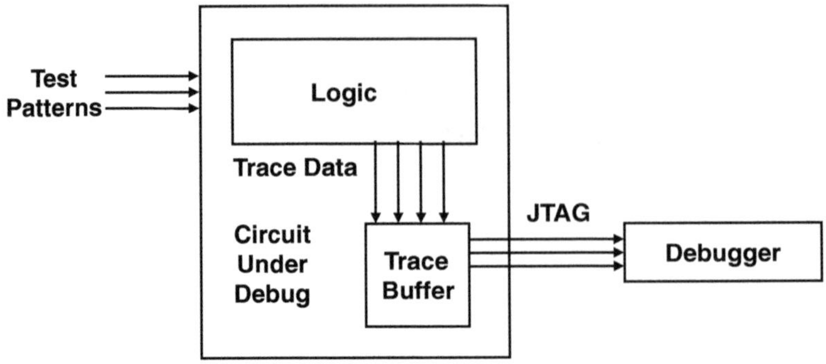

Fig. 3.1 Overview of trace buffer based post-silicon debug

internal signal states for a certain number of cycles. The trace buffer is regulated by an external controller, which dictates when the trace buffer should start operating and when it will stop. Once the trace buffer is full, the contents are transferred to some offline debugger via *JTAG* interface. The entire process is illustrated in Fig. 3.1.

It should be recalled that since the trace buffer is part of debug hardware, its capacity is restricted. The trace buffer is generally a block ram [2] which has two parameters—depth and width. The depth indicates the number of cycles over which the trace data is stored, while the width denotes the number of signals which are stored every cycle. The trace buffer is designed before a chip is manufactured. The different stages of trace buffer design, including specifying the area, determining the depth as well as selecting the signals for tracing are determined during design time. The status of trace buffer modeling during a chip design flow is presented in Fig. 3.2.

As can be seen in Fig. 3.2, trace signal selection can be performed at various stages of the design, during specification, during design or during synthesis depending on the logic and the user. Signal selection is an important issue because the performance of a debug mechanism depends on the signals selected. The trace buffer capacity is extremely limited. For example, a typical large $ISCAS'89$ benchmark will have around 1500 flip-flops. If a trace buffer has a width of 32, choosing 32 signals out of 1500 means $\binom{1500}{32}$ combinations, which is about 10^{69}. A trial and error method of signal selection from these huge number of combinations is impractical. Random signal selections result in inferior performance as demonstrated in [3]. Thus, efficient algorithms need to be developed for efficient trace signal selection so that debug objectives are attained.

3.2 Related Work

The bugs that evade the functional verification phase needs to be identified as soon as possible [4] before the chip is shipped to the customer. Failure to detect these bugs in time leads to increasing respin cost as well as production delay [5, 6]. Typically,

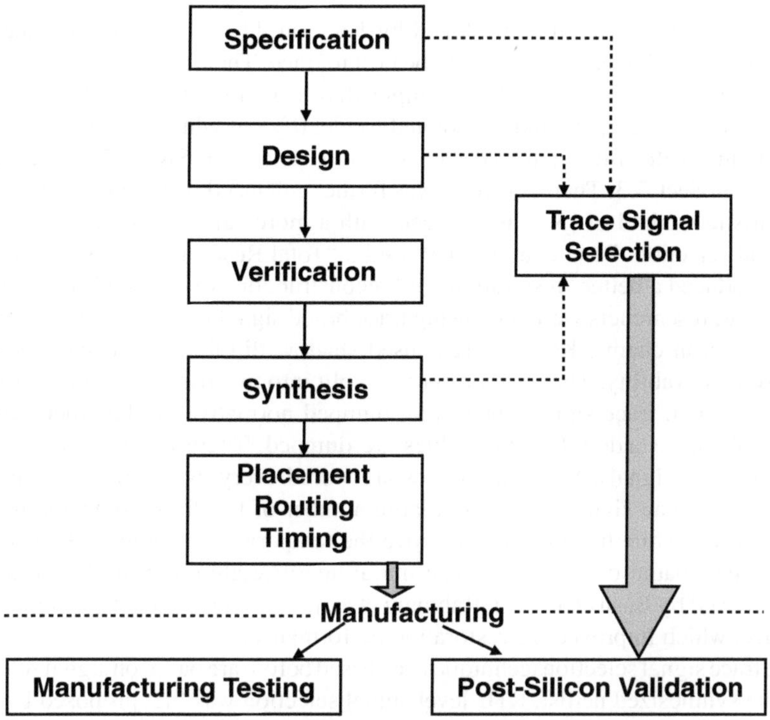

Fig. 3.2 Trace signal selection during design flow

design companies used in-house debug methodologies to detect them [3]. However, with increase in design size, there is an enhancing focus on scalable post-silicon validation methodology [7]. It has been reported [8] that the industry spends around 50% of its overall design cost in post-silicon debug.

Over the years, researchers have tried to develop various methods for post-silicon validation and debug. Formal analysis for post-silicon debug was recommended by De Paula et al. [9]. The primary problem with this approach was that it was not applicable to designs with a large number of gates due to state space explosion. It can be questioned that since both post-silicon debug and manufacturing testing are performed at the same level, as seen in Fig. 3.2, why not leverage manufacturing testing techniques like scan chains which have been a stable debug technique for a long time. However, scan chains require to stop the circuit functionality and hence, are not applicable [10], specifically, when the functional errors are drastically apart [11]. Although shadow flip-flops and double buffering can deal with this problem, they come with a large area overhead [12].

Trace buffer based silicon debug methodologies have been universally adopted to this end [13–15]. Once the signal states of a circuit are learned, they can be analyzed to recover the erroneous state using techniques like failure propagation tracing [16] or latch divergence analysis [17]. An automatic state restoration based trace signal selection technique was first developed by Ko et al. [1, 3]. They used a heuristic method to decide which signals would be best to trace. Their objective was to reconstruct the untraced signal states. For example, if, in a circuit of 1500 flip-flops, only 32 are traced every cycle, the authors wanted to identify out which 32 will reconstruct a maximum of the 1468 untraced signals. More details on Signal Restoration are explained in Sect. 3.3. Their approach was further enhanced by Liu et al. [18], where the heuristic method of [3] was replaced with a more substantial theoretical basis for signal selection. Basu et al. [2, 19] used a "Total Restoration" based technique which produced a better "restoration" or "reconstruction" of untraced signal states.

Till now, researchers were only using trace based signal restoration. In an attempt to include scan chains, Ko et al. [20] used shadow flip-flops with scan chains to improve observability. The trace buffer was split into two parts—trace and scan. In the "trace" part, trace signal states were dumped normally. On the other hand, in the "scan" part, shadow flip-flop values are dumped. There is a tradeoff between trace and scan signals. While the traces are stored every cycle, the scans are not. However, the scan signals combine a large number of flip-flops compared to trace signals. Thus, while trace signals improve the temporal observability, scan signals improve the spatial observability. The initial heuristic algorithm by Ko et al. [20] was improved by Basu et al. [21]. Rahmani et al. [22, 23] proposed fine-grained scan and trace, which improved the restoration performance.

The trace signal selection techniques discussed before are based on signal selection from the synthesized netlist. RTL-level signal selection was first proposed by Basu et al. [19]. This was subsequently improved by Kumar et al. [24]. Layout-aware trace signal selection was developed by Thakyal et al. [25]. Instead of statically selecting a set of signals for tracing, researchers came up with the notion of multiplexed trace signal selection [26–28], where the signals to be traced are selected depending on the mode of the design. Signal selection using formal analysis was proposed by Kumar et al. [29].

3.3 Signal Restoration

In this section, we explain signal restoration, which has been the conventional basis for signal selection [2, 3, 18]. In post-silicon debug, unknown signal states can be reconstructed from the traced signal states in two ways—forward and backward restoration. Forward restoration deals with the restoration of signals from input to output, that is, knowledge of input states can help in restoring the value of the output. Backward restoration, on the other hand, deals with reconstructing the input from the output. Forward and backward restoration can be illustrated using the example in Fig. 3.3. We use a 2-input *AND* gate to explain the restoration process. Forward restoration is presented in Fig. 3.3a. When one input of a gate has the controlling input value for the particular gate, the output will reflect the input. Controlling input of a gate represents the input signal, which the output can directly follow, regardless of the other gates' inputs. For example, in an *AND* gate, 0 is the controlling input, since if any input of the gate has a value of 0, the output will also have 0, irrespective of the total number of inputs to the gate as well as the values of the other inputs. Similarly, 1 is the controlling input of a *OR* gate. An *XOR* gate will not have any controlling inputs. It can be seen from Fig. 3.3a that forward restoration is performed in either of two ways:

1. When one input of the gate is controlling input.
2. When values of all the inputs of the gate is known.

For the first case, an input value of 0 for the signal a in Fig. 3.3a directly translates to an output value of 0 for c. For the second case, when both b and c have values of 1 in Fig. 3.3a, it can be derived that c will also be 1.

Figure 3.3b shows backward restoration. The simplest way to implement backward restoration is when the output is noncontrolling value of a gate. In this case, all the inputs need to be of the same value irrespective of the number of inputs. For example, consider the *AND* gate in Fig. 3.3b. When the output is 1, it is easily inferred that the inputs must also be 1. This inference is irrespective of the number of inputs. For example, even if the *AND* gate had been of 20 inputs, an output value of 1 indicates all the inputs also to be 1. The lower figure in Fig. 3.3b shows another form of backward restoration. In this case, the output c has a controlling value 0, which indicates at least one of the two inputs should be 0. If one of the inputs' value is known ($b = 1$ here), and that value is a noncontrolling value, it can be inferred that the other input should be controlling value, i.e., 0. For a m input gate, it can be stated that if the output is a controlling value and $m - 1$ input values are known and they are all noncontrolling, the mth input can be restored. However, this technique will not hold when one of the $m - 1$ inputs has a controlling value as demonstrated in Fig. 3.3c. In this example, the output c is 0, like Fig. 3.3b. However, the input b has a value of 0. There is no way to realize what a can be. In this case, backward restoration will fail. Thus, backward restoration succeeds only in the following situations:

- When the output of the gate is a noncontrolling value.
- When the output is controlling value, and all but one input's values are known and all of them are noncontrolling.

Although we have illustrated the signal reconstruction using a 2-input *AND* gate, the restoration procedure can also be expressed in an identical manner for other types of logic gates as well as with more inputs.

The signal reconstruction procedure is illustrated using a simple circuit shown in Fig. 3.4. Let us assume that the trace buffer width is 2, that is, states of two signals can be recorded. We restore the other signal states by application of the signal selection methods presented in [1, 18]. The results are shown in Table 3.1. The 'X's represent those states which cannot be determined. The selected signals are shown in shades. For both [1, 18], the signals selected are C and F, in that order. **Restoration ratio**, which is a popular metric for estimation of signal restorability is defined as

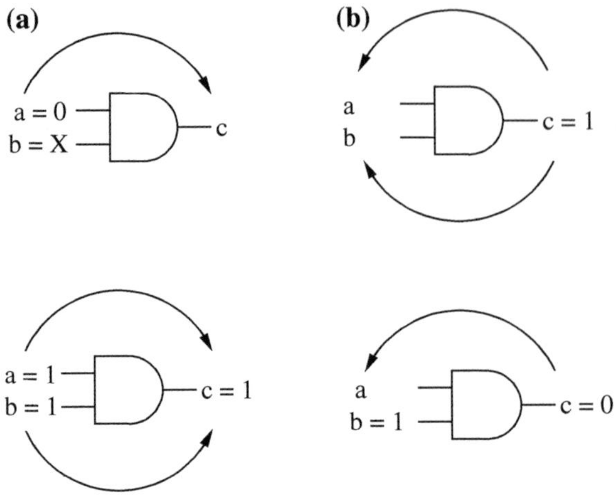

Fig. 3.3 Signal restoration

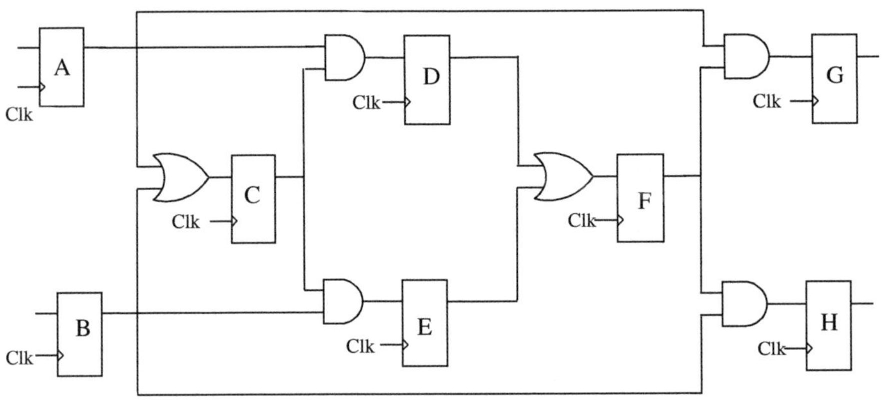

Fig. 3.4 Example circuit

$$\textit{Restoration Ratio} = \frac{\textit{number of states restored} + \textit{number of states traced}}{\textit{number of states traced}}. \quad (3.1)$$

Let us determine the number of restored states in Table 3.1. If we consider the row corresponding to signal A, two entries have value 0, while the rest have value X (non-restored state). Thus, two states are known. Similarly, two states are known for the row corresponding to signal B. Since signal C is traced, all the states are known (no X in the row). For signal D, three entries in the row have value 0, hence three states are reconstructed. Computing in this manner, a total of 26 states are reconstructed. Out of them, 10 entries (corresponding to signals C and F) are traced states. Therefore, $\textit{Restoration Ratio} = \frac{26}{10} = \mathbf{2.6}$.

Table 3.1 Restored signals using two selected signals

Signal	Cycle 1	Cycle 2	Cycle 3	Cycle 4	Cycle 5
A	X	0	X	0	X
B	X	0	X	0	X
C	1	1	0	1	0
D	X	X	0	0	0
E	X	X	0	0	0
F	0	1	1	0	0
G	X	0	0	X	0
H	X	0	0	X	0

Popular signal selection approaches [1, 18], which utilize **partial restorability**[1] are not able to provide best possible signal reconstruction. In this chapter, we will discuss **total restorability**[2] based signal selection algorithm proposed by Basu et al. [19] that outperform the previous approaches in both signal selection time and the quality of the selected signals. First, we will propose gate-level signal selection algorithm in Sect. 3.4 and then RTL-level signal selection in Sect. 3.5.

3.4 Gate-Level Signal Selection (GSS)

Algorithm 1 shows the gate-level signal selection procedure (GSS) developed by [19] that has five important steps, each of which is described in the remainder of this section.

Algorithm 1: Gate-level Signal Selection

Input: Circuit, Trace Buffer
Output: List of selected signals S (initially empty)
1: Compute the node values.
2: Find the state element with highest value and add to S.
3: Create Initial Region.
while *trace buffer is not full* **do**
  4: Recompute the node values of state elements.
  5: Compute region growth by finding the state element with highest value not in S and add to S.
end
return S

[1]**Partial Restorability** of a signal refers to the probability that the signal value can be reconstructed using known values of some other traced signals.

[2]**Total Restorability** of a group of signals refer to the fact that the signal states can be completely restored, that is, it is a special case of partial restorability with restorability value of 100%.

3.4.1 Computation of Edge Values

An edge between two flip-flops is defined a path between the two which contains only combinational elements, i.e., there cannot be any flip-flops in an edge. Therefore, not all flip-flops can be connected by an edge. An edge is not directed. The same edge can be used for forward as well as backward computation. For example, in Fig. 3.4, an edge between the two flip-flops A and C, that passes through an OR gate, can be used for forward computation from A to C and backward computation from C to A. An edge is used to find the influence of one flip-flop on another. For example, the edge between A and C can be used to find the influence of A on C (forward) and C on A (backward). There can be two cases here, independent or dependent, which we will describe below.

3.4.2 Independent Signals

To explain independent edges, we reconsider Fig. 3.4. The OR gate in front of flip-flop C is driven by signals A and B, which do not have any interdependency, that is, input of B is not driven by A and vice versa. Hence, the edges AC and BC are independent. The edge value calculation for an independent edge using the generic example is shown in Fig. 3.5, which has two flip-flops K and L.

The input of flip-flop L corresponds to the output of the combinational gate G_n. For each of the combinational gates in the path, four probabilities are defined: ${P^I}_{0,N}$, ${P^I}_{1,N}$, ${P^O}_{0,N}$ and ${P^O}_{1,N}$, where, ${P^I}_{k,N}$ indicates the probability that a node N (gate or flip-flop) has an input value of 'k' when another node is controlling it ($k \in \{0, 1\}$). ${P^O}_{k,N}$ indicates the probability that a node N (gate or flip-flop) has an output value of 'k' when another node is controlling it ($k \in \{0, 1\}$). The output of flip-flop K can influence the output of G_1 in two cases:

1. Output of K is a controlling value for G_1.
2. All the inputs to G_1 are complement of the controlling value.

The computations are shown by assuming G_1 to be a 2-input AND gate. P_{G_1} is defined as the overall probability of K controlling G_1. According to [30],

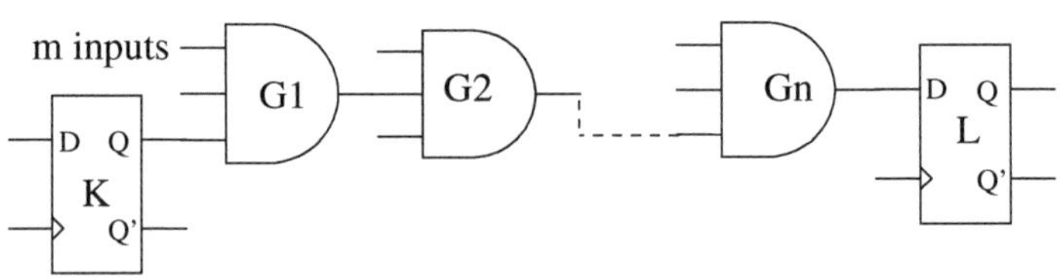

Fig. 3.5 Example circuit with n gates

$$P_{G_1} = P^O{}_{1,G_1} + P^O{}_{0,G_1} \tag{3.2}$$

Let P_{condm,G_1} be the probability that the output of G_1 follows the output of K ($m \in \{0, 1\}$). That is, P_{cond0,G_1} will be the probability of output of G_1 being 0 when output of K is also 0. For simplicity of calculation, in this example, let us assume $P^I{}_{0,G_1} = P^I{}_{1,G_1} = 0.5$ (that is, the inputs of G_1 have an equal probability of being 0 or 1).

$$P^O{}_{0/1,G_1} = P_{cond0/1,G_1} \times P^I{}_{0/1,G_1} \tag{3.3}$$

Now, for a 2-input AND gate, P_{cond0,G_1} is 1, since 0 is the controlling input. Therefore, we obtain $P^O{}_{0,G_1} = 0.5$. Similarly, since 1 is the noncontrolling input, P_{cond1,G_1} is 0.5, which gives $P^O{}_{1,G_1} = 0.25$. From Eq. 3.2, it can be seen that $P_{G1} = 0.75$. Therefore, the probability that G_1 will follow the value of K is 0.75. In the next step, the probability how G_1 controls G_2 is found out. This is multiplied by the probability P_{G1} to derive how K controls P_{G2}. In this way, the probabilities are computed along the path till L is reached. It has to be remembered that while computing P_{G2}, the input probabilities will not be 0.5 like G_1. The values of $P^I{}_{0,G_2}$ and $P^I{}_{1,G_2}$ would be $P^O{}_{0,G_1}$ and $P^O{}_{1,G_1}$ obtained from Eq. 3.3. For example if G_2 is also a 2-input AND gate, applying Eq. 3.3, we obtain, $P^O{}_{0,G_2} = 0.5$, and $P^O{}_{1,G_2} = 0.125$. Therefore, we get $P_{G2} = 0.625$.

If there are n combinational gates between K and L, we get

$$P^O{}_{0/1,G_n} = \prod_{1 \le i \le n} (P_{cond0/1,G_i}) \times P^I{}_{0/1,G_1} \tag{3.4}$$

The final probability P_{G_n} is obtained using Eq. 3.2.

These computations can be used to obtain an edge value of the circuit in Fig. 3.4. To compute the value of edge AC, the OR gate in between the two signals is named as G. As before, assuming equal probabilities of 0 and 1 at the input, $P^I{}_{0,G} = P^I{}_{1,G} = 0.5$. Since it is an OR gate, $P_{cond0,G} = 0.5$ and $P_{cond1,G} = 1$. Therefore, Eq. 3.3 can be used to obtain $P^O{}_{0,G} = 0.25$ and $P^O{}_{1,G} = 0.5$. Equation 3.2 can now be used to obtain $P_G = 0.75$, which is the value of the edge AC.

3.4.3 Dependent Signals

In Fig. 3.5, we saw that there is only one path connecting the flip-flops K and L. We would like to observe what happens when there is a re-convergent fan-out, i.e., multiple paths emanating from K converge at L. This case is called **Dependent Signals**. We need to determine the probability of a gate output influencing an m-input gate, if it affects l inputs ($m \ge l \ge 2$) of the gate. An example circuit similar to Fig. 3.5 is shown in Fig. 3.6 to calculate the edge value in case of dependent signals. None of the works prior to [19] like [1, 18] considered re-convergent fan outs.

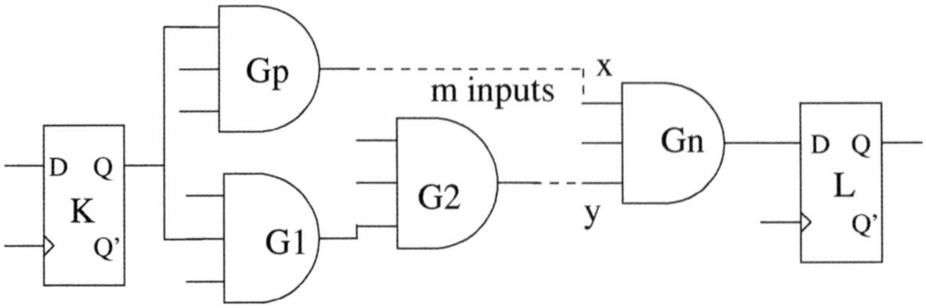

Fig. 3.6 Example circuit

In Fig. 3.6, two inputs (*x*,*y*) of the *m* input gate G_n are affected by flip-flop *K*. We assume the paths *Kx* and *Ky* are independent. If they are dependent, the method described below can be applied recursively. The procedure to resolve this is by combining them to a single independent edge, so that the computations are easier as in Sect. 3.4.2. Let us assume that G_n is an AND gate. For an AND gate, since 0 is the controlling value, having any of the inputs of the gate G_n as 0 will ensure a 0 being propagated to the output. Therefore

$$P^I{}_{0,G_n} = P^O{}_{0,x} + P^O{}_{0,y} - P^O{}_{0,x\&y} \tag{3.5}$$

$P^O{}_{0,x\&y}$ subtracts the probability when both are 0, since it is being computed twice. Similarly, since 1 is the noncontrolling input, we get

$$P^I{}_{1,G_n} = P^O{}_{1,x\&y} \tag{3.6}$$

where $P^O{}_{1,x\&y}$ is the probability when both *x* and *y* are '1'. Let $P_{cond0/1,x/y}$ be the probabilities that *x*(*y*) is 0(1) when the output of *K* is 0(1). $P^O{}_{0/1,x\&y}$ can be defined as

$$P^O{}_{0/1,x\&,y} = (P_{cond0/1,x} \times P_{cond0/1,y}) \times P^O_{0/1,K}$$

With the help of Eq. 3.3, this can be reduced to

$$P^O{}_{0/1,x\&y} = \frac{P^O{}_{0/1,x} \times P^O{}_{0/1,y}}{P^O{}_{0/1,K}} \tag{3.7}$$

Equation 3.4 can be used to obtain the values $P^O{}_{0/1,x/y}$ and Eqs. 3.5 and 3.6 to obtain the values of $P^I{}_{0/1,G_n}$. The final P_{G_n} can be obtained using Eqs. 3.2 and 3.3, and the information on the number of inputs to the gate G_n, which in turn, is the value of the edge *KL*.

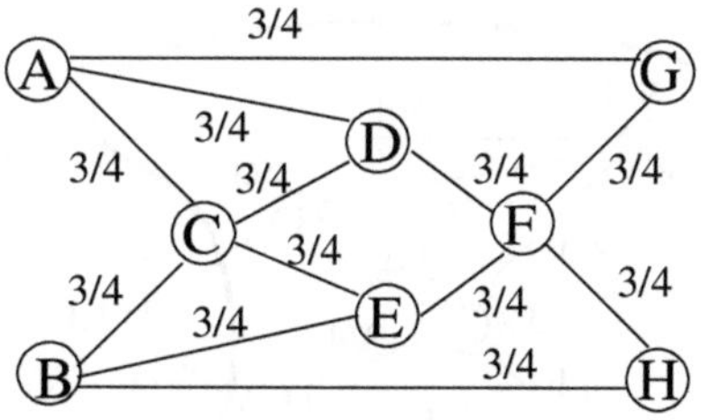

Fig. 3.7 Graphical representation of example circuit

3.4.3.1 Example

Using the calculations described in Sects. 3.4.2 and 3.4.3, we can determine the edge values of the circuit in Fig. 3.4. A graphical representation of the circuit is shown in Fig. 3.7. Each gate (flip-flop or combinational gate) is represented as a node while the path between two gates is represented as an edge.

There are no dependent edges in this example circuit. Since all the edges have only one two-input gate in between them, they have values are $\frac{3}{4}$ (obtained from Sect. 3.4.2). This graph will be used hereon to explain the total restorability based signal selection algorithm.

3.4.4 Initial Value Computation for State Elements

The value of a state element is defined as the sum of all the edges attached with it, in both forward and backward direction. For example, in Fig. 3.7, the value of flip-flop *C* is the sum of the edge weights connected with it, that is, *CA*, *CB*, *CD* and *CE*. A "threshold" in order to prevent combinational loops inside the circuit during edge value computation, which was used by [1] as well. This computation is independent of sequential loops in the design.

3.4.5 Initial Region Creation

A region is a collection of state elements, such that each state element is connected to at least another one in the region. It is not necessary that a region is fully connected, i.e., all state elements are connected to each other. Please note, by "connected", we mean a flip-flop having an edge with another. In Fig. 3.7, the flip-flops *A*, *B*, *C*, *D*, and *E* form a region. The first state element selected for tracing is the one with the highest value, based on the calculations in Sect. 3.4.4. It is added to a list called "known". Now, all state elements which have an edge with the recently selected element are added to the region.

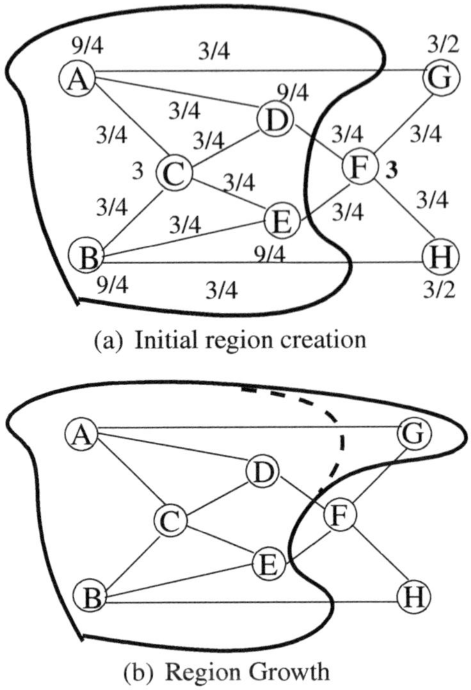

Fig. 3.8 Region creation and growth

Figure 3.8a is used to show region creation and growth. The values of edges and nodes are shown along each edge and node respectively. For example, node A has 3 edges AC, AD, and AG, each having a value $\frac{3}{4}$. Therefore, the value for A is $\frac{3}{4} + \frac{3}{4} + \frac{3}{4} = \frac{9}{4}$. The flip flop with the highest value is C. All the nodes which have an edge from C are included in the region, which is represented by the spline in Fig. 3.8a.

3.4.6 *Recomputation of Node Values*

As we have seen in Sect. 3.4.5, the first state element in Fig. 3.7 to be traced is already known (C in the previous example). However, other trace signals need to be found out till the trace buffer width is reached. To select the subsequent signals, the value of each node is recomputed. The flip-flop whose value is being recomputed may have an edge to an flip-flop inside the region as well as one outside the region. Edges to state elements inside the region are given higher weight. In order to utilize the knowledge from signals that are already in the region, thus increasing their restorability values and therefore, aiming for total restorability of those signals, edges to state elements inside the region are given higher weight during recomputation.

3.4.7 Region Growth

To select the next trace signal, the flip-flop with the highest value and not in the list "known" is determined. If two flip-flops have the same value, the one with the higher forward restoration is traced. This is because, when all the input signals are known, forward restoration is guaranteed to succeed, but not backward restoration as seen in Sect. 3.3. For example in Fig. 3.8a, the next state element to be traced is A, which is included in the list "known". If the trace buffer width has reached, calculations will stop. Otherwise, the region grows by adding all state elements having an edge to the recently selected node. As shown in Fig. 3.8b, in this case G is added since G is the only node connected to A and not in the region. The dotted line in Fig. 3.8b indicates the original region, while the spline indicates the new region. The whole process of recomputation and region growth is iterated till the trace buffer width is reached.

3.4.8 Complexity Analysis

Let V be the number of nodes in the circuit and E be the number of edges in the circuit. Let N be the width of the trace buffer, that is, N signals are traced every cycle. Edge value computation takes $O(E)$ time, while flip-flop value computations takes $O(V)$ time. Therefore, to select N signals, the time required is $O(NV)$. Since edge value computation is done only once and node value computation is done each time a signal is selected for tracing, the overall time complexity of our algorithm is $O(E+NV)$.

3.4.9 Motivational Example

The total restorability based method (described in Sect. 3.4) is used for selecting signals in the circuit in Fig. 3.4. The first signal that traced is C. Note that this was the same signal that was selected in Table 3.1. The second signal selected is A, based on total restorability computations. Tracing A along with C guarantees to reconstruct D every cycle. In Table 3.1, F is selected as the second traced signal. C and F together do not provide any such guarantees. The results are shown in Table 3.2. It can be seen that the restoration ratio using this approach is **3.2**, which is better than the one in Table 3.1 (2.6).

Table 3.2 Restored signals using our method

Signal	Cycle 1	Cycle 2	Cycle 3	Cycle 4	Cycle 5
A	0	0	0	0	1
B	1	0	1	0	X
C	1	1	0	1	0
D	X	0	0	0	0
E	X	1	0	0	0
F	X	X	1	0	0
G	X	0	0	0	0
H	X	X	0	0	0

3.5 RTL-level Signal Selection

In this section, signal selection at RTL-level is described. Signal reconstruction at RTL-level is shown using the Verilog design in Fig. 3.9a. There are three register variables (flip-flops) a, b and c as well as two input signals d and e. $m1$, $m2$, and $m3$ are three other signals in the design. All registers are initialized to 0. Flip-flop a is the concatenated value of d and c and 8-bits long. The result of *AND* between a and $m1$ and the *AND* between $m2$ and $m3$ are *OR*ed to derive b, which is also 8-bits long. c is the sum of e along with a stretch of all 1s. Both c and e are 7-bits long. Signal d is of 1-bit length. Let us consider that signal a is traced. We will now explain how that reconstructs signals in other cycles. Since a is part of b's input, state of b can be reconstructed from a. c and d are both inputs to a, both of which can be reconstructed using backward restoration. Finally, from value of c, state of e can be reconstructed using backward restoration. Thus, tracing of a helps to restore states of four other signals (though not in the same cycle).

Algorithm 2 shows our signal selection procedure that has six important steps. In the first step, a control data flow graph ($CDFG$) is generated to model the entire system. Since in the RTL-level, each register variable represents multiple state elements, we use the register variables for signal selection purpose. However, the trace buffer width refers to the total number of state elements represented by these register variables. For example, the register variable $[7:0]a$ will represent 8 state elements, and therefore selection of variable a implies that 8 trace buffer locations are needed. The relationship between the different register variables is obtained from the $CDFG$. These relations are used to produce the total restorability values for the variables. The register variable with the highest value is chosen for tracing. Once a variable is chosen for tracing, all the other variable values are recomputed in the same manner as in Algorithm 1. The Steps 4–6 are continued until the trace buffer is full. The remainder of this section describes each of the steps in detail.

```
always @ (posedge clk or negedge reset)
begin
if (!reset)
 begin
  a <= 7'b0;   b <= 7'b0; c <= 7'b0;
end
else
 begin
  a <= {d, c}; b <= (a & m1) || (m2 &
m3); c <= e + 7'b1;
 end
end
```

(a) RTL Verilog example

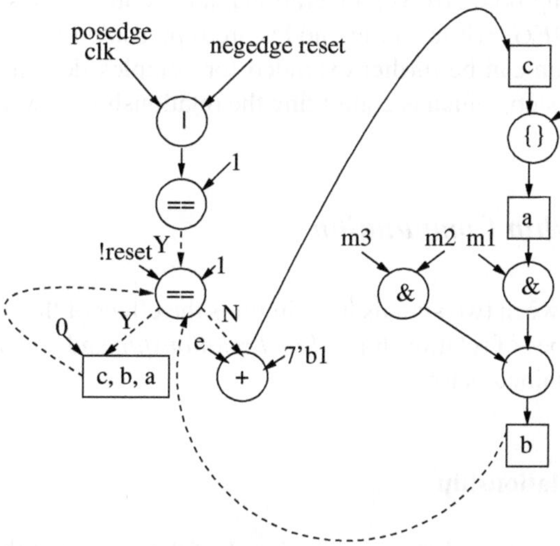

(b) CDFG of Verilog code

Fig. 3.9 Verilog code and CDFG

3.5.1 CDFG Generation

First, a Control Data Flow Graph (*CDFG*) of the RTL-model is generated using any standard *HDL* parser. The one used here is open source Icarus Verilog parser [31] for

Algorithm 2: RTL-level Signal Selection

Input: RTL description of design, No. of trace entries
Output: List of selected signals S (initially empty)
1: Develop the *CDFG* of the RTL description.
2: Find the *relationship* between the register variables.
3: Find the initial values of the register variables.
while *trace buffer is not full* **do**
  4: Find the register variable with the highest value.
  5: Add all corresponding state elements to the list S.
  6: Recompute values for all the register variables.
end
return S

the Verilog circuits. The *CDFG* format is similar to the one developed by Mohanty et al. [32].

The *CDFG* representation of the Verilog code in Fig. 3.9a is shown in Fig. 3.9b. The *CDFG* represents both the movement of control signals (using dotted arrow) and data values (using bold arrows). Operational and control nodes are represented by circles in the *CDFG*, while storage nodes are represented by boxes. This basic *CDFG* representation can be further extended for complex design. This *CDFG* is the input to the next step, which is computing the relationship between the elements.

3.5.2 Relationship Computation

The relationship between two signals in a circuit is the effect of that signal on others. There can be two types of relationships: *direct relationship* and *conditional relationship*, which are explained below:

3.5.2.1 Direct Relationship

There is a direct relationship between two signals if they occur on the same line of a signal assignment. For example, in Fig. 3.9a, the signal pairs a has a direct relationship with b. Direct relationship can be either forward or backward relationship. Forward relationship deals with the propagation of values from the right-hand side of the assignment to the left-hand side. Backward relationship on the other hand deals with the reverse, that is from left-hand to right-hand side of the assignment. In Fig. 3.9b, a has a forward relationship on b, while b has a backward relationship on a. To show relationship computation, let us consider a signal assignment statement looks like

$$y <= x_1 \; OP_1 \; x_2 \; OP_2 \; x_3 \; OP_3 \; .... \; x_n$$

where *OP* may represent any operation (eg., AND, OR, etc.). We want to find out the relationship of each of the n signals on the right hand side of the assignment statement on y. Each of these signals is assumed to be k bits long and independent. If each of the OP_i's is *AND* gates, the assignment statement can be rewritten as

$$y <= x_1 \,\&\, x_2 \,\&\ldots\ldots\ldots\ldots\&\, x_n$$

It should be remembered that similar computations can be extended to different types of gates OP_i, like *OR*, *XOR*, etc. The relationship of y on x_i is computed as a probability that the value of y will follow the value of x_i, which is similar to the independent edge value computation in Sect. 3.4.2. The relationship of y and x, $P^y{}_{0/1,x_i}$ is equivalent to $P_{cond0/1,y}$ in Eq. 3.3. Assuming all OP_is to be *AND* gates, y follows x_i completely when either all the k bits of x_i are 0 (P_0), or when all the x_i's have their k bits as 1 (P_1)(Since 0 is the controlling input of AND gate). The relationship of y on x_i is given by

$$P^y{}_{0,x_i} = \frac{2^{k(n-1)}}{2^{kn}} \tag{3.8}$$

$$P^y{}_{1,x_i} = \frac{1}{2^{kn}} \tag{3.9}$$

According to [30] and again, assuming for simplicity that 0 or 1 can occur with 50% probability, we get,

$$P^y{}_{x_i} = \frac{2^{k(n-1)}+1}{2^{kn}} \tag{3.10}$$

These calculations are when all the *OP*s are AND operations. Similar equations can be written for other operations as well. For example, if *OP* was an *OR* operation, the equations will be rewritten as

$$P^y{}_{1,x_i} = \frac{1}{2^{kn}} \tag{3.11}$$

$$P^y{}_{0,x_i} = \frac{2^{k(n-1)}}{2^{kn}} \tag{3.12}$$

Let us see how can we apply these computations to the example in Fig. 3.9b. The dependency of signal b on signals a and c in Fig. 3.9b is shown in Fig. 3.10 (assuming all signals are independent). Two new variables $g1$ and $g2$ have been introduced as

$$g1 = a \,\&\, m1$$
$$g2 = m2 \,\&\, m3$$
$$b = g1 \mid g2$$

We first find the relation of a on $g1$, and then $g1$ on b, to find the complete relationship between a and b. From Eqs. 3.8 and 3.9,

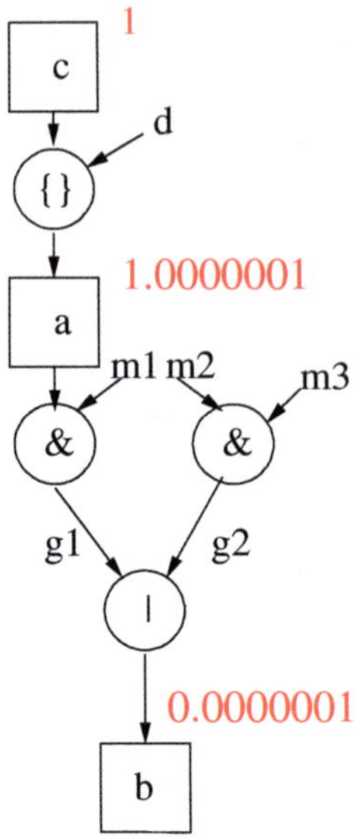

Fig. 3.10 A portion of the CDFG in Fig. 3.9b

$$P^{g1}{}_{0,a} = \frac{2^8}{2^{16}}$$
$$P^{g1}{}_{1,a} = \frac{1}{2^{16}}$$

To find the relationship of $g1$ on b, using Eqs. 3.11 and 3.12,

$$P^{b}{}_{1,g1} = \frac{1}{2^{16}}$$
$$P^{b}{}_{0,g1} = \frac{2^8}{2^{16}}$$

Thus, combining all these equations, we get the relation between a and b as:

$$P^{b}{}_{1,a} = \frac{1}{2^{16}} \times \frac{2^8}{2^{16}}$$
$$P^{b}{}_{0,a} = \frac{1}{2^{16}} \times \frac{2^8}{2^{16}}$$

Finally, using Eq. 3.4, we get,

$$P^{b}{}_{a} = \frac{1}{2^{23}} = 0.0000001$$

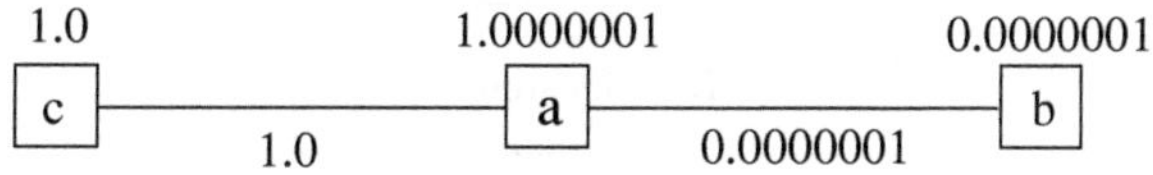

Fig. 3.11 Simplified version of Fig. 3.10

The value of $P^a{}_c$ is 1.0, since a and c have direct concatenated relationship. Figure 3.11 A simplified version of Fig. 3.10 along with the edge values is shown in Fig. 3.11. As before, the node values are computed by adding the edge values and node values (on top of nodes).

In this section, we have described the relationship when the signals are independent. Edge and node computations of dependent signals are of the same form as in Sect. 3.4.3.

3.5.2.2 Conditional Relationship

Non-assignment dependencies are computed using conditional relationship. For example, in the following code, the symbol x has a conditional dependence on m and n.

$$if\,(m\ or\ n)\ \ x <= y;$$

In this case, the dependency value is $\frac{3}{4}$, as obtained from Eqs. 3.2 and 3.3 in Sect. 3.4.2, since there are only 2 variables m and n in the *if* condition. In this way, conditional dependency is measured for the entire circuit.

3.5.3 Signal Selection

After the edge and node values are computed, signals are selected using a procedure similar to the gate-level signal selection using region growth and recomputation similar to Algorithm 1, until the trace buffer is full.

3.6 Experiments

In this section, the total restorability-based signal selection technique is first compared with partial restorability-based signal selection techniques [1, 18]. Finally, impact of RTL-level signal selection techniques are presented.

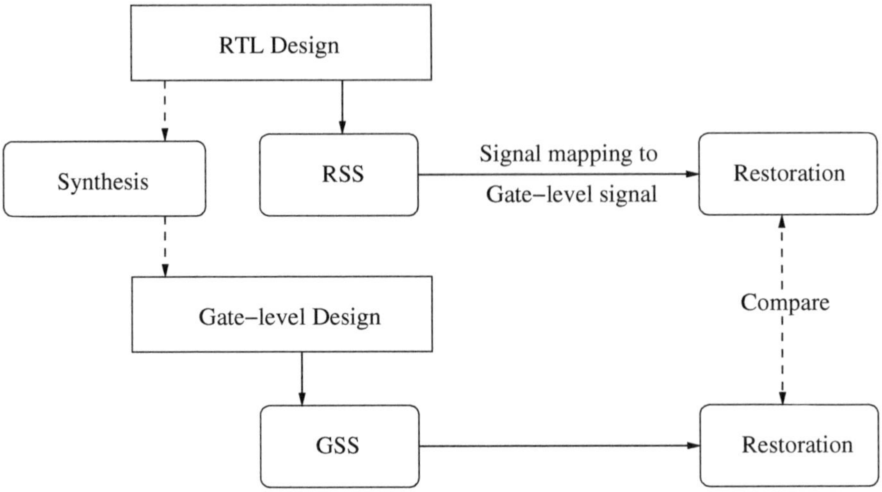

Fig. 3.12 Overview of our experiments to verify RSS

3.6.1 Experimental Setup

First, the gate-level signal selection approach (GSS) is compared with partial restorability based signal selection techniques (used by [1, 18]), by application on the ISCAS'89 benchmark circuits. A simulator, similar to the one described by [18] has been used for this purpose.

For verifying the RTL-level signal selection technique (RSS), Verilog circuits from Opencores website are used. To ensure the performance of RSS, gate-level signal selection algorithm (GSS) is also applied for these circuits. The circuits are first synthesized using Synopsys Design Compiler to gate-level netlist before GSS is applied. Overview of the approach is shown in Fig. 3.12. A comparison of restoration performance for both RSS and GSS reveals that the restoration ratio obtained by both is almost similar, as discussed in Sect. 3.6.3.

3.6.2 Results on Gate-Level Signal Selection (GSS)

Comparison of the Table 3.3 compares the performance of the total restorability-based approach [19] with the partial restorability based approach proposed by Ko et al. [1]. The three largest ISCAS '89 benchmark circuits are used for this purpose. The trace buffer width is selected as 32. The improvement is defined as the ratio between the restoration ratio of the two approaches. In Table 3.3, random inputs refer to the case when all inputs are driven randomly, including the control inputs. On the other hand, "deterministic inputs" refer to when the control inputs are driven with specified

Table 3.3 Comparison with Ko et al.

	Restoration ratio with random inputs			Restoration ratio with deterministic inputs		
Circuit	[1]	[19]	Improvement	[1]	[19]	Improvement
s38584	38	42	1.1	6	20	3.33
s38417	9	16	1.8	9	16	1.8
s35932	48	50	1.04	25	35	1.4

Table 3.4 Comparison with Liu et al. with deterministic inputs

	Restoration ratio		
Circuit	[18]	[19]	Improvement
s38584	9	20	2.22
s38417	14	16	1.14
s35932	22	35	1.6

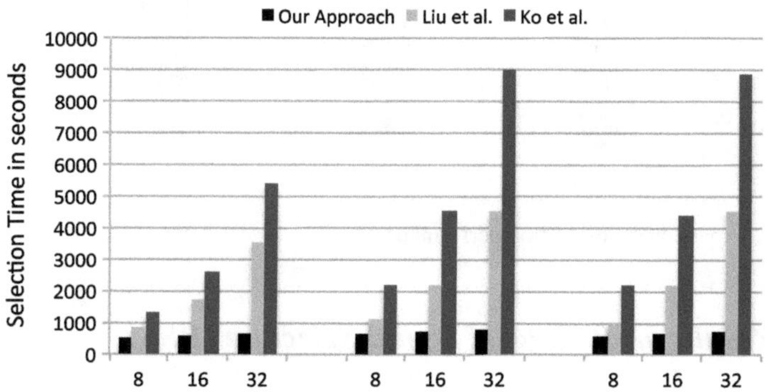

Fig. 3.13 Comparison of signal selection time

inputs to ensure the circuit does not fall into a reset state. All the other inputs are driven randomly. It is seen that using random inputs, the improvement of [19] over [1] is moderate, 31% on average. On the other hand, for the deterministic inputs, the improvement is significant, 117%. Since, real-life applications are represented using deterministic inputs as stated by [18], performance of the signal selection algorithms in those cases are important.

Table 3.4 compares the restoration ratio of total restorability based method [19] with the approach proposed by Liu et al. [18] for the three largest ISCAS'89 benchmarks using a trace buffer width of 32 is used. In this case, only deterministic inputs are chosen. An average improvement of 65% is observed for all the benchmarks.

The signal selection time of the total restorability-based approach [19] is compared with the partial restorability based approaches [1, 18]. The results are shown in Fig. 3.13. For each of the benchmarks, the trace buffer width is varied in steps of 8, 16 and 32. As seen, [19] takes up to 90% less signal selection time compared to [1, 18].

Table 3.5 RTL-level versus gate-level signal selection

Circuit	Memory size reduction	Speedup
Total CPU	8.1	697
Wishbourne LCD controller	22.81	1923
dmx512 tranceiver	191.24	733
OPB onewire	3.22	3600
Simple RS232 Uart	3.8	500

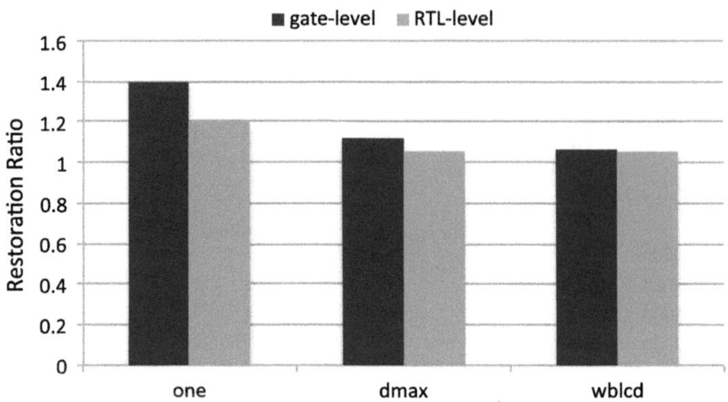

Fig. 3.14 Comparison of restoration performance

3.6.3 Results on RTL-level Signal Selection (RSS)

RTL-level signal selection is both time and memory efficient compared to GSS. For 5 Opencore circuits, the corresponding memory size reduction and speedup obtained by using RSS, when compared with GSS, is shown in Table 3.5. As can be seen, a speedup of up to 3600 as well as memory size reduction of up to 191 is obtained using RSS.

Table 3.5 has shown that RSS provides higher speedup and memory size reduction compared to GSS. However, we are yet to check whether this comes at any penalty on restoration performance. To evaluate the restoration performance of these two methods, we use three opencore benchmarks (*OPB onewire, dmx512 transceiver and Wishbourne LCD controller*). The benchmarks are driven using deterministic inputs. The results are shown in Fig. 3.14. As can be seen, the restoration performance of both GSS and RSS are almost similar, with a minimal penalty for RSS.

3.7 Conclusion

Post-silicon validation is extremely important to discover bugs that have escaped the pre-silicon verification phase. Limited observability is a major bottleneck for post-silicon validation. Trace buffer based signal selection helps to alleviate this problem. In this chapter, a total restorability-based signal selection algorithm has been discussed which provides better restoration performance compared to other structural signal selection algorithms. An RTL-level signal selection algorithm is also discussed which reduces the memory size as well as the signal selection time.

References

1. H.F. Ko, N. Nicolici, Algorithms for state restoration and trace-signal selection for data acquisition in silicon debug. IEEE Trans. Comput. Aided Design Integr. Circuits Syst. **28**(2), 285–297 (2009)
2. K. Basu, P. Mishra, Efficient trace data compression using statically selected dictionary, in *2011 IEEE 29th VLSI Test Symposium (VTS)* (IEEE, 2011), pp. 14–19
3. H.F. Ko, N. Nicolici, Automated trace signals identification and state restoration for improving observability in post-silicon validation, in *Proceedings of the Conference on Design, Automation and Test in Europe* (ACM, 2008), pp. 1298–1303
4. B. Vermeulen, S.K. Goel, Design for debug: catching design errors in digital chips. IEEE Design Test **19**(3), 37–45 (2002)
5. M. Abramovici, P. Bradley, K. Dwarakanath, P. Levin, G. Memmi, D. Miller, A reconfigurable design-for-debug infrastructure for SoCs, in *Proceedings of the 43rd Annual Design Automation Conference* (ACM, 2006), pp. 7–12
6. A.B. Hopkins, K.D. McDonald-Maier, Debug support for complex systems on-chip: a review. IEE Proc. Comput. Digit. Tech. **153**(4), 197–207 (2006)
7. A. Khoche, D. Conti, TRP in action: embedded instrumentations in FPGA, in *Proceedings of the 24th IEEE VLSI Test Symposium* (2006)
8. A. Nahir, A. Ziv, R. Galivanche, A. Hu, M. Abramovici, A. Camilleri, B. Bentley, H. Foster, V. Bertacco, S. Kapoor, Bridging pre-silicon verification and post-silicon validation, in *Proceedings of the 47th Design Automation Conference* (ACM, 2010), pp. 94–95
9. F.M. De Paula, M. Gort, A.J. Hu, S.J. Wilton, J. Yang, Backspace: formal analysis for post-silicon debug, in *Proceedings of the 2008 International Conference on Formal Methods in Computer-Aided Design* (IEEE Press, 2008), p. 5
10. G.-J. Van Rootselaar, B. Vermeulen, Silicon debug: scan chains alone are not enough, in *International Test Conference, 1999. Proceedings* (IEEE, 1999), pp. 892–902
11. D.D. Josephson, The manic depression of microprocessor debug, in *International Test Conference, 2002. Proceedings* (IEEE, 2002), pp. 657–663
12. D. Josephson, B. Gottlieb, The crazy mixed up world of silicon debug [IC validation], in *Proceedings of the IEEE 2004 Custom Integrated Circuits Conference, 2004* (IEEE, 2004), pp. 665–670
13. G.M. Uhler, R. Thekkath, Trace control block implementation and method, U.S. Patent 7,055,070, 30 May 2006
14. S. Tang, Q. Xu, A multi-core debug platform for NoC-based systems, in *Design, Automation & Test in Europe Conference & Exhibition, 2007. DATE'07* (IEEE, 2007), pp. 1–6
15. C. MacNamee, D. Heffernan, Emerging on-ship debugging techniques for real-time embedded systems. Comput. Control Eng. J. **11**(6), 295–303 (2000)

16. O. Caty, P. Dahlgren, I. Bayraktaroglu, Microprocessor silicon debug based on failure propagation tracing, in *ITC 2005, IEEE International Test Conference, 2005. Proceedings* (IEEE, 2005), 10 pp.
17. P. Dahlgren, P. Dickinson, I. Parulkar, Latch divergency in microprocessor failure analysis, in *null* (Citeseer, 2003), p. 755
18. X. Liu, Q. Xu, Trace signal selection for visibility enhancement in post-silicon validation, in *Proceedings of the Conference on Design, Automation and Test in Europe* (European Design and Automation Association, 2009), pp. 1338–1343
19. K. Basu, P. Mishra, Rats: restoration-aware trace signal selection for post-silicon validation. IEEE Trans. Very Large Scale Integr. (VLSI) Syst. **21**(4), 605–613 (2013)
20. H.F. Ko, N. Nicolici, Combining scan and trace buffers for enhancing real-time observability in post-silicon debugging, in *2010 15th IEEE European Test Symposium (ETS)* (IEEE, 2010), pp. 62–67
21. K. Basu, P. Mishra, P. Patra, Efficient combination of trace and scan signals for post silicon validation and debug, in *2011 IEEE International Test Conference (ITC)* (IEEE, 2011), pp. 1–8
22. K. Rahmani, P. Mishra, Efficient signal selection using fine-grained combination of scan and trace buffers, in *2013 26th International Conference on VLSI Design and 2013 12th International Conference on Embedded Systems (VLSID)* (IEEE, 2013), pp. 308–313
23. K. Rahmani, S. Proch, P. Mishra, Efficient selection of trace and scan signals for post-silicon debug. IEEE Trans. Very Large Scale Integr. (VLSI) Syst. **24**(1), 313–323 (2016)
24. B. Kumar, K. Basu, M. Fujita, V. Singh, RTL level trace signal selection and coverage estimation during post-silicon validation, in *2017 IEEE International High Level Design Validation and Test Workshop (HLDVT)* (IEEE, 2017), pp. 59–66
25. P. Thakyal, P. Mishra, Layout-aware selection of trace signals for post-silicon debug, in *2014 IEEE Computer Society Annual Symposium on VLSI (ISVLSI)* (IEEE, 2014), pp. 326–331
26. S. Prabhakar, M.S. Hsiao, Multiplexed trace signal selection using non-trivial implication-based correlation, in *2010 11th International Symposium on Quality Electronic Design (ISQED)* (IEEE, 2010), pp. 697–704
27. X. Liu, Q. Xu, On multiplexed signal tracing for post-silicon validation. IEEE Trans. Comput. Aided Design Integr. Circuits Syst. **32**(5), 748–759 (2013)
28. K. Basu, P. Mishra, P. Patra, A. Nahir, A. Adir, Dynamic selection of trace signals for post-silicon debug, in *2013 14th International Workshop on Microprocessor Test and Verification (MTV)* (IEEE, 2013), pp. 62–67
29. B. Kumar, K. Basu, A. Jindal, B. Pandey, M. Fujita, A formal perspective on effective post-silicon debug and trace signal selection, in *International Symposium on VLSI Design and Test* (Springer, Berlin, 2017), pp. 753–766
30. E. Taylor, J. Han, J. Fortes, Towards accurate and efficient reliability modeling of nanoelectronic circuits, in *Sixth IEEE Conference on Nanotechnology, 2006. IEEE-NANO 2006*, vol. 1 (IEEE, 2006), pp. 395–398
31. S. Williams, Icarus verilog (2006)
32. S.P. Mohanty, N. Ranganathan, E. Kougianos, P. Patra, *Low-Power High-Level Synthesis for Nanoscale CMOS Circuits* (Springer Science & Business Media, Berlin, 2008)

Chapter 4
Simulation-Based Signal Selection

Debapriya Chatterjee and Valeria Bertacco

4.1 Introduction

Shrinking transistor sizes with each new generation of CMOS technology nodes have allowed modern digital integrated chips (ICs) to include more and more logic, and thus have become increasingly complex. At the same time, immense market competition has been rapidly shrinking the time-to-market for new IC products. This phenomenon has greatly impacted the verification flow of digital designs. Traditionally, functional bugs in a design have been identified through the extensive use of simulation and formal verification techniques in the pre-silicon phase. However, shorter design cycles, growing design complexity, and limited validation capabilities have lead to situations where functional bugs may be missed or overlooked, particularly when they manifest deep in the design's state space. Hence, initial silicon prototypes often contain functional logic bugs, even if they clear manufacturing testing. As a direct consequence, post-silicon debug, which involves the detection and investigation of these hard-to-find bugs, has become a crucial component of business success.

The fundamental challenge in post-silicon debug lies in the very limited visibility of internal design signals. Physical probing tools [11] have limited capabilities and are primarily used for I/O ports. Reuse of *design for test* (DFT) circuit structures, such as internal scan chains, for post-silicon debug has been widely adopted in the industry [15]. Though scan chains can capture all or a subset of internal state elements, and thus increase signal observability, it may take several thousand clock cycles to dump

D. Chatterjee (✉)
IBM Systems, 12430 Metric Boulevard, Apt 3307, Austin, TX 78758, USA
e-mail: dchatte@us.ibm.com

V. Bertacco
University of Michigan, 4645 BBB, 2260 Hayward St, Ann Arbor, MI 48109-2121, USA
e-mail: valeria@umich.edu

P. Mishra and F. Farahmandi (eds.), *Post-Silicon Validation and Debug*,
https://doi.org/10.1007/978-3-319-98116-1_4

out one state snapshot and, in most cases, the circuit's execution must be suspended until the completion of this process.

To facilitate post-silicon debug, *design for debug* (DFD) structures such as embedded logic analyzers (ELAs), have been proposed [1] and have found widespread use in both the field-programmable and the integrated circuit logic industry [2, 3, 16]. The main component in an ELA is the trace buffer. A trace buffer consists of a mix of trigger units and sampling units. Programmable trigger units are used to specify an event that should trigger the ELA. The trigger can either prompt or stop the logging of signals. Sampling units are used to log the values of a small set of signals (trace signals) over a specified number of clock cycles into trace buffers. The number of signals traced is known as the *width* of the trace buffer, while the length of the tracing interval is called *depth*. Trace buffers are implemented with on-chip embedded memories [16] and data acquisition can be performed during normal chip operation by setting up the relevant trigger event. Subsequently, the sampled data is transferred off-chip for postprocessing debug analysis. Note that DFD structures must maintain a low area overhead profile, since they do not provide added features to a design. As a result, only a very small number of signals can be traced, in comparison to those available in the design.

For trace buffers to be effective, designers must carefully select for tracing those signals that yield the most debug information. Through a judicious choice of trace signals, one can even reconstruct data for state elements that are not traced. As an example, for microprocessor designs, it is common practice to trace pipeline control signals, so that the values of other data registers can be inferred during post-analysis. This approach is not effective for a generic circuit, however, because it leverages architectural knowledge of the design. Hence, the need for generalized solutions in this domain is growing.

Even though the inferred information does not necessarily boost the number of design errors identified, it still increases internal signal visibility and has the potential of providing valuable debugging information. Because bugs tend to occur in unexpected design regions and configurations, it is not always possible to predict the most critical signals to trace. Ideally, we would like a mechanism, which allows to reconstruct almost all internal signals from the tracing of just a handful of signals, so as to offer pre-silicon quality observability during post-silicon debug.

Recent research addressing these challenges [7] has shown that many un-traced signals and state elements can be inferred from a small number of traced state elements by forward and backward implication, even in arbitrary logic. Ko and Nicolici [7] were the first to propose an automated trace signal selection method that attempts to maximize the number of non-traced states restored from a given number of traced state elements. The quality of the trace signal selection was quantified by the state restoration ratio (SRR), that is, the ratio of the number of state values restored over the state values traced, for a given time interval. This measure has been adopted by subsequent researchers for quantitative comparison. Further research [4, 10, 12] has proposed several automated trace signal selection methods based on a range of different heuristics, which estimate the state restoration capabilities of a group of signals. These solutions share a common structure: (i) a metric to estimate the state restoration capability of a set of state elements and (ii) the use of the metric in the process of selecting which set of signals should be part of the traced set. The majority of solutions to date have proposed a greedy selection process.

In this chapter, we explore whether the estimate of the state restoration capability of a set of signals can be gathered by actually simulating the restoration process on the circuit over a small number of cycles, and measuring the corresponding restoration ratio. In this process, we also observe that greedy selection processes suffer from diminishing returns: as the number of trace signals increases, the incremental restoration amount diminishes. The chapter presents a solution to remedy this trend through an elimination-based process that replaces previous greedy solutions.

4.2 Related Work

Automatic trace buffer signal selection is a relatively new research area. One of the earliest solutions for this problem [6] considered only the reconstruction of data at the combinational logic nodes of the circuit. Ko and Nicolici [7] introduced an efficient algorithm to perform state restoration as a post-analysis process on recorded trace buffer data. This work formally defined the state restoration ratio (SRR) as the ratio of the number of restored state elements to the number of traced state elements. They also introduced the first trace signal selection algorithm, striving to maximize the amount of restored state. Subsequent research in this area has strived for to improve the SRR with automatic signal selection solutions [4, 10, 12].

As mentioned earlier, these solutions share a common structure, with a metric to estimate the restoration capacity of a certain set of state elements and a selection algorithm to decide which ones to trace, guided by the estimator metric. These solutions primarily differ in the way estimation is performed. There are three variants of estimation metrics explored in recent literature: (a) probability based (b) simulation-based and (c) hybrid approaches that combine both aspects. In this chapter, we will focus on simulation-based metrics.

Both [7, 10] leverage a probabilistic metric: the steady-state probability of the value a flip-flop's output is estimated assuming uniform random distribution of 0 and 1 logic values at the primary inputs. Given these assumptions, and using the knowledge of the traced signal values, a probabilistic model of the *visibility* of 0 and 1 values at the other circuit nodes can be generated. This probabilistic model leverages the circuit topology and logic functionality of individual gates, and the estimation process performs forward and backward propagation of probability values across logic gates. The final state restoration capacity estimate is then expressed as a sum of the predicted visibility of 0 and 1 values at the state elements of the circuit. The probabilistic model presented in [7] lacks theoretical basis and it is then improved on in [10]. In contrast, [4] considers only the restoration probability along paths connecting flip-flops. The probability that a flip-flop output value controls the input value of another flip-flop is computed and called *direct restorability* of the corresponding path. The selection algorithm grows a region of flip-flops in a greedy fashion based on this metric, while an adjustment mechanism accounts for flip-flops that are already selected in the region and updates the path's probabilities accordingly. An earlier chapter explains the details of probability-based estimation. In Sect. 4.3 of this chapter, we analyze the accuracy of such probabilistic metrics. We explore why probability-based estimates often differ from the actual restoration, as obtained by logic simulation of the system. Indeed, based on this analysis, we then progress to evaluate the accuracy of a restoration ratio estimate obtained by gathering simulation data over a short period of time.

As a logical successor to probability and simulation-based metrics, hybrid methods that combine probability and simulation-based estimation have appeared in the literature as well. Some examples of this approach include [5, 9]. These methods combine the best aspects of both families of estimation metrics: they use a probability-based metric (inaccurate but fast to compute) to quickly arrive at a short list of traced elements, which is further refined using simulation-based analysis (accurate but slower to compute). A later chapter in this manuscript covers one such hybrid solution.

A different line of research [13, 17] suggests that not all state elements or signals are equally relevant for debugging purposes. Hence, instead of striving to maximize the state restoration ratio, the authors of those works focus on maximizing restorability of a specified subset of signals, while minimizing the impact to other flip-flops. In particular, the algorithm in [13] uses a probabilistic estimation metric analogous to [10], and follows a pareto-optimal selection process. It is possible to extend any metric-based selection methods to this problem by assigning higher weight coefficients to this specified subset during the selection process.

4.3 Background and Motivation

An ideal post-silicon debugging solution would enable pre-silicon quality observability, i.e., every signal value is observable at each cycle, with little design effort and area overhead. However, a practical goal is to attain partial observability by tracing a small set of signals and use them to find the root cause of the bug. Several solutions have suggested automatic signal selection algorithms to determine which state elements, if traced, allows maximum restoration. An intuitive measure for evaluating restoration quality is the state restoration ratio, defined as $SRR = \frac{N_{traced}+N_{restored}}{N_{traced}}$, where N_{traced} is the number of traced state elements and $N_{restored}$ is the number of restored ones during the time window dictated by the trace buffer's depth. Automated signal selection strives to maximize SRR.

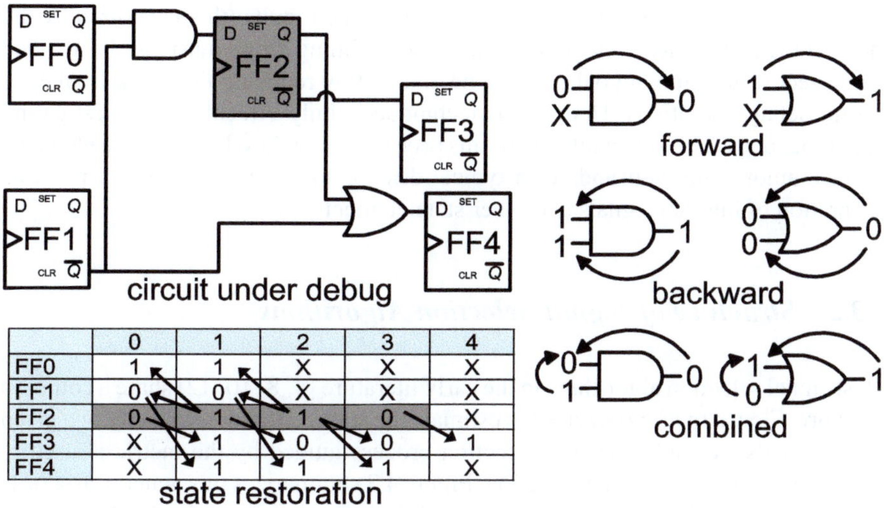

Fig. 4.1 Example of state restoration process. The circuit shown at the top left is the circuit under debug, with flip-flop FF2 traced for 4 clock cycles (shown in gray). The table below lists the values of all flip-flops, whether traced, restored, or unknown (X). Forward inference and backward justification through the logic gates (shown with forward and backward arrows in the table) allows to restore several flip-flop values that were not traced. The elementary rules of forward inference, backward justification and combined inference are shown for two types of logic gates on the right side of the figure

4.3.1 State Restoration Process

The state restoration process relies on the property that if a controlling value is known for at least one input of a logic gate, the output can be inferred without the knowledge of other inputs. This property is used for forward inference of signal values in case of partial knowledge. Similarly, if a noncontrolled value is observed at the output of a gate, all inputs can be inferred to hold the noncontrolling value for that type of gate, enabling backward justification. Combined inferences leveraging knowledge of both inputs and output are also possible (see Fig. 4.1. Repeated application of these simple operations for all gates of a circuit leads to value reconstruction for state elements beside those traced. This process is used in the post-analysis of the data obtained from trace buffers to restore non-traced signals.

Figure 4.1 illustrates this process with an example inspired by [7]. In this example, the flip-flop FF2 is traced over four clock cycles; additional values at other flip-flops can be inferred as shown in the table in the lower part of the figure. In this particular example, the state restoration ratio is computed as: $SRR = 15/4 = 3.75$ ($N_{traced} = 4$, $N_{restored} = 11$). An efficient bit-parallel algorithm to perform this restoration process is introduced in [7] and has been used in subsequent research. It is important to note that the forward inference and backward justification operations are correct only if the logic functions of the gates in the circuit conform to the structural netlist, with no stuck-at-faults or other such faults (this is assured by the fact that the IC has cleared manufacturing tests). Timing errors must also be avoided for correct restoration, a goal that can be attained by reducing the clock frequency during debug operations. Hence, this technique is only effective for investigating functional bugs. The key challenge of this process is how to select the state elements to trace among the thousands of a typical design, so to achieve the best possible restoration of internal signals and other state elements.

4.3.2 Structure of Signal Selection Algorithms

Most signal selection algorithms in the early literature [4, 8, 10, 12] share a common structure. First, a metric is devised to estimate the state restoration capacity of a given set of signals; second, a greedy selection process guided by the metric is used to converge to a locally optimal selection. Figure 4.2 summarizes this general structure.

For this general algorithm to be successful, the capacity metric should have the following properties: (i) it should be proportional to the actual average SRR that can be obtained with the given set of signals over many runs, (ii) it should be as computationally inexpensive as possible, since several such computations will be needed in the final selection process. The first criterion is especially important for the greedy selection process to be successful, since it guides the successive greedy choices towards the optimal subset. The greedy selection process starts off with the signal that promises the maximum capacity, and then expand the set, one signal at a time, by evaluating the restoration capacity of all possible candidate sets generated

```
Input: circuit, width of trace buffer w,
restoration capacity metric fC(...)
Output: selected flip-flop set T

while (|T| < w)  {
  maximum visibility maxV = 0
  for (each unselected flip-flop s in circuit){
    T = T ∪ {s}
    visibility V = fC(T)
    T = T \ {s}
    if(V > maxV){
      selected = s
      maxV = V }
  }
  T = T ∪ {selected}
}
```

Fig. 4.2 General structure of greedy selection algorithms. A set T is being expanded one state element at a time by evaluating the gains in visibility for obtained by including that element over others. The state element providing the maximum gain is then added to the traced set permanently

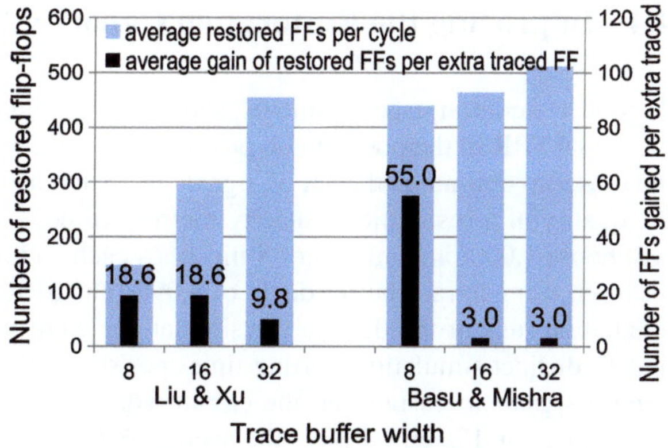

Fig. 4.3 Diminishing returns in restored flip-flops when increasing trace buffer size are observed for two prior greedy algorithm solutions. The plots correspond to circuit s38417

by including one more signal. Unfortunately, as we discuss below, this approach leads to diminishing restoration returns.

4.3.3 *The Problem of Diminishing Returns with Greedy Selection*

The greedy selection process suffers from another critical problem with regards to the quality of the chosen signals. Figure 4.3 plots the average number of restored flip-flops per cycle for 3 different trace buffer widths (8, 16 and 32) for the ISCAS89 benchmark circuit s38417. Alongside, we also plot the average number of restored flip-flops attained when adding each new traced flip-flop (FF). The plots correspond to the data reported by Liu and Xu [10] and by Basu and Mishra [4].

Note that in the result obtained by Liu and Xu, an increase of the observed FFs from 8 to 16 corresponds to an increase in the number of restored FFs from 149 to 298, leading to a $(298 - 149)/(16 - 8) = 18.62$ gain per added new trace signal, as shown by the inner dark bar corresponding to width 16. However, when the traced signals increase from 16 to 32, the rate of gain is much lower (9.8). This effect is even more pronounced in the results by Basu and Mishra [4], where a much better initial restoration is obtained, but as the number of trace signals is doubled, the improvement is minute. This behavior results from the inaccuracy of the estimation metric, as well as the very nature of the greedy selection. Indeed, the restoration obtained by greedy selection algorithm plateaus when a large number of flip-flops are traced. This trend is due to the selection of $2n$ flip-flops being constrained by the prior choice of the first n flip-flops. In contrast, the best possible set of $2n$ flip-flops might not even include some of the first n flip-flops. Hence, we propose an alternative approach that applies greedy selections backward: i.e., we start off with the set of all FFs, and then we iteratively reduce this set until we obtain a set of the desired cardinality. In the following section, we outline an algorithm based on this approach.

4.4 Improving the Restoration Capacity Metric

A good restoration capacity metric should have a high degree of correlation with the actual SRR in the post-silicon post-analysis, since the more accurate the metric, the closer to optimal is the set of signals that such metric will identify. To evaluate the quality of a restoration capacity metric, we devised the following experiment: we choose 1,000 random sets of 8 flip-flops each and measured the average SRR in each set, using a trace buffer depth of 4,096. The SRR was computed based on 100 simulation runs of the design; the simulations were rooted with 10 random seeds, and 10 distinct simulation starting times per seed. We also asserted the appropriate control signals to ensure that the circuit would operate in its normal mode during the simulation. Figure 4.4 plots the average SRR versus the estimated one obtained with the Liu and Xu's restoration capacity estimation metric. Data is shown using a scatter plot to highlight the correlation of the metric with the actual measured SRR.

As can be noted from the figure, although the metric has positive correlation with measured SRR, the extent of the correlation is poor, as indicated by the small correlation coefficient (R). The fundamental reason behind this pattern lies in the lossy information compaction of probability-based restorability estimates. For example, consider the two input AND gate of Fig. 4.5, where the only knowledge available is that the restoration probability of the value 1 (V_1) at the inputs is 0.5. A probability-based estimation scheme will infer the restoration probability of the value 1 at the output to be $0.5 \times 0.5 = 0.25$. However, if the actual restored values for the two inputs over 6 successive clock cycles are $1X1X1X$ and $X1X1X1$, compatible with

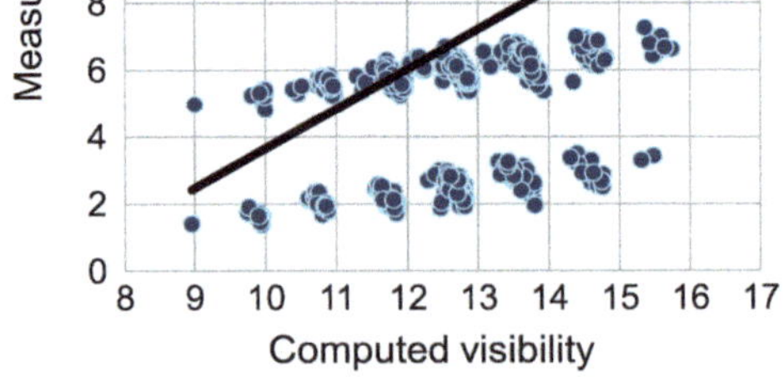

Fig. 4.4 Correlation of the Liu and Xu probability-based restoration capacity metric with measured SRR for s35932. The metric has a positive but poor correlation with measured SRR. We also report a linear regression fit of the data and the square of the correlation coefficient. Data points in the lower right corner represents selection of flip-flops that have a high estimated value of state visibility but rather poor measured SRR. The inclusion of these flip-flops can drive the greedy selection algorithm to sub-optimal selections

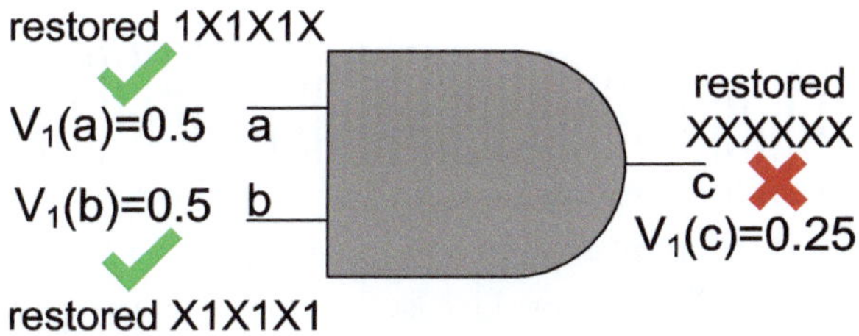

Fig. 4.5 Example of a misleading restoration probability estimate. No restoration is achieved over 6 cycles while probability-based estimates suggest a 25% restoration ratio

the estimated restoration probability, we cannot restore the output for any cycle. This limitation is common to all probability-based estimates and it results from the compaction of information over several cycles into a single measure. It could be avoided if we had a conditional probability distribution of each signal's restorability, given the value of other signals, an infeasible level of accuracy in practice. In conclusion, the example shows that restoration probability estimates are not reliable, and often do not correlate well with actual restoration.

Keeping the ideal characteristics of a restoration capacity metric in mind, we investigated whether a new metric could be derived from the simulation data itself. Indeed, a better estimate of SRR for a given group of signals and trace depth can be obtained by performing a large number of simulations while randomizing input values and the starting point for tracing; then performing the restoration process for the circuit; and finally averaging the SRR values from each individual simulation. This corresponds to estimating the SRR for the group of trace signals by Monte Carlo simulation, and unfortunately it is a very compute intensive process for typical trace buffer sizes and depths. In contrast, as we indicated earlier, individual restoration capacity estimations should be kept fairly simple, due to the large number of estimations required for a selection to converge to a final set.

A key insight in our search for an accurate SRR estimator is that the estimate of state restoration capacity metric does not need to match the SRR exactly, but only needs to be highly correlated with it, so that it guides us to the same group of traced signals. A common method of reducing effort in simulation-based estimations is to perform several short simulations and average their outcomes. Specifically, we could use a shorter trace buffer depth. This observation led us to a study of SRR sensitivity to trace buffer depth. The results for one selection of 8 flip-flops for circuit s35932 circuit are shown in Fig. 4.6. In the figure, we plot the SRR estimate computed over several trace buffer depths, three different random starting points of tracing and three different random input value selections per starting point.

The main conclusion that can be derived from the study is that the SRR obtained from a certain group of traced signals is fairly insensitive to the trace buffer depth: indeed it can be noticed from the figure that the SRR variation is negligible beyond a trace buffer size of 64. We observed a similar behavior for all other ISCAS circuits, as well as when using a larger set of random samples. Intuitive reasoning suggests

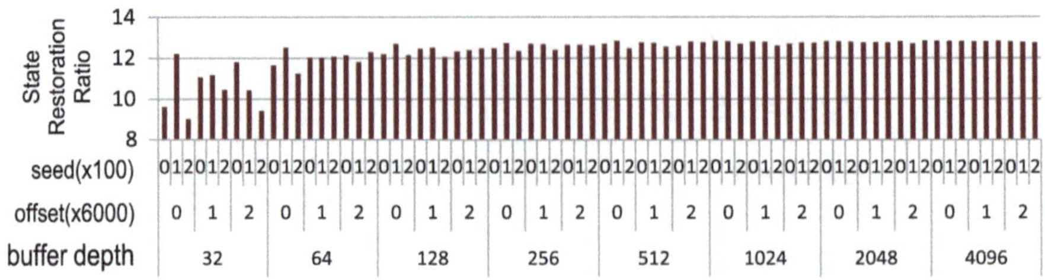

Fig. 4.6 Impact of trace buffer size on SRR. Analysis on s35932 over 3 random starting points of tracing and 3 random sets of input values per starting point indicates that SRR for a fixed set of signals is fairly insensitive to trace buffer sizes beyond 64

SRR is relatively insensitive to trace buffer depth beyond a certain size, since most circuits tend to stay in a small fraction of possible states, and each occurrence of such states has similar restoration behavior. We conclude then that SRR measurements over simulated restorations on small trace buffer sizes (~64) provide an accurate estimation of restoration capacity.

To further validate our hypothesis that short trace buffer sizes are sufficient for accurate SRR estimation we performed the previous correlation study using our new estimation metric, as shown in Fig. 4.7. The SRR estimate is computed using a fast mock simulation with a trace buffer size of 64 and only one random set of inputs and starting time for tracing. This is the setup for estimations that we used in the rest of the chapter. We conclude that the SRR measurements over simulated restorations on small trace buffer sizes (~64) provide a reliable estimate of restoration capacity. The plots of Fig. 4.7, obtained for s38417 and s35932, clearly indicate a very high

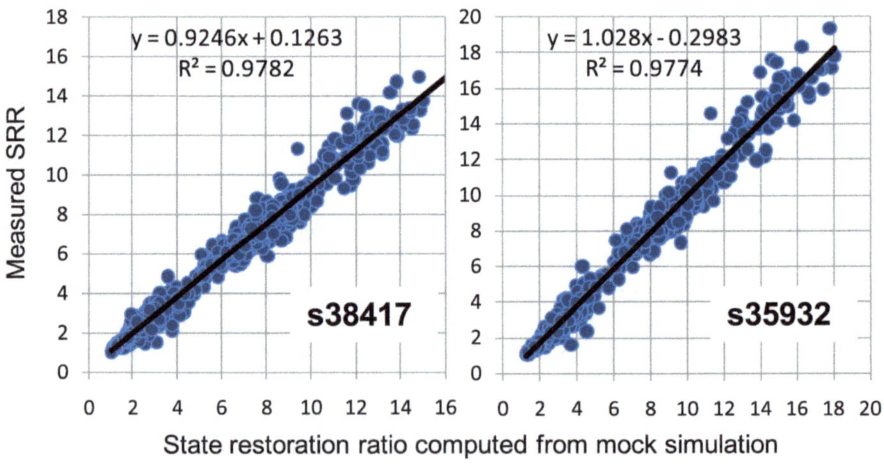

Fig. 4.7 Correlation of our simulation-based restoration capacity metric with observed SRR using a mock simulation trace depth of 64 for s38417 and s35932. The proposed metric bears strong positive correlation with the observed SRR as indicated by the high value of the correlation coefficient

correlation between our estimation metric and the observed SRR. The simulation-based capacity estimation evidently shows an extremely high degree of linear correlation with the observed SRR. Similarly, strong correlations were observed for other ISCAS circuits as well. These results confirm the viability of an SRR estimator based on mock simulation of restoration over a small trace buffer size. We expect that greater buffer sizes and averaging over more simulation with different random input values and starting points would further improve the accuracy of the estimate, although with smaller returns.

4.5 Selection Algorithm Design

The problem of selecting an optimal set of flip-flops can be thought of as the problem of retaining the maximum amount of information in the unrolled circuit graph. In our algorithm, we start off including all flip-flops in the circuit, which will restore almost all signals and states, and then we try to reduce this set by removing flip-flops incrementally. This process will ensure that early selections do not limit the quality of the final pool, as discussed in Sect. 4.3.3. The flip-flops that contribute least to restoration of others should be eliminated first. The process terminates when we are left with a set of flip-flops of the desired cardinality. During each step of the algorithm, we use the proposed simulation-based estimator to evaluate the restoration capacity of the candidate set of flip-flops. If elimination of two or more candidate flip-flops results in the same restoration estimate, we break the tie by comparing the total number of signals restored. If a tie still exists, then we consider the number of connected flip-flops via a forward or backward path in the circuit graph: flip-flops with fewer connections will get eliminated, if a tie still remains it will be broken by random choice.

Our algorithm is illustrated in Fig. 4.8: the schematic represents each elimination step of the algorithm when operating on a circuit with 5 flip-flops and a target trace

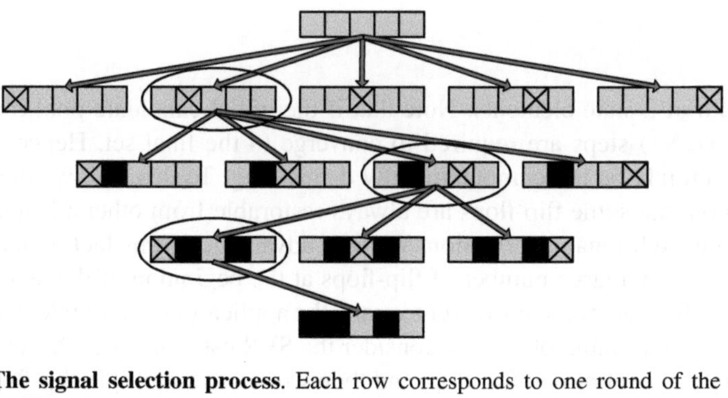

Fig. 4.8 The signal selection process. Each row corresponds to one round of the algorithm; the flip-flop (FF) whose elimination leads to maximum retention of restored states according to the estimation metric is removed in next round. The black squares correspond to FFs previously eliminated, while crosses indicate the FF being evaluated for elimination. In this example, there are a total of 5 FFs and a trace buffer width of 2, so 3 FFs must be eliminated

```
Input: circuit, width of trace buffer w,
mock simulation based SRR estimator fSRR(...)
Output: selected flip-flop set T
Parameter: step-size d, pruning termination parameter PT

while (V > PT) {
  for(each flip-flop s in T){
    T = T \ {s}
    visibility V = fSRR(T) × |T|
    restoration capacity without s RCW[s] = V
    T = T ∪ {s}
  }
  T = T−{s | RCW[s] is within top d values }
  V = fSRR(T) × |T|
} //end of pruning

while (|T| > w) {
  maximum visibility maxV = 0
  for each s ∈ T{
    T = T \ {s}
    visibility V = fSRR(T) × |T|
    T = T ∪ {s}
    if(V > maxV){
      selected = s
      maxV = V
    }
  }
  T = T \ {selected}
}
```

Fig. 4.9 Pseudo-code for our proposed algorithm. A set T of traced state elements is being constructed by one elimination at a time of least restoration information carrying state element in it. Pruning is used to eliminate multiple such elements in one iteration to quickly arrive at a short list

buffer width of 2 state elements. Note that if the initial candidate pool includes N flip-flops, $O(N^2)$ steps are required to converge to the final set. Hence, for large circuits this might be too computationally demanding. To this end, we noticed that it is common that some flip-flops are always restorable from others; hence they do not carry any additional information. We take advantage of this fact by using a fast pruning phase on a large number of flip-flops at the beginning of the algorithm, so as to reduce this size of the initial set to make the application of an $O(N^2)$ algorithm feasible. For the pruning phase, we consider the SRR estimate of each candidate set obtained by removal of one flip-flop, and then we remove multiple flip-flops in one single step, all characterized by a small contribution to restoration capacity. As shown in the pseudo-code of Fig. 4.9, we consider all possible eliminations in sorted order of SRR estimate values (stored in $RCW[]$ vector). The flip-flops whose elimination lead to the top SRR estimate values are selected to be in the elimination set. The size of the set is a parameter called step size d, set to 50 in our experiments. To limit the extent to which this coarse grain pruning is applied, we specify a pruning termination parameter PT such that, if the average number of restored flip-flops in the mock simulation drops below PT, the coarse grain pruning phase ends. This parameter establishes a trade-off between quality of selection and computational cost of the algorithm. In our experiments, we set $PT = 95\%$.

4.6 Experimental Results

We evaluated the trace signal selection quality of the proposed algorithm by comparing the SRR obtained on six ISCAS89 benchmark circuits, against that obtained by several probability-based metric solutions [4, 8, 10, 12]. In addition, we also experimented with three control path blocks taken from the OpenSparc processor core design [14], synthesized from their RTL description. Relevant circuit characteristics are presented in Table 4.1. The benchmarks were resynthesized using Synopsys Design Compiler targeting the GTECH gate library to have the same level of optimizations as performed on industrial netlists. The synthesis tool performs automatic removal of certain redundant flip-flops in the designs under evaluation.

We used an X-simulator that we developed in house to compute the simulation-based estimation metric and to measure the final SRR obtained by applying our proposed algorithm. The X-simulator takes a design along with traced values, and

Table 4.1 Benchmark circuits used to evaluate our signal selection algorithm contains a mix of ISCAS89 and OpenSPARC circuit blocks

Circuit	Flip-flops before synthesis	Flip-flops after synthesis	Gates after synthesis
s5378	179	164	1,058
s9234	211	145	920
s15850	534	524	3,619
s38584	1,426	1,426	12,560
s38417	1,636	1,564	10,564
s35932	1,728	1,728	4,981
Sparc MMU	–	262	1,977
Sparc EXU	–	327	2,168
Sparc IFU	–	2,755	19,912

it restores all possible values of non-traced signals and states. The 3-input or larger gates are internally de-composed into elementary 2-input gates in the X-simulator for efficient computation, a transformation that has no other consequence since the trace signals are only flip-flop values. We implemented our X-simulator using the efficient event-driven bit-parallel propagation technique described in [8]. All the experiments were run on a quad core Intel processor running at 2.4 GHz. The width of the bit-parallel operations in the restoration process was extended to 64 bits from 32 bits described in [8], to better utilize the 64-bit integer arithmetic capabilities of the processor. Further extension to wider vector operations harnessing the vector arithmetic capabilities of modern processors are also possible. In our specific case, this improvement led to much better performance in the estimation phases, since the trace buffer depth was also chosen as 64 cycles.

Each design was forced to operate in its normal functional mode during tracing by forcing fixed values at the relevant control inputs, including reset, while assigning random values to all other inputs. This setup is referred as "deterministic random" in literature [4, 8]. This restriction at the inputs is critical for evaluation of trace signal selection quality. If control inputs are allowed to toggle, the circuit might intermittently enter the reset state and the reset signal itself might be traced, leading to a large amount of state restoration. However, during debug, this scenario is unlikely to happen and the circuit will operate in the functional mode most of the time, so the state restoration ratio obtained when control signals are allowed to toggle is not representative of actual restoration capacity of the trace signals. This issue has been pointed out in [8, 10].

4.6.1 Restoration Quality

Table 4.2 compares the state restoration ratio obtained by several other probability-based metric solutions against our proposed technique. As in [4, 10], the trace buffer widths used in the experiments are 8, 16 and 32, while its depth is kept at 4,096 cycles. The corresponding SRR for each solution (wherever known) is reported. The percentage improvement of SRR obtained by our proposed algorithm over the best reported value is indicated in last column. Each restoration ratio is averaged over 100 simulations, using 10 different random seeds (to generate random values at non-control primary inputs), and 10 different starting points past from the initial reset state, per seed. For certain buffer sizes, especially in smaller circuits, the SRR obtained by our solution is not better than some of the competing solutions. This is primarily due to the fact that our optimized ISCAS89 circuits have fewer flip-flops. Hence, even though our technique actually restores a higher fraction of the flip-flops, the reported SRR of other solutions has the advantage of including the restoration of redundant flip-flops. For example, for s9234 at a buffer size of 32, our algorithm restores $4.18 \times 32 = 134$ (approx.) flip-flops on average, per cycle,

Table 4.2 Comparison of state restoration ratio with no input knowledge. The table compares our solution against some of the probability-based metric solutions, computing restoration only based on traced state elements. The last column reports change over the best value obtained by competing solutions

Circuit	Trace width	Ko and Nicolici [8]	Liu and Xu [10]	Basu and Mishra [4]	Proposed solution	Improv. (%) over best
s5378	8	–	*14.67*	–	*13.24*	*−9.75*
	16	–	*8.99*	–	*7.83*	*−12.93*
	32	–	*4.72*	–	*4.89*	**+3.60**
s9234	8	–	*4.76*	–	*10.68*	**+24.36**
	16	–	*7.18*	–	*7.16*	*−0.27*
	32	–	*4.67*	–	*4.18*	*−10.49*
s15850	8	–	19.93	–	39.54	**+98.39**
	16	–	24.22	–	24.85	**+2.60**
	32	–	13.30	–	13.60	**+2.25**
s38584	8	19.00	19.23	78.00	84.10	**+7.82**
	16	10.56	13.96	40.00	47.04	**+17.60**
	32	6.32	8.68	20.00	26.97	**+34.85**
s38417	8	19.62	18.63	55.00	45.21	−17.80
	16	11.22	18.62	29.00	30.77	**+6.10**
	32	6.73	14.20	16.00	20.25	**+26.56**
s35932	8	41.45	64.00	95.00	96.12	**+1.17**
	16	39.31	38.13	60.00	67.45	**+12.41**
	32	24.76	21.06	35.00	43.23	**+23.51**

Table 4.3 SRR for OpenSparc blocks, using only traced state elements. These blocks are representative of typical microprocessor design blocks

Circuit	Trace width		
	8	16	32
Sparc MMU	12.22	8.03	4.67
Sparc EXU	4.53	3.46	4.02
Sparc IFU	99.10	62.01	35.67

out of total 145, which is 92% of all flip-flops, whereas the best reported solution only restores $4.67 \times 32 = 149$(approx.) out of 211 flip-flops, corresponding to 70%. For larger circuits, which better represent practical post-silicon debug situations, our solution achieves an improvement of up to 34.85% (for s38584) in the SRR.

We report the SRR obtained for the OpenSparc blocks in Table 4.3. The primary inputs were driven by the trace recorded during the execution of a functional test

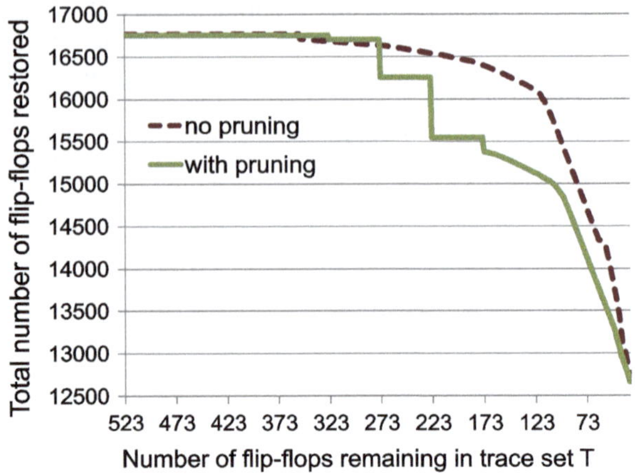

Fig. 4.10 The effect of the pruning phase in the trace signal selection algorithm for s15850

from the OpenSparc regression suite. The trace buffer depth is kept at 4,096 cycles for these designs as well.

4.6.2 *Effect of Pruning*

We studied the effect of the pruning optimization (discussed in Sect. 4.5) in our elimination-based algorithm. The effect of pruning is shown in Fig. 4.10. This data corresponds to execution of the proposed algorithm for circuit s15850, when the $f_{SRR}()$ metric is based on a simulation with a trace buffer depth of 32 (instead of the usual 64, for purposes of visible fine granularity), and a trace buffer width also of 32. Hence, the algorithm terminates when the traced set reaches 32. A total of $524 \times 32 = 16,768$ flip-flop values (s15850 has 524 flip-flops, refer Table 4.1) are present in the simulation window for the estimator metric. The y-axis plots the value of $f_{SRR}(T) \times |T| \times 32$ during each iteration in the execution of our signal selection algorithm. Note that the no-pruning line is smooth as only one flip-flop is removed per iteration, and the total number of restored flip-flops in the mock simulation gradually decreases. On the other hand, pruning uses a step size(d) of 50 flip-flops; hence, during the pruning phase, the total number of restored flip-flops drops as a step function. In this example pruning termination (PT) was set at 93% i.e. $16,768 \times 0.93 = 15,594$, a value by which the set is reduced to a size of approximately 200. Note that the quality of pruning is only slightly worse than the exact version (the with-pruning line ends slightly lower than the no-pruning line). Thus, pruning trades-off some accuracy for faster execution.

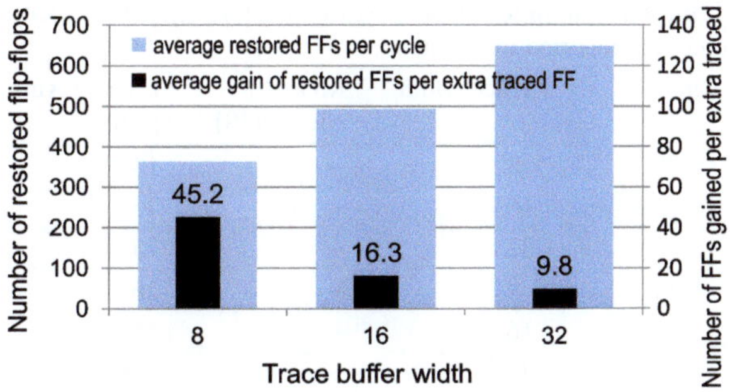

Fig. 4.11 Restored flip-flops versus trace buffer size for circuit s38417. A moderately steady rate of increase of the number of restored flip-flops with increasing trace buffer size is observed for our proposed solution

4.6.3 Restoration Gain on Additional Traced Signals

Diminishing gain with additional traced flip-flops was pointed out as a shortcoming of the greedy algorithms in Sect. 4.3.3. Our proposed algorithm alleviates this issue to a large extent. Figure 4.11 plots the same information as Fig. 4.3 but using our algorithm. It can be noticed that we restore on average more flip-flops than other solutions for buffer sizes of 16 and 32. Moreover, far more steady gain in the number of restored flip-flops per additional traced signal is observed, compared to Basu and Mishra [4], the best probability-based metric solution so far in terms of total restoration. Similar trends are observed for other benchmarks as well.

4.6.4 Algorithm Execution Performance

Trace signal selection is performed only once during the design phase of the circuit blocks to be included in the signal list for the ELA. Hence, the run-time of the selection algorithms is less important than the quality of the selected signals. However, if an inordinate amount of time is needed for even moderately sized circuit blocks, performance would be an issue. In our algorithm, the pruning phase was designed specifically for this reason. A comparison of the execution time of few other solutions in literature and our solution is presented in Table 4.4. Note that, the execution performance of the proposed algorithm is often worse for small designs, this is due to the large number of simulations needed in our algorithm. However, these simulations are for computation of the estimation metric, and they are independent of each other during each iteration of the selection algorithm. A possible way to improve the algorithm's performance, if necessary, is to further leverage the vector capabilities of general purpose processors or pattern parallelism of GPU platforms, where the same execution is applied on different data sets.

Table 4.4 Comparison of execution performance for the algorithms considered. All execution times are reported in seconds

Circuit	Trace width	Ko and Nicolici [8]	Liu and Xu [10]	Basu and Mishra [4]	Proposed solution
s5378	8	–	14	–	656
	16	–	36	–	634
	32	–	75	–	600
s9234	8	–	26	–	456
	16	–	75	–	441
	32	–	148	–	433
s15850	8	–	298	–	3,877
	16	–	764	–	3,823
	32	–	1,656	–	3,781
s38584	8	34,440	388	1,200	18,143
	16	73,500	802	2,600	18,091
	32	149,580	2,826	5,500	18,003
s38417	8	28,200	2,319	2,200	24,943
	16	69,060	5,285	4,500	24,819
	32	149,940	11,732	9,100	24,734
s35932	8	31,440	1,407	2,200	19,857
	16	68,700	5,251	4,400	19,832
	32	142,800	10,496	8,900	19,801

4.7 Conclusion

In this chapter, we have presented a simulation-based metric-guided trace signal selection algorithm that strives to maximize state restoration ratio. We introduced a mock simulation based restoration capacity metric which provides more accurate guidance to the selection algorithm than only probability-based metrics. It also achieves better restoration trends per additional traced signal while restoring a higher number of states on average, compared to greedy incremental set of trace signal construction algorithms.

References

1. M. Abramovici, P. Bradley, K. Dwarakanath, P. Levin, G. Memmi, D. Miller, A reconfigurable design-for-debug infrastructure for SoCs, in *Proceedings of the DAC* (ACM, New York, 2006), pp. 7–12
2. Altera Verification Tool: SignalTap II Embedded Logic Analyzer (2006), http://www.altera.com/products/software/products/quartus2/verification/signaltap2/sig-index.html

3. ARM limited: Embedded Trace Macrocells (2007), http://www.arm.com/products/solutions/ETM.html
4. K. Basu, P. Mishra, Efficient trace signal selection for post silicon validation and debug, in *Proceedings of the VLSI design* (2011), pp. 352–357
5. K. Basu, P. Mishra, Rats: Restoration-aware trace signal selection for post-silicon validation. IEEE Trans. VLSI Syst. **21**(4), 605–613 (2013)
6. Y.C. Hsu, F. Tsai, W. Jong, Y.T. Chang, Visibility enhancement for silicon debug, in *Proceedings of the DAC* (2006), pp. 13–18
7. H.F. Ko, N. Nicolici, Automated trace signals identification and state restoration for improving observability in post-silicon validation, in *Proceedings of the DATE* (2008), pp. 1298–1303
8. H.F. Ko, N. Nicolici, Algorithms for state restoration and trace-signal selection for data acquisition in silicon debug. IEEE Trans. CAD **28**(2), 285–297 (2009)
9. M. Li, A. Davoodi, A hybrid approach for fast and accurate trace signal selection for post-silicon debug. IEEE Trans. CAD **33**(7), 1081–1094 (2014)
10. X. Liu, Q. Xu, Trace signal selection for visibility enhancement in post-silicon validation, in *Proceedings of the DATE* (2009), pp. 1338–1343
11. N. Nataraj, T. Lundquist, K. Shah, Fault localization using time resolved photon emission and STIL waveforms, in *Proceedings of the ITC* (2003), pp. 254–263
12. S. Prabhakar, M. Hsiao, Using non-trivial logic implications for trace buffer-based silicon debug, in *Proceedings of the ATS* (2009), pp. 131–136
13. H. Shojaei, A. Davoodi, Trace signal selection to enhance timing and logic visibility in post-silicon validation, in *Proceedings of the ICCAD* (2010), pp. 168–172
14. Sun Microsystems OpenSPARC, http://www.opensparc.net/
15. B. Vermeulen, T. Waayers, S. Bakker, IEEE 1149.1-compliant access architecture for multiple core debug on digital system chips, in *Proceedings of the ITC* (2002), pp. 55–63
16. Xilinx Verification Tool: ChipScope Pro (2006), http://www.xilinx.com/ise/optional_prod/cspro.html
17. J.S. Yang, N.A. Touba, Automated selection of signals to observe for efficient silicon debug, in *Proceedings of the VTS* (2009), pp. 79–84

Chapter 5
Hybrid Signal Selection

Azadeh Davoodi

5.1 Introduction

Trace buffers are specialized hardware added on-chip that can facilitate the process of diagnosis and debug of Integrated Circuits. Once the chip is fabricated, trace buffers can be used to track the values stored in a limited number of internal state elements, which are otherwise inaccessible, during the actual operation of the chip, and within a specific "capture window" [1]. These internal state elements which are referred to as the trace signals are design-specific and selected prior to fabrication.

During the operation of the chip, firing of a trigger signal starts the capture process from the trace signals. The values of the trace signals within the capture window are then stored in the trace buffer. Due to the limited size of the trace buffer, there is a limitation on the number of trace signals and the size of the capture window, which correspond to bandwidth and depth of the trace buffer, respectively. See Fig. 5.1.

Various automated algorithms have been proposed which can analyze an arbitrary design and select up to a fixed number of trace signals. For a given design, the trace signals should be selected to maximize the observability to the remaining, inaccessible signals on the chip as much as possible. Majority of existing algorithms measure the quality of selected trace signals using a State Restoration Ratio (SRR) metric. We will review the definition of SRR shortly but typically, a higher SRR indicates doing better in recovering the observability of the remaining inaccessible signals inside the chip.

Many algorithms for trace signal selection end up following one of these two alternatives for deciding the signals to be traced: either approximate the value of SRR quickly, typically using some analytical metric, or use a much more accurate but lengthy simulation process to evaluate SRR for a considered set of trace signal candidates. The former (known as metric-based selection) allows more aggressive exploration of design space for searching among the trace signal candidates but optimization decisions are made based on a rough approximation of SRR [2–6]. On the contrary, the latter of algorithms (known as simulation-based selection) are

A. Davoodi (✉)
Department of Electrical and Computer Engineering, University of Wisconsin, Madison, WI 53706, USA
e-mail: adavoodi@wisc.edu

P. Mishra and F. Farahmandi (eds.), *Post-Silicon Validation and Debug*,
https://doi.org/10.1007/978-3-319-98116-1_5

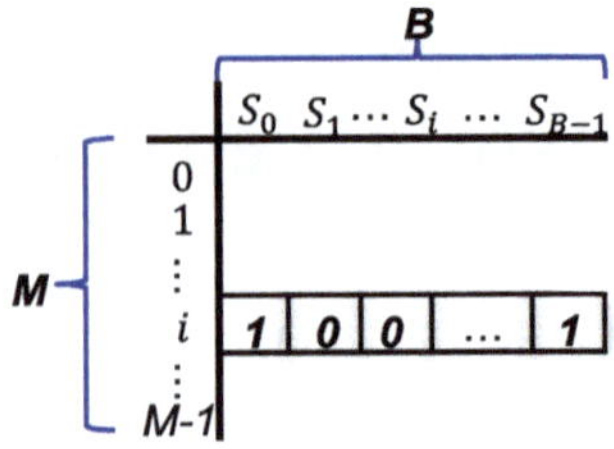

Fig. 5.1 In a trace buffer, the bandwidth (denoted by B) corresponds to the number of traced signals. The size of the capture window corresponds to the depth of the buffer and is denoted by M

naturally constrained by how in-depth they can explore the design space because of the tradeoff with the execution time of the algorithm to generate the trace signals [7, 8]. For example, a typical simulation-based algorithm follows a simple design exploration process which selects the trace signals one at a time in a sequential manner. At each step, the signal that maximizes the SRR for the currently selected subset is greedily picked as the next trace signal. This requires evaluating the SRR for each trace candidate at each step. As a result, the majority of the execution runtime of the algorithm is spent on the SRR evaluation of the trace candidates.

Other algorithms were also been proposed prior to the above two classes of algorithms including those based on Integer Linear Programming [9] and circuit satisfiability [10]. However, these earlier techniques suffered from runtime scalability with an increase in the design size.

In this chapter, we give an overview of the class of *hybrid* trace signal selection algorithms which aim to bridge the gap between the metric-based and simulation-based algorithms. Hybrid algorithms try to achieve the best of both worlds by only occasionally incorporating a simulation-based SRR evaluation, while frequently relying on metric-based evaluation of SRR to explore the search space more aggressively. Specifically, we will discuss the algorithm proposed in [11]. We first give an overview of the basics including elaboration of the restoration process of inaccessible signals using the traced ones which is known as the "X-Simulation" process. Next, we review the formal definition of SRR and definition of the trace signal selection problem considered in this chapter.

5.2 Overview of Basics

Trace buffer is an on-chip buffer of size $B \times M$ as shown in Fig. 5.1, where B is the bandwidth of the buffer and M is its depth and equal to the number of clock cycles when the trace signals are captured. We refer to $B \times M$ as the "capture window".

To restore untraced signals on-chip using the traced ones, a procedure known as X-Simulation is typically used. Figure 5.2 explains it using an example. Here, flip-flop 2 is the only trace signal and is used to restore as many as the remaining signals in a capture window of 4 clock cycles. As expected the values of flip-flop 2 are known at each of the 4 clock cycles because it is traced.

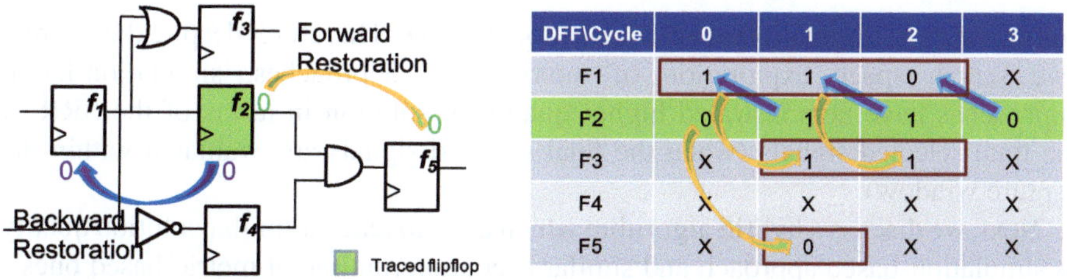

Fig. 5.2 Overview of the signal restoration from trace signals via the X-Simulation process

During the X-Simulation process, forward and backward restoration are used iteratively at each cycle of the capture window to restore as many remaining signals as possible, until no more signals can be restored. In this example, first, the values of flip flop 1 can be restored in cycles 0, 1, 2 via a backward restoration. This restoration is possible, obviously due to the specific circuitry connecting flip flops 1 and 2 to each other (which is a direct connection in our simple example). Next, given these restored signals, the values of flip flops 3 and 5 can be restored in some of the clock cycles. The process terminates at this point because no more signals can be recovered. Overall 7 signals are restored with this capture window.

The quality of the state restoration process is measured by the State Restoration Ratio (SRR) metric which is given by the following equation.

$$SRR = \frac{B \times M + \#restored\ signals}{B \times M} \tag{5.1}$$

For the above example, we have $SRR = \frac{4+6}{4} = 2.5$. For a trace buffer of size $B \times M$, the *trace signal selection problem* aims to select B state elements (alternatively called flip-flops in this chapter), such that the SRR metric is maximized as much as possible.

In simulation-based approach for trace signal selection, trace signals are selected (or alternatively non-traced signals are eliminated) iteratively. At each iteration, X-Simulation is performed to evaluate SRR for each signal and the one with highest SRR is selected for tracing (or the one with lowest SRR is eliminated). This process requires many rounds of X-Simulation to evaluate each candidate signal that has not been selected yet, and needs to be repeated at each iteration. Therefore, to accelerate the process, X-Simulation is performed in a window that is significantly smaller than the capture window. Specifically, each time X-Simulation may be done within an "observation window" of size $B \times N$ with $N << M$. Despite this acceleration, the runtime of simulation-based trace selection is much higher than metric-based techniques, especially for the variation that is based on elimination of signals because B is typically much smaller than the number of signals in the design.

Alternatively, metric-based approach for trace signal selection completely relies on (typically probabilistic) metrics to approximate SRR and select the trace signals. As a result, they are much faster and can explore the search space more aggressively. However approximation of SRR using metrics can be significantly less accurate compared to X-Simulation in an observation window. Therefore, despite the aggressive search space exploration of metric-based approaches, simulation-based approaches have demonstrated higher quality of solution in terms of the SRR of the final selected signals (when the final selected signals are evaluated within the capture window).

Next, we discuss a hybrid algorithm which aims to reach a similar solution quality as simulation-based approach and similar execution runtime of metric-based ones.

5.3 Algorithm for Hybrid Signal Selection

Inspired by simulation-based algorithms, the hybrid approach relies on X-Simulation (but within an observation window much smaller than the capture window), yet this is done significantly fewer times during the course of signal selection. Also inspired by metric-based algorithms, the hybrid approach relies on metrics (though a new set of them designed for a hybrid exploration) to drive the search space exploration more efficiently during the times that X-Simulation is not used.

The goal of the defined metrics in the hybrid approach is to identify the top candidates for tracing at each iteration such that these top candidates are significantly fewer than the total number of candidates for tracing. This allows X-Simulation to be used only for these few top candidates to select the next trace signal in a much more runtime-efficient manner than a purely simulation-based approach.

5.3.1 High-Level Overview

Figure 5.3 gives an overview of the hybrid algorithm in [11]. First, the metrics are initialized and then the trace signals are selected sequentially until B target signals are selected.

Every time a trace signal is selected, the metrics need to be updated to allow more accurate selection of the trace signal in the next iteration. The selection of the trace signal follows one of the two complementary methods.

Method (i) is applied first and is driven by use of metrics *and* few X-Simulations. (In fact, some of the metrics themselves require performing a few number of X-Simulation.) Specifically, method (i) first uses the metrics to narrow down a very small set of promising trace signal candidates in that iteration from the set of all unselected signals up to that point. Then, it performs X-Simulation to evaluate each promising candidate and select the one with maximum SRR to be the next trace signal.

The process of selecting the promising candidates using method (i) ignores a set of flip-flops in the design which are of type "island" flip-flops and will be defined later. The goal of method (ii) is specifically to consider adding an island flip-flop as the next trace signal to ensure all flip-flops are considered for tracing during the course

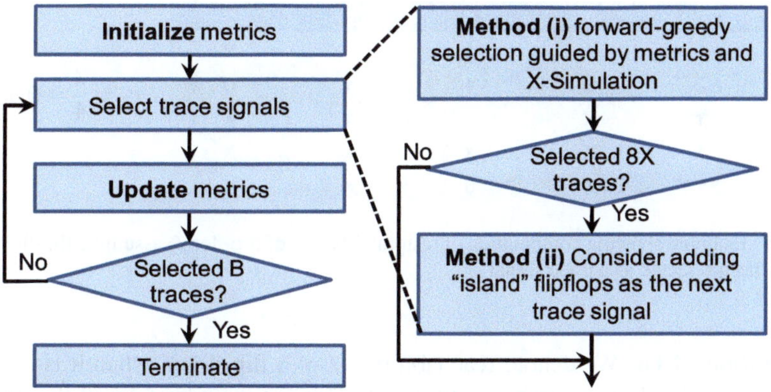

Fig. 5.3 Overview of a hybrid trace signal selection algorithm

Fig. 5.4 Overview and dependency of the metrics used by the hybrid algorithm

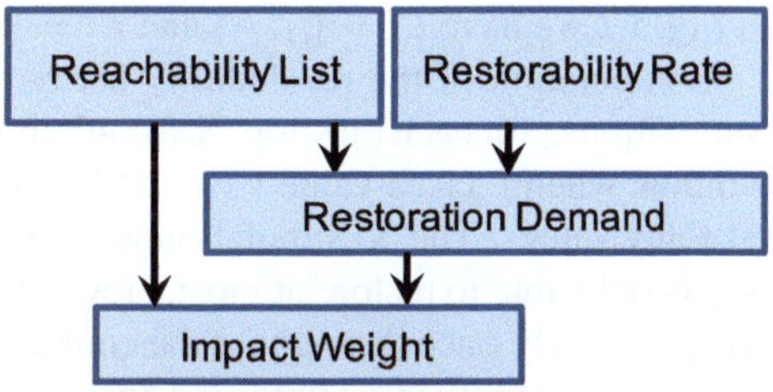

of the algorithm. Therefore, the two methods are complementary to each other. In practice, it is found that method (i) is more effective because the number of island flip-flops is small relative to the non-island ones. Therefore, the hybrid algorithm is tuned to consider selecting one trace signal using method (ii) only after selecting 8 trace signals using method (i) as shown in the figure. The algorithm terminates as soon as the target B signals are selected.

5.3.2 Metrics for Hybrid Evaluation of SRR

The metrics discussed in this section aim to identify a *small* number of top trace signal candidates, relative to the total number of signals that need to be evaluated. This is done at each iteration of the hybrid algorithm and use of metrics allows the identification of top candidates to be done significantly faster than pure use of X-Simulation. Figure 5.4 shows the four metrics and how they depend on each other. This implies that computation and any update of the metrics should be done based on the order imposed by this dependency. The final metric which is used to identify a small number of top candidates at each iteration is the "impact weight". Next, we discuss these metrics one by one in the order shown in Fig. 5.4.

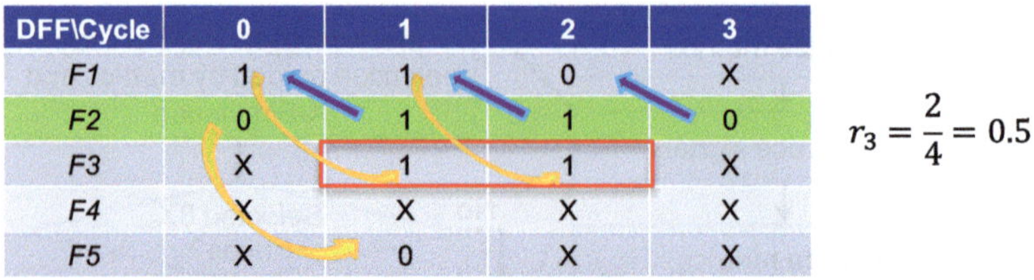

Fig. 5.5 Example showing computation of restorability rate of flip-flop f_3 assuming flip-flop f_2 has already been selected as a trace signal

Reachability List: We denote reachability list of a flip-flop f when it is assigned value v by L_f^v, where v can be set to 0 or 1. This list includes a set of the flipflops which can be directly restored by f when it takes value v. By direct restoration, we mean that it is done without help of any other flip-flop. For example, for the circuit in Fig. 5.2 we have $L_2^0 = \{f_1, f_5\}$ and $L_2^1 = \{f_1, f_3\}$.

Computation of the reachability list requires performing a small number of X-Simulations.; for each flip-flop, X-Simulation is performed within a small observation window when f takes value v (0 and 1 separately), while all other flip-flops do not take any values. The X-Simulation is done very fast per flip-flop because each flip-flop is only able to restore at most a few other flip-flops by itself. Reachability list is computed only once throughout the course of the hybrid algorithm when the metrics are computed for the first time, and does not depend on how trace signals are selected. As a result, it takes only a negligible portion of the overall execution runtime of the algorithm.

Restorability Rate: For a flip-flop f, we denote a restorability rate by r_f. This metric is computed at each iteration of the hybrid algorithm for any flip-flop that has not been selected so far for tracing. It reflects the probability that f can be restored using the trace signals selected so far. Computation of this metric also requires a *small* number of X-Simulations; first, at each iteration, all r_f values are computed using an observation window of 64 cycles instead of the entire capture window.

Figure 5.5 shows an example. It illustrates the computation of the restorability rate and reachability list of flip-flop f_3 in the example circuit of Fig. 5.2.

For simplicity, the example is shown for a capture window of only 4 cycles. Assuming flip-flop f_2 has been selected up to that iteration as a trace signal, the likelihood of restoring f_3 is $r_3 = 0.5$. This is because f_3 can be restored in 2 of 4 the cycles of the capture window.

Restoration Demand: This metric is computed at each iteration of the algorithm between flip-flops i and f, where f is a candidate flip-flop considered for tracing and i is a flip-flop that belongs to one of the reachability lists of f (i.e., $i \in L_f^0$ or $i \in L_f^1$). Furthermore, i should not have been selected so far for tracing up to that iteration of the hybrid algorithm but other signals may have already been selected for tracing up to that iteration.

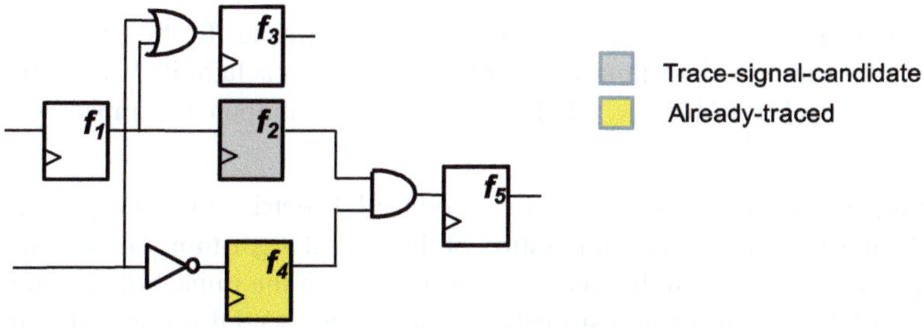

Fig. 5.6 Example showing the definition of the restoration demand

For example, in the circuit shown in Fig. 5.6, assume flip-flop f_4 has already been selected for tracing and flip-flop f_2 is a candidate considered for tracing and i is flip-flop f_3, which obviously falls in the reachability list of f_3.

The restoration demand metric aims to compute the remaining restoration needed by i from f such that i can be fully restored. This is done for two cases when f is assigned a value v of 0 or 1. This metric is given by the following equation.

$$d^v_{i,f} \approx \min(1 - r_i, a^v_f) \tag{5.2}$$

where r_i is restorability rate of i which was introduced before, and a^v_f is probability that f takes value v.

In this equation, the first argument $(1 - r_i)$ is the remaining restoration needed for i to be fully recovered. The second argument computes an upper bound on the restoration that f can offer i.

Note that the above equation is just a fast approximation of the demand of flip-flop i from f such that i can be restored. The computation of this metric is fast because r_i has already been computed. The quantity a^v_f is computed only once at the beginning of the algorithm (as a preprocessing step) by performing standard simulation and calculating the likelihood that each flip-flop may take a value of 0 or 1. In the example of Fig. 5.6, we have $d^1_{3,2} = \{1 - r_3, a^1_3\}$.

Impact Weight: This is the final metric that is computed at each iteration, for each untraced flip-flop. Once the impact weights are computed, the top 5% of flip-flops with the highest impact weight are selected as set of top candidates. Recall, X-Simulation will then be used for these top 5% flip-flops and the one with the highest value of SRR will be selected as the trace signal in that iteration.

We denote the impact weight of untraced flip-flop f with w_f which is defined by the following equation:

$$w_f = \sum_{v=0,1} \sum_{\forall i \in L^v_f} d^v_{i,f} \tag{5.3}$$

Basically, for flip-flop f this metric is sum of the restoration demands of all flip-flops i, which fall in its reachability list. A higher impact weight obviously reflects that selecting f as the next trace signal will help more with the restoration of the remaining untraced flip-flops.

For the example of Fig. 5.2, assuming f_2 is an untraced flip-flop in the current iteration of the hybrid algorithm, we have the following reachability lists for this flip-flop: $L_2^0 = \{f_1, f_5\}$ and $L_2^1 = \{f_1, f_3\}$. Therefore, we can compute the impact weight of f_2 as $w_2 = d_{1,2}^0 + d_{5,2}^0 + d_{1,2}^1 + d_{3,2}^1$.

Summary: Overall, in this section we introduced 4 metrics to identify a small set of candidates for tracing at each iteration of the hybrid algorithm. The metrics were built upon each other with the final one approximating the impact of each untraced flip-flop as far as its ability to restore the remaining untraced flip-flops. This allowed quick identification of a significantly smaller number of untraced flip-flops (i.e., only 5% of the total) at each iteration of the algorithm so X-Simulation can then be afforded to accurately evaluate these top candidates and select the one with the highest SRR as the trace signal in that iteration.

We note the above metrics from [11] only serve as an example of how a hybrid algorithm can be created to bridge the gap between a purely metric-based and simulation-based trace signal section algorithms. For example, some extensions to further improve these metrics are offered in [12].

5.3.3 *Island Flip-flops*

Recall in Fig. 5.3, method (i) is first used to select the top candidates. The process of selection of the top candidates using the impact weight metric does not allow consideration of a class of flip-flops which we define as island flip-flops. We define flip-flop f to be an island flip-flop if both its reachability lists are empty so we have $L_f^0 = L_f^1 = \emptyset$. These types of flip-flops do not have any impact weight because the impact weight metric of a flip-flop is computed by summing the demands of the flip-flops in its reachability list, and the reachability list is computed when all other flip-flops are ignored. However, it is possible that the selection of an island flip-flop as a trace signal improves the quality of the final solution due to the cumulative impact of other trace signals. Note the definition of the reachability list only considers the case when one flip-flop is assigned a value at each time so it ignores the cumulative impact of other flip-flops.

Method (ii) in the hybrid algorithm is, therefore, designed to specifically address the above shortcoming. First, all island flip-flops are identified. Next, X-Simulation is performed to measure the SRR of each island flip-flop and the one with highest SRR is selected as the next trace signal. As shown in Fig. 5.3, method (ii) kicks in every time method (i) selects 8 additional trace signals because the number of island flip-flops is a very small percentage of the total number of flip-flops in practice.

5.4 Conclusions

We discussed the pros and cons of simulation-based and metric-based trace selection algorithms with respect to two important evaluation metrics, namely execution run-time of the algorithm and quality of the solution measured in terms of SRR. We then discussed how a hybrid approach can be designed to take advantage of the strengths of these two classes of algorithms so the process of signal selection is driven by more accurate assessment of SRR without sacrifice in the runtime of the algorithm using many rounds of time-consuming X-Simulation.

We focused on a case study of one hybrid algorithm. The hybrid realization relied on carefully designed metrics to identify a very small set of top candidates (relative to the total number of candidate signals) at each iteration of the algorithm. It then used X-Simulation on this small set to accurately evaluate the SRR corresponding to each one and select the best one as the next trace signal in that iteration. Other hybrid realizations may also be possible which may define other metrics to identify the top candidates, or rely on an entirely different algorithmic process to select the trace signals.

References

1. M. Abramovici, P. Bradley, K.N. Dwarakanath, P. Levin, G. Memmi, D. Miller, A reconfigurable design-for-debug infrastructure for SoCs, in *Design Automation Conference* (2006), pp. 7–12
2. H.F. Ko, N. Nicolici, Algorithms for state restoration and trace-signal selection for data acquisition in silicon debug. IEEE Trans. Comput. Aided Des. Integr. Circuits Syst. **28**(2), 285–297 (2009)
3. X. Liu, Q. Xu, On signal selection for visibility enhancement in trace-based post-silicon validation. IEEE Trans. Comput. Aided Des. Integr. Circuits Syst. **31**(8), 1263–1274 (2012)
4. H. Shojaei, A. Davoodi, Trace signal selection to enhance timing and logic visibility in post-silicon validation, in *International Conference on Computer-Aided Design* (2010), pp. 168–172
5. E. Hung, S.J.E. Wilton, Scalable signal selection for post-silicon debug. IEEE Trans. Very Large Scale Integr. VLSI Syst. **21**(6), 1103–1115 (2012)
6. K. Basu, P. Mishra, RATS: restoration-aware trace signal selection for post-silicon validation. IEEE Trans. VLSI Syst. **21**(4), 605–613 (2013)
7. D. Chatterjee, C. McCarter, V. Bertacco, Simulation-based signal selection for state restoration in silicon debug, in *International Conference on Computer-Aided Design* (2011), pp. 595–601
8. K. Zhao, J. Bian, Pruning-based trace signal selection algorithm for data acquisition in post-silicon validation. Inst. Electron. Inf. Commun. Eng. Trans. **95**(A(6)), 1030–1040 (2012)
9. J.-S. Yang, N.A. Touba, Efficient trace signal selection for silicon debug by error transmission analysis. IEEE Trans. Comput. Aided Des. Integr. Circuits Syst. **31**(3), 442–446 (2012)
10. Y.-S. Yang, A.G. Veneris, N. Nicolici, Automating data analysis and acquisition setup in a silicon debug environment. IEEE Trans. Very Large Scale Integr. VLSI Syst. **20**(6), 1118–1131 (2012)
11. M. Li, A. Davoodi, A hybrid approach for fast and accurate trace signal selection for post-silicon debug, in *Design, Automation, and Test in Europe* (2013), pp. 485–490
12. M. Li, A. Davoodi, A hybrid approach for fast and accurate trace signal selection for post-silicon debug. IEEE Trans. Comput. Aided Des. Integr. Circuits Syst. **33**(7), 1081–1094 (2014)

Chapter 6
Post-Silicon Signal Selection Using Machine Learning

Alif Ahmed, Kamran Rahmani and Prabhat Mishra

6.1 Introduction

The goal of post-silicon validation is to ensure that the fabricated, pre-production silicon functions correctly while running actual applications under on-field operating conditions. Post-silicon validation is a complex activity performed under an aggressive schedule, accounting for more than 50% of the overall validation cost of a modern integrated circuit [1]. A fundamental constraint in post-silicon validation is *limited observability*: limitations in the number of output pins, coupled with restrictions imposed by area and power constraints on internal trace buffer sizes, imply that only a few hundred signals among the millions can be traced during a silicon execution. Furthermore, in order for a signal to be observed, the design must be instrumented a priori with appropriate hardware that routes the signal to an observable point. It is, therefore, crucial to develop techniques for identifying trace signals that maximize design visibility and debug information under the constraints imposed by the post-silicon observability restrictions.

Post-silicon trace signal selection in current industrial practice is primarily manual and guided by the designer's experience and insight, often with no objective techniques for qualifying the observability quality of a selected set of signals. Critical observability holes manifest themselves only during silicon debug, typically in the form of inadequacy of the set of traced signals for diagnosis or localization of a bug. However, this is too late for the redesign of the debug infrastructure or selec-

A. Ahmed (✉) · K. Rahmani · P. Mishra
Department of Computer and Information Science and Engineering, University of Florida, Gainesville, FL, USA
e-mail: alifahmed@ufl.edu

K. Rahmani
e-mail: kamran@ufl.edu

P. Mishra
e-mail: prabhat@ufl.edu

P. Mishra and F. Farahmandi (eds.), *Post-Silicon Validation and Debug*,
https://doi.org/10.1007/978-3-319-98116-1_6

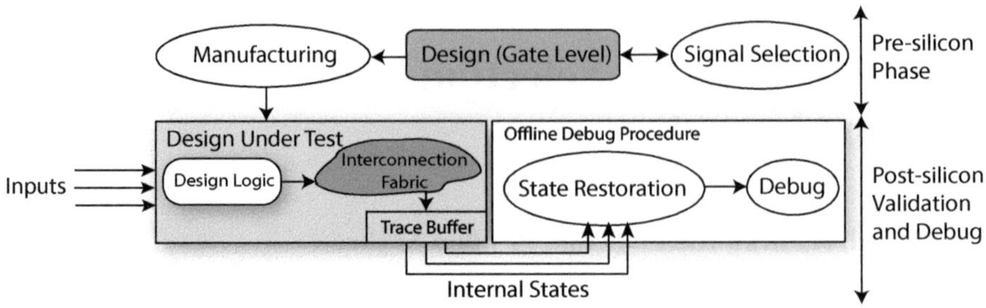

Fig. 6.1 Signal selection and restoration flow. Signals for trace buffer is selected in pre-silicon stage. Signal restoration for debugging is done post-silicon [8]

tion of new trace signals (with associated routing hardware), which would require a significant hardware change. Thus one has to contend with costly escapes, complex workarounds, and in many cases, more silicon re-spins.

There has been significant recent research to address the above issue by developing algorithms for signal selection through automatic analysis of pre-silicon designs, and by restoring internal signals during post-silicon debug (Fig. 6.1) [2–12]. The focus is to identify a set of signals S that maximizes *state restorability*, the set of states that can be reconstructed based on the observation of the signals in S. A common class of signal selection techniques involves defining a metric based on the design structure, which is then used in a (typically greedy) selection process to evaluate a candidate signal set [13–15]. These approaches are fast but provide a relatively low value of state restoration. Recent work on simulation-based signal selection [11] provides superior restoration quality but incurs prohibitive computation overhead. A hybrid signal selection approach [12] has been proposed which incorporated a combination of metric-based and simulation-based signal selection approaches. However, using less simulation to save selection time sacrifices the restoration performance.

Application of machine learning for signal selection is emerging as a promising technique [7, 8]. These have shown to provide excellent restoration performance with a small runtime overhead. This chapter will go into the details of two such methods. The first one is learning-based signal selection technique [8]. This method performs $O(n)$ simulations on the design to train a representative model. Here, n is the number of flip-flops in the design. After training is completed, trained model is used for further exploration. Different selection algorithms are applied on the model, and signals that give best restoration performance is selected for trace buffer. This method is faster than simulation- based techniques [11], which use $O(n^2)$ simulations. Section 6.4 describes this method in detail.

The second method is feature-based signal selection technique [7]. For large industrial designs, running even $O(n)$ simulations as described by learning-based method may be infeasible. Feature-based signal selection avoids this issue by simulating small designs rather than actual ones. These small designs should have similar characteristics as actual design. The model trained by this technique is then subjected to same steps as learning-based technique. Even training with such small designs can provide very accurate predictions for selecting profitable signals. Section 6.5 describes this technique in detail. Section 6.6 compares different signal selection techniques in terms of signal selection time and restoration quality.

6.2 Background

The goal of a signal selection algorithm is to construct a set of w flip-flops (out of total N flip-flips in the circuit) so that the total restoration r_m is maximized. Here, w is the width of the trace buffer. The selected signals can be mapped to an *input vector* $v = \langle f_1, f_2, ..., f_N \rangle$, where $f_i = 1$ if the i-th flip-flop is selected, 0 otherwise. Therefore, the signal selection problem can be formulated as the following constrained optimization problem:

$$\begin{aligned} &\text{maximize} \quad r_m(v) \\ &\text{under constraint} \quad \sum_{k=1}^{N} f_k = w \end{aligned} \tag{6.1}$$

The term *restoration ratio* is used to quantify the restoration ability of signal selection techniques. It entails inferring values of untraced signal states from a sequence of traced signals sampled over a period of time. It is defined as

$$\text{Restoration ratio} = \frac{\text{number of states traced and restored}}{\text{number of states traced}}$$

Signal values are recovered using forward and backward restoration. Forward restoration means inferring values of output signals from inputs. Figure 6.2a gives some examples of forward restoration, like in case of OR gates, if any input is 1, the output will also be 1. On the other hand, backward restoration means inferring values of inputs from output. Some example of backward restoration is given in Fig. 6.2b. Like, for an AND gate, if output is 1, both inputs can be inferred as 1. Also, in case of an OR gate, if output is 0, then both inputs are also 0.

6.3 Exploration Strategies

In a large design, there can be billions of signals. Efficient exploration of this huge search space is crucial to select the most profitable signals while keeping the runtime low. In this section, we will discuss three common strategies that are adopted by various signal selection approaches. Figure 6.3 provides an overview of these exploration strategies.

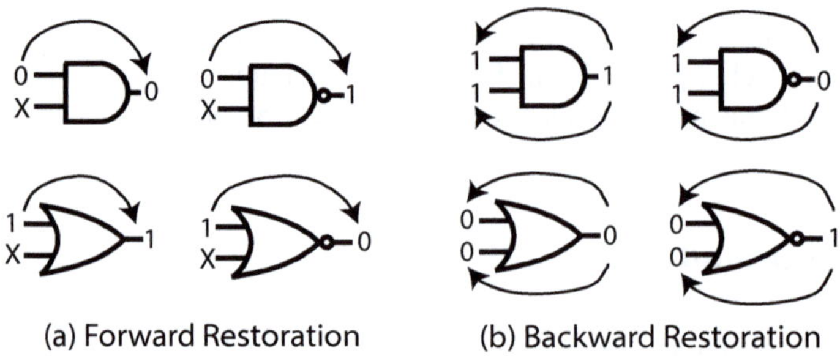

Fig. 6.2 Examples of signal restoration for various logic gates [8]. **a** Forward restoration: For AND gate, if any input is 0, output can be inferred to 0, etc. **b** Backward restoration: For AND gate, if output is 1, then both inputs are 1, etc.

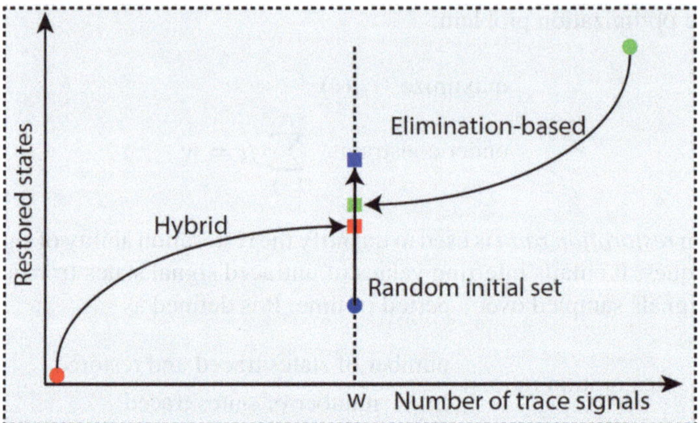

Fig. 6.3 Signal selection strategies. Elimination-based starts with all signals, and removes the least beneficial one. Hybrid (augmentation) starts with no signals, and selects the most beneficial one. Random initial set starts with random w signals. It removes least profitable signal and adds most profitable one [8]

6.3.1 Elimination-Based

The first strategy is introduced by Chatterjee et al. and used in the simulation-based approach [11]. This strategy uses method of elimination— starts with all flip-flops and stops when the number of remaining flip-flops is equal to the trace buffer width w. Algorithm 1 outlines the steps involved. First, all the flip-flops are selected as part of the candidate signals set (i.e., they are set to 1 in v). In each iteration of the algorithm, a signal with minimum impact on the restoration ratio of the candidate signals vector is removed by setting its value to 0 in v. This process stops when the number of remaining flip-flops in the candidate signals vector is equal to the trace buffer width (w). The final selected signals in v are returned as the output for the algorithm.

Algorithm 1 Elimination-based Signal Selection

```
procedure ELIMINATION(circuit, w, m)
    Create initial vector of v =< 1, 1, ..., 1 >, |v| = N
    remainedSignals = N
    while remainedSignals > w do
        maxRestorability = −∞
        maxIndex = −1
        for i = 1; i <= N; i ++ do
            if v[i] = 1 then
                v[i] = 0
                if r_m(v) > maxRestorability then
                    maxRestorability = r_m(v)
                    maxIndex = i
                v[i] = 1
        v[maxIndex] = 0
        remainedSignals = remainedSignals − 1
    return v
```

Algorithm 2 Augmentation-based Signal Selection

```
procedure AUGMENTATION(circuit, w, m)
    Create initial vector of v =< 0, 0, ..., 0 >, |v| = N
    for selected = 1; selected <= w; selected ++ do
        maxRestorability = −∞
        maxIndex = −1
        for i = 1; i <= N; i ++ do
            if v[i] = 0 then
                v[i] = 1
                if r_m(v) > maxRestorability then
                    maxRestorability = r_m(v)
                    maxIndex = i
                v[i] = 0
        v[maxIndex] = 1
    return v
```

6.3.2 Augmentation-Based

Another way of selecting w signals can be a method of augmentation. Algorithm 2 outlines the steps involved in selecting signals using this augmentation technique. It is similar to the approach described by Li et al. [12]. In this technique, in each iteration, we add the most beneficial signal to the candidate signals set, instead of removing the least beneficial one. This process stops when the total number of selected signals is equal to trace buffer width w. The final vector of selected signals v is returned as the algorithm output.

Algorithm 3 Random Initial Set Signal Selection

procedure RANDOMINITIALSET($circuit, \hat{r}_m(v), w$)
    Create selected signals set S
    Create initial vector of $v =< 0, 0, ..., 0 >, |v| = N \times p$
    Randomly set w elements of v to 1
    Set v_{new} as v
    do
        Set v as v_{new}
        Find signal with minimum effect on $\hat{r}_m(v_{new})$ and set it to 0
        Find signal with maximum effect on $\hat{r}_m(v_{new})$ and set it to 1
    while $\hat{r}_m(v_{new}) > \hat{r}_m(v)$
    for $i = 1; i <= N \times p; i++$ **do**
        if $v[i] = 1$ **then**
            Add i to S
    return S

6.3.3 Random Initial Set

This exploration strategy is used based on random selection. Algorithm 3 outlines the steps involved. Initially, w signals are selected at random. Then in each iteration, the least beneficial signal is removed and the most profitable one is added. This process is continued until removing a signal and adding back another one does not improve the predicted restoration in $\hat{r}_m(v)$.

6.4 Learning-Based Signal Selection

This section discusses the approach proposed by Rahmani et al. [8]. This is one of the first machine learning based approaches ever used to address signal selection problem. The main advantage of this approach over other techniques is execution speed. While simulation-based methods use $O(n^2)$ simulations [11], this method uses $O(n)$—which is comparable to hybrid [12] approach. Furthermore, this method demonstrates higher restoration ratio compared to hybrid approach.

Figure 6.4 presents the overview of learning-based signal selection technique. It can be divided into several steps.The first step involves creating a model of the design by running mock simulations. The second step involves running all three exploration strategies (discussed in Sect. 6.3) on top of the trained model. Finally, the most profitable one among them will be selected for trace buffer. The following sections will discuss these steps in detail.

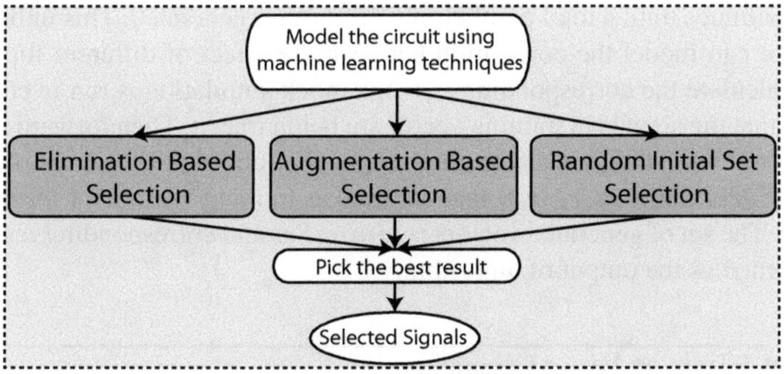

Fig. 6.4 Overview of learning-based signal selection technique [8]

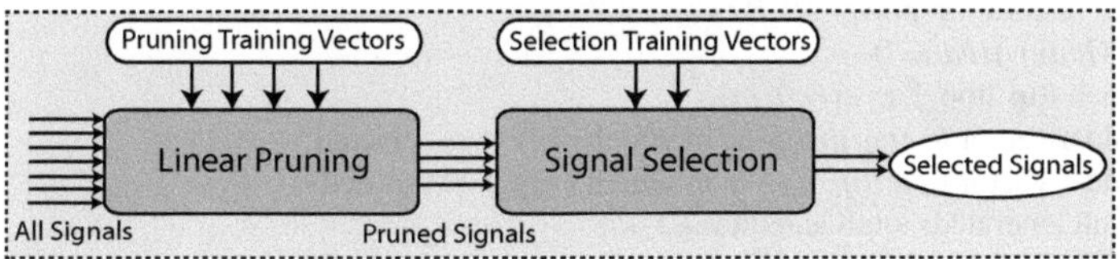

Fig. 6.5 A quick linear model is used to eliminate most of the non-beneficial flip-flops. A more accurate nonlinear model is used for the remaining flip-flops [8]

6.4.1 Selection Model Training

In order to increase the accuracy of the prediction while simultaneously reducing the runtime of modeling/prediction in large circuits, a two-step modeling scheme is used by the authors. Figure 6.5 illustrates the framework. In the first step, a linear model is applied to eliminate less important flip-flops and reduce the size of feature vector. Although the accuracy of linear modeling is low, it is fast and can be used to quickly prune out the non-beneficial signals and determine top candidates using simple calculations. In the second step, a nonlinear regression is applied on the reduced set to produce a finer model of the remaining flip-flops. The reduced number enables to use a more accurate nonlinear model with fewer training vectors for selecting the final set of signals.

6.4.1.1 Generating Training Vectors

Algorithm 4 outlines the pseudo-code for training vector generation used in both pruning and final model. To consider the effect of each flip-flop on total restorability, two vectors are generated. First, a vector in which only a particular flip-flop is selected, and second a vector in which all the flip-flops are selected except that particular flip-flop. In addition, to include the vectors with different number of flip-flops, $N-1$ vectors with $2, 3, \ldots, N$ randomly chosen flip-flops are generated. This process continues until a total number of t vectors are generated. This unbiased random vector can model the correlation between the effect of different flip-flops. In order to calculate the corresponding $r_m(v)$, a mock simulation is run over m cycles assuming that the signals in training vector are being traced. Then forward and backward restoration techniques are applied to get the total number of restored states. Finally, we get t pairs $\langle v_i, r_m(v_i)\rangle$ that are used as training vectors for the regression technique. The set of generated vectors $trainingSet$ and corresponding restorability R are returned as the output of algorithm.

Algorithm 4 Training Vector Generation

```
procedure GENERATEVECTORS(circuit, m, t)
    Create training vectors set trainingSet
    Create restoration power set R
    totalGenerated = 0
    for each flip-flop f in circuit do
        Add a vector to trainingSet in which only f is selected
        Add a vector to trainingSet in which only f is omitted
        totalGenerated=totalGenerated+2
    for i = 2; i <= N; i++ do
        Add a vector to trainingSet in which exactly i random flip-flops are chosen
        totalGenerated++
    while totalGenerated < t do
        length= a random number between 1 and N
        randomVector=a vector in which exactly length random flip-flops are chosen
        if randomVector ∉ trainingSet then
            Add randomVector to trainingSet
            totalGenerated++
    for each vector trainingVector in trainingSet do
        R(v) = Restoration power of trainingVector using a mock simulation followed by a
    restoration process over m cycles
    return trainingSet, R
```

6.4.1.2 Linear Pruning

Linear pruning is used to quickly eliminate most of the non-beneficial flip-flops (in term of restorability effectiveness). The authors have used support vector regression with linear kernel, which is the simplest form of this well-known model. However, any other linear regression technique can be used as well. Given the training set $\langle v_i, r_m(v_i)\rangle$, the support vector regression solution is a set of j support vectors which is used for predicting new vectors. Denoting the predicted $r_m(v)$ as $\hat{r}_m(v)$, we have the following equation:

$$\hat{r}_m(v) = \hat{w}_0 + \sum_{k=1}^{j} \alpha_k k(v_k, v) \tag{6.2}$$

In Eq. 6.2, v is the vector whose restorability we wish to predict, v_k is the kth support vector, and α_k is the corresponding coefficient. In addition, $k(v_k, v)$ is the output of the kernel function used in support vector regression. In linear mode, the kernel function is of the form $k(v_k, v) = v_k^T.v$, where v_k^T is the transpose of v_k. Then Eq. 6.2 can be rewritten as follows.

$$\hat{r}_m(v) = \hat{w}_0 + \sum_{k=1}^{j} \alpha_k v_k^T.v \tag{6.3}$$

$$\Rightarrow \hat{r}_m(v) = \hat{w}_0 + \hat{w}^T.v \ (where\ \hat{w} = \sum_{k=1}^{j} \alpha_k v_k) \tag{6.4}$$

Equation 6.4 illustrates the simplified version of the prediction formula when a linear kernel is used. In fact, the model is a simple hyperplane which has the minimum error among all the hyperplanes over the training set. Although this linear model may not be the best fit for the nonlinear function $r_m(v)$, it can be used to quickly detect and eliminate non-beneficial flip-flops as those will get a smaller coefficient in the $\hat{w}$ vector.

Algorithm 5 outlines the linear pruning process. First, a set of training vectors are generated followed by a linear modeling using support vector regression. Next, the weight vector $\hat{w}$ of predicted function is calculated as illustrated in Eq. 6.4. The flip-flops with most effect on restorability have largest values in corresponding index of weight vector. Therefore, the index of $p \times N$ largest values in weight vectors are kept as most useful flip-flops in terms of restorability and the rest is removed. Here, N is the number of flip-flops in the circuit and p is the pruning factor. Smaller p means less features in the next step which leads to a more accurate and faster nonlinear model. However, due to lower accuracy of linear model, lower value of p will also increase the chance of eliminating a useful flip-flop by mistake. The output of the process is the preserved flip-flops set S.

The linear model has a higher prediction error; however, this is compensated by selecting a bigger set (compared to the buffer width) of the top signals in the pruning phase. A more accurate nonlinear model will be used in next step, which enables picking the most profitable signals from this set with a more fine-grained selection. To illustrate the fact that top signals are not removed in the linear pruning, Fig. 6.6 shows how many of the 32 top signals are kept when we keep reducing the p value for benchmark *S38417*. As we can see, most of the profitable signals are left even for $p = 0.05$.

6.4.1.3 Accurate Model Building

The reduced number of flip-flops in feature vector enables creating a more accurate nonlinear model of the circuit with significantly less number of training vectors. The effective number of required training vectors in this step is reduced by $1 - p$, where

Algorithm 5 Linear Pruning Algorithm

1: **procedure** LINEARPRUNING($circuit, m, t, p$)
2: Create selected features set S
3: $trainVectors$ =GenerateVectors($circuit, m, t$)
4: Model $\hat{r}_m(v)$ using support vector regression with $trainVectors$ and linear kernel
5: Calculate the weight vector $\hat{w} = \sum_{k=1}^{j} \alpha_k v_k$
6: $S =$ the index of top $p \times N$ values in vector $\hat{w}$
7: **return** S

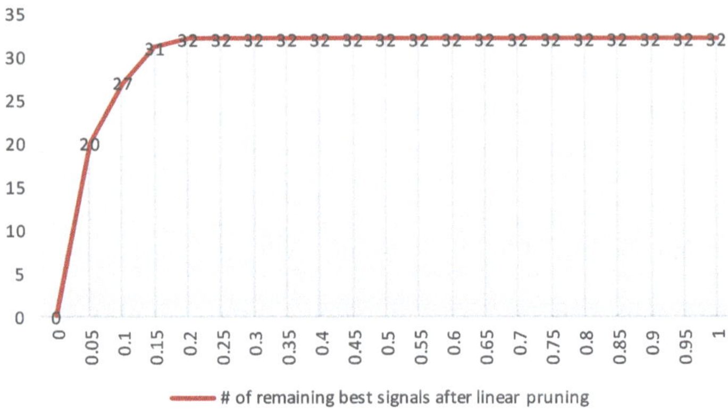

Fig. 6.6 Number of profitable signals remaining after linear pruning (out of 32 signals selected when linear pruning is not applied or $p = 1$) for different values of p in *S38417* benchmark [8]

p is the pruning factor. There are several nonlinear models available to use, each of which can be a good fit in a specific domain. Mean Prediction Error (MPE) can be used to measure the quality of a model on a test vector set of size n, is defined as below.

$$MeanPredictionError = 1/n * \sum_{k=1}^{n} |\hat{r}_m(v_k) - r_m(v_k)| \tag{6.5}$$

Algorithm 6 outlines the accurate model building process after the pruning. First, a set of $t_{selection}$ training vectors is generated for training. In order to find the best nonlinear model in the *candidateModels* set, a quick training is followed by a MPE calculation on a small set of vectors randomly selected from the bigger training vectors set. It should be noted that different set of vectors are used for quick training and testing (MPE calculation); this makes the model selection process unbiased and yields a better result for new input vectors. After choosing the best model with minimum MPE, the model is retrained with all the training vectors.

Prediction accuracy can largely vary depending on the algorithm used to train the model. Figure 6.7 illustrates the real versus predicted restoration states for different algorithms in *S38584* benchmark. Authors found that model trained with *cubist*

Algorithm 6 Accurate Model Building Algorithm

procedure SELECTFINALMODEL($circuit, m, t_{selection}, w, candidateModels$)
    $trainVectors$ = GenerateVectors($circuit, m, t_{selection}$)
    $quickTrainVectors$ =random 10% of trainVectors
    $quickTestVectors$ =random 10% of trainVectors, exclusive with quickTrainVectors
    for each model $model$ in $candidateModels$ **do**
        Model $\hat{r}_m(v)$ with pruned features using $model$ and $quickTrainVectors$
        Calculate MPE for $\hat{r}_m(v)$ on $quickTestVectors$
    $bestModel$ = model with minimum MPE
    $result$ = Model $\hat{r}_m(v)$ with pruned features using $bestModel$ and $trainVectors$
    return result

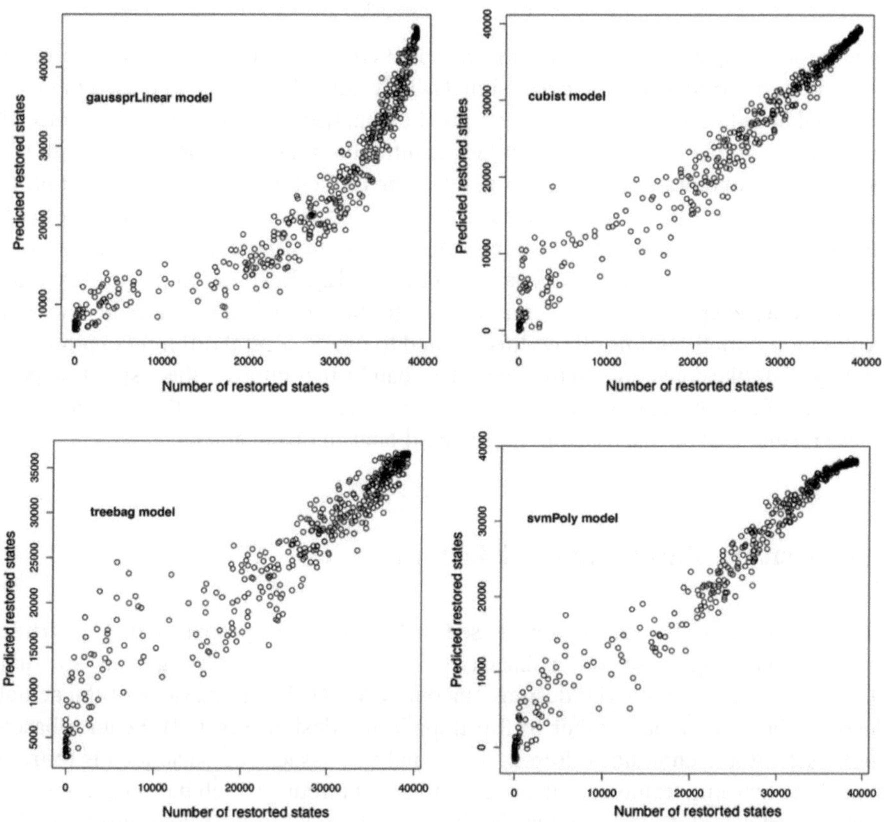

Fig. 6.7 Real versus predicted restoration states for different models in *S38584* benchmark. Each random vector represents a set of randomly selected trace signals and is showed as a circle in the graph [8]

provided the most accurate prediction. It should be noted that the MPE is bigger for larger benchmarks in *cubist*; however, it still maintains the relative relationship between the restoration values. In other words, the percentage of error ($|Predicted - Actual|/Actual$) will not grow linearly as the actual restoration absolute value grows in larger benchmarks. In fact, the predicted values in *cubist* match the real values in most of the cases [8]. This enables high-quality signal selection without any further real simulation.

6.4.2 Selection Using Trained Model

In previous section, we discussed how a model can be trained to accurately predict restoration performance of selected signals. This trained model can be used in place of actual simulations. It can provide a good estimation of which signals to trace if combined with the exploration strategies mentioned in Sect. 6.3. Rahmani et al. [8] used all three exploration strategies on top of the trained model to get the most profitable signals for tracing, as depicted in Fig. 6.4. As simulations on actual designs are performed only during the training stage and not during exploration, it has significantly less runtime overhead compared to [11, 12]. Restoration performance on the other hand depends on the accuracy of the trained model and the effectiveness of exploration. Small runtime allows this method to run on several different exploration strategies, unlike conventional methods. This can further improve the restoration performance. Tables 6.1 and 6.2 compares learning-based technique with other methods. These results will be discussed in more detail later in this chapter.

6.5 Feature-Based Signal Selection

This section describes the feature-based signal selection proposed in [7]. While learning-based signal selection discussed in Sect. 6.4 is promising, it is still not applicable on large industrial designs since it needs $O(N)$ simulations of the actual design, where N is the number of flip-flops in the design (Fig. 6.8). Feature-based signal selection technique addresses this scalability issue. The basic idea is to train a machine learning framework using a small set of circuits which has a similar characteristic as actual designs, and apply the trained model to the bigger circuit under test. This strategy avoids the need for simulating large industry-scale designs.

6.5.1 Overview

Figure 6.9 shows the overview of feature-based signal selection approach. First, a set of small training circuits are chosen to build the selection model. For each training

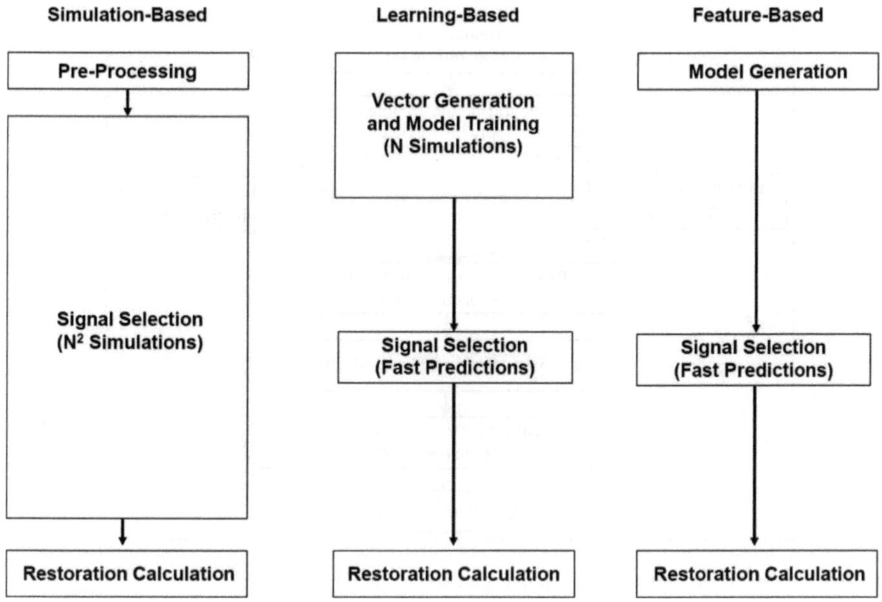

Fig. 6.8 Runtime distribution of different signal selection techniques

circuit, a modified version of both elimination-based [11] and augmentation-based [12] approaches are applied and then the best result is selected. A set of training vectors is then generated and added to the training vectors set. Next, this training set is used to create a selection model using different machine learning regression techniques and pick the one with best accuracy. In this step, a model is trained with the criteria of a good candidate signal and the relation with its properties. This model can then be used to select trace signals on any related design under signal selection without expensive mock simulations involved. The selection model training is a one-time process. Once it is done the model can be used to select trace signal on any other related circuits, as long as it shares same properties like connectivity and coding guideline with the training circuits.

6.5.2 *Model Training*

The trained model should be generic enough so that it can be applied to the circuit under test and accurate enough to produce high quality result. Before going into the details of modeling technique, a few useful terms are defined below for a circuit with N flip-flops and a mock simulation window of m.

- $\mathbf{Fan-out_g(x)}$ for flip-flop f is defined as the number of gates of type g (and, or, etc.) connected to its output.

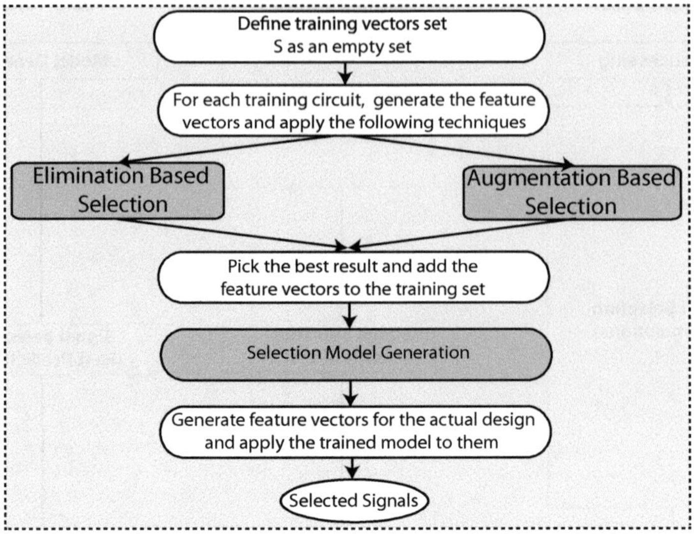

Fig. 6.9 Overview of feature-based signal selection technique [7]

- **Fan − in$_g$(x)** for flip-flop f is defined as the number of gates of type g (and, or, etc.) connected to its inputs.
- **Connectivity(x)** for flip-flop f is defined as the number of flip-flops connected to it through other combinational gates in both backward and forward directions.
- **InputDistance(x)** for flip-flop f is defined as the minimum distance of the flip-flop from the primary input signals (in terms of number of gates).
- **OutputDistance(x)** for flip-flop f is defined as the minimum distance of the flip-flop from the primary output signals (in terms of number of gates).
- **ZeroProbability(x)** for flip-flop f is defined as the percentage of 0 values for the flip-flop in a mock simulation over m cycles.
- **SingleRestoration(x)** for flip-flop f is defined as the number of restored states in a mock simulation/restoration process over m cycle if f is the only trace signal.
- **SelectionOrder$_g$(x)** for flip-flop f is defined as the sequence number of selection when technique g is applied. This number is 1 for the first (best) selected signal and N for the last selected signal.
- **Rank(g(x))** for flip-flop f is defined as the number of flip-flops with $g(x) <= g(f)$ divided by N. In other words, it is the normalized relative rank of applying function g to flip-flop f compared to the other flip-flops. This value would be 1 and $1/N$ for the flip-flops with the maximum and minimum value of g(f), respectively.

6.5.2.1 Feature Selection

Features that have a high correlation with structural properties of a signal and its performance in state restoration should be chosen for training. In addition, it should be independent of the circuit size and structure. This is crucial as we want to train the model using a set of small circuits and apply the learning to the bigger circuit under test. Lastly, the generation of feature vectors should not be computationally expensive. Keeping these in mind, authors of [7] have suggested the following feature vectors:

- Rank(Fan-out$_g$(f)) for all gates of type g in the circuit.
- Rank(Fan-in$_g$(f) for all gates of type g circuit.
- Rank(Connectivity(f)) for flip-flop f.
- Rank(InputDistance(f) for flip-flop f.
- Rank(OutputDistance(f)) for flip-flop f.
- Rank(ZeroProbability(f)) for flip-flop f.
- Rank(SingleRestoration(f)) for flip-flop f.

It can be seen that the chosen features are mostly based on the circuit structure and fast to evaluate. In addition, rank function is applied to all the features to make them relative values instead of absolute values. This makes the features independent of the circuit size and number of gates. Intuitively, the feature vector captures the fan-in and fan-out of the signal, its relative position and depth in the circuit, and its impact on restoring its neighbors when selected as a trace signal. There is a high correlation between these features and the restoration performance of a flip-flop [7].

6.5.2.2 Generating Training Vector

Generating training vector is the same as learning-based method. Detail algorithm of this process is presented in Sect. 6.4.1.1.

6.5.2.3 Model Building

In this step, a selection model is built by applying simulation- based techniques on a small set of training circuits. Intuitively, the model learns the criteria of a good trace signal and the relation with its feature vector described before. Algorithm 7 outlines the steps involved in creating the selection model from a set of small training circuits. This algorithm is similar to learning-based model building explained in Sect. 6.4.1.3. MPE is used to calculate prediction error for both methods. However, there is one small difference. In feature-based technique, the authors have not used Random Initial Set exploration strategy. Only augmentation and elimination based exploration techniques are applied and the one with better average restoration ratio is selected. Details of these strategies have been discussed in Sect. 6.3.

Next, for each flip-flop f in the circuit, a pair of feature vector v and selection order rank r is added to the training vectors set. Choosing selection order rank helps to normalize the training data across all the circuits and makes it independent of number of flip-flops in the circuit. Different regression techniques (like SVM, linear modeling, etc.) is then applied to the training vectors and the best one is returned.

Algorithm 7 Model Generation Algorithm

procedure MODELGENERATION(trainingCircuits, regressionModels, m)
    Create training vectors set trainingVectors
    for each circuit c in trainingCircuits **do**
        n = number of flip-flops in c
        apply AugmentationBased(c,n,m) and EliminationBased(c,1,m) to c and pick the best one as g
        for each flip-flop f in c **do**
            Add $< v, r >$ to trainingVectors where v is the feature vector for the flip-flop and r is Rank(SelectionOrder$_g$(f))
    apply all the models in regressionModels to trainingVectors
    return model m with minimum MPE

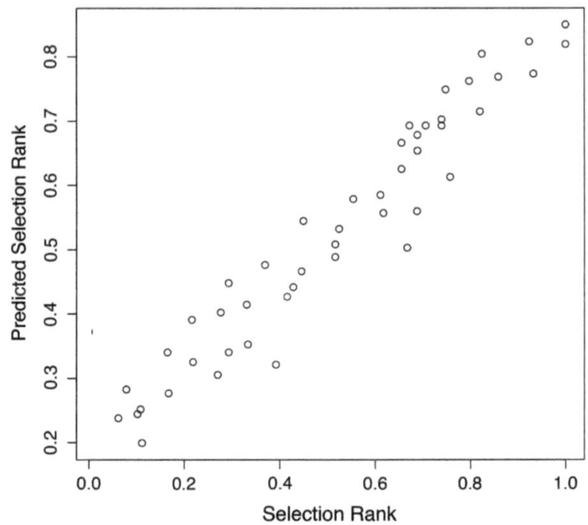

Fig. 6.10 Actual versus predicted values of selection ranks in s38584 benchmark. Model is trained using small ISCAS'89 benchmarks [7]

6.5.3 Signal Selection Process

Once the selection model trained as described in the previous section, it can be used on any circuit for selecting trace signals. Authors of [7] have used the smallest circuits in ISCAS'89 benchmarks[1] to train the model using *cubist* regression technique. Figure 6.10 shows the actual versus predicted selection ranks (using the trained model) on a set of flip-flops in s38584 benchmarks (in this example, s38584 is assumed as the actual design under signal selection). It can be seen that there is a high correlation between the real and predicted values. This is significant as it enables us to select high-quality trace signals in the circuit under test using very fast predication to generate the selection ranks based on the feature vectors, instead of expensive mock simulations.

[1]s1494, s1488, s713, s1238, s1196, and s838.

Algorithm 8 outlines the signal selection technique. First, feature vectors for all the flip-flops in the circuit are generated. Then, these vectors are used to predict the selection sequence rank for the flip-flops with the selection model *m*. Finally, the top *w* (trace buffer width) flip-flops with the highest value of predicted selection sequence rank is returned as the result.

Algorithm 8 Signal Selection Algorithm

```
procedure SIGNALSELECTION(circuit, m, w)
    Initialize predictionMap as an empty map
    for each flip-flop f in circuit do
        v = feature vector of f
        r = m(v), the predicted value of selection sequence rank of f using model m
        Add < f, r > to predictionMap
    Sort predictionMap based on r values
    return top w flip-flops with the highest values of r
```

6.6 Comparison of Different Signal Selection Techniques

This section compares different signal selection techniques in terms of restoration quality and signal selection time.

6.6.1 Restoration Quality

Table 6.1 presents the restoration ratios different state-of-the-art techniques using different ISCAS'89 and ITC'99 benchmarks [7]. The ones shown as "N/A" means that the technique was not able to finish within 24 hours of runtime. The improvement in restoration performance for feature-based technique proposed in [7] is up to 135.4% in *s38584* and 8.8% on average. Compared to Chatterjee et al. [11], feature-based technique [7] ran the elimination on training circuits without any pruning which reduces the chance of removing effective flip-flops prior to selection itself. Similarly, Li et al. [12] incorporated simulations for only top 5% of the candidate flip-flops, which sacrifices the precision of the selection process. In addition, building the selection model using small training circuits allows [7] to use both elimination-based and augmentation-based techniques at the same time and pick the best one for each circuit. In addition, model in [7] is trained using the best result of simulation-based techniques on a set of training circuits (instead of just one), which provides a more globally optimized selection model. Although the model is trained using small circuits in ISCAS'89 benchmarks, it still outperforms [16] in ITC'99 benchmarks (b15 and b17). This shows that the feature vector and selection model is generic

Table 6.1 Comparison of restoration ratios between different signal selection methods [7]

Circuit	#Flip-flops	Buffer width	Simulation based [11]	Hybrid [12]	Learning based [16]	Feature based [7]
s5378	179	8	13.41	**14.35**	14.20	14.13
		16	7.35	8.36	8.40	**8.92**
		32	4.47	4.99	4.93	**5.12**
s9234	228	8.0	13.98	9.25	15.33	**15.82**
		16	8.30	6.13	8.76	**9.10**
		32	4.46	4.38	4.84	**5.11**
s15850	597	8	26.33	21.90	44.03	**45.12**
		16	19.89	14.78	23.13	**24.37**
		32	13.19	10.88	**13.92**	13.82
s13207	669	8	35.52	33.60	47.18	**49.30**
		16	20.13	23.22	29.00	**31.21**
		32	11.25	13.64	15.42	**16.13**
s38584	1452	8	N/A	27.00	54.25	**127.72**
		16	N/A	13.97	69.03	**79.09**
		32	N/A	7.50	43.66	**44.02**
s38417	1636	8	N/A	37.71	52.33	**53.27**
		16	N/A	23.80	**27.12**	26.97
		32	N/A	11.83	16.73	**17.10**
s35932	1728	8	132.00	144.00	186.80	**186.90**
		16	67.45	72.00	**93.60**	93.42
		32	34.63	36.00	46.98	**47.15**
b15	449	8	5.99	N/A	6.15	**7.18**
		16	3.56	N/A	4.83	**4.98**
		32	**34.63**	N/A	3.31	3.46
b17	1415	8	N/A	N/A	14.12	**14.43**
		16	N/A	N/A	13.19	**13.31**
		32	N/A	N/A	7.93	**8.77**
b18	3320	8	N/A	N/A	N/A	**25.12**
		16	N/A	N/A	N/A	**21.60**
		32	N/A	N/A	N/A	**12.49**
b19	6642	8	N/A	N/A	N/A	**32.00**
		16	N/A	N/A	N/A	**24.64**
		32	N/A	N/A	N/A	**18.11**

Table 6.2 Runtime comparison of existing selection approaches [7]

Circuit	Buffer width	Simulation based [11]	Hybrid [12]	Learning based [16]	Feature based [7]
s5378	8	00:01:53	00:00:08	00:01:46	00:00:11
	16	00:01:52	00:00:10	00:01:52	00:00:11
	32	00:01:48	00:00:16	00:02:09	00:00:11
s9234	8	00:08:52	00:00:32	00:00:10	00:00:11
	16	00:08:43	00:00:40	00:00:10	00:00:11
	32	00:08:10	00:00:50	00:00:10	00:00:11
s15850	8	03:44:12	00:05:20	00:04:20	00:00:13
	16	03:44:04	00:06:00	00:04:35	00:00:13
	32	03:43:39	00:06:36	00:05:04	00:00:13
s13207	8	01:21:41	00:01:36	00:03:45	00:00:13
	16	01:21:35	00:02:00	00:04:01	00:00:13
	32	01:21:13	00:02:40	00:04:12	00:00:13
s38584	8	N/A	00:05:28	00:16:52	00:00:36
	16	N/A	00:06:06	00:17:09	00:00:36
	32	N/A	00:09:02	00:17:35	00:00:36
s38417	8	N/A	00:22:42	00:20:23	00:00:39
	16	N/A	00:33:04	00:21:07	00:00:39
	32	N/A	00:34:28	00:23:55	00:00:39
s35932	8	11:39:36	00:04:28	00:16:49	00:00:37
	16	11:39:09	00:05:56	00:17:33	00:00:37
	32	11:38:01	00:08:38	00:18:21	00:00:37
b15	8	06:12:09	N/A	00:06:49	00:00:12
	16	06:09:55	N/A	00:07:03	00:00:12
	32	06:06:40	N/A	00:07:11	00:00:12
b17	8	N/A	N/A	00:19:10	00:00:35
	16	N/A	N/A	00:20:30	00:00:35
	32	N/A	N/A	00:21:40	00:00:35
b18	8	N/A	N/A	N/A	00:06:11
	16	N/A	N/A	N/A	00:06:11
	32	N/A	N/A	N/A	00:06:11
b19	8	N/A	N/A	N/A	00:21:09
	16	N/A	N/A	N/A	00:21:09
	32	N/A	N/A	N/A	00:21:09

and can be applied to the designs from the same domain with similar characteristic and standard cell library. This enables training the model with a small set of related circuit with same coding guideline or sub-components of SoCs and use it for signal selection in the larger industrial circuits.

6.6.2 Signal Selection Time

Table 6.2 presents the runtime of different approaches. Results were collected from an Octa-Core AMD Opteron 6378 (1400 MHz) machine with 188GB of memory [7]. The runtime for is calculated as the summation of required time for generating training vectors on the training circuits, modeling, generating the feature vectors for the circuit under test, and the signal selection process itself. The reported runtime format is "hour:minute:second". Like the previous table, "N/A" means that the technique was not able to finish within 24 hours. As we can see, feature-based technique shows significant speed-up compared to others. The speed-up is up to 37X in *s38417* and *b17* for buffer width of 32 and 17.6X on average. This is because [7] needs mock simulations only on a set of small training circuits. Once the model is created, there is no need for any simulations on the circuit under test as the selection process just uses fast predictions instead of actual simulations.

6.7 Conclusion

This chapter described two machine learning based approaches toward solving signal selection problem. State-of-the-art signal selection methods require running a large number of simulations on the design. This often amounts to infeasible signal selection time for actual industrial scale designs. The basic idea of these machine learning based techniques is to replace costly simulations with lightweight prediction models. The first method discussed in this chapter only performs a small number of simulations on actual design to build the model. Subsequently, the model is subjected to more thorough exploration strategies for profitable signal selection. The second method further improves runtime by training the model with small designs instead of actual large designs. Finally, the chapter compared the restoration performance and runtime of the signal selection methods.

References

1. P. Mishra, R. Morad, A. Ziv, S. Ray, Post-silicon validation in the soc era: a tutorial introduction. IEEE Des. Test **34**(3), 68–92 (2017)
2. K. Rahmani, P. Mishra, Efficient signal selection using fine-grained combination of scan and trace buffers, in *2013 26th International Conference on VLSI Design and 2013 12th International Conference on Embedded Systems (VLSID)*(IEEE, New York, 2013), pp. 308–313
3. K. Rahmani, P. Mishra, S.Ray, Efficient trace signal selection using augmentation and ILP techniques, in *2014 15th International Symposium on Quality Electronic Design (ISQED)* (IEEE, New York, 2014), pp. 148–155
4. P. Thakyal, P. Mishra, Layout-aware signal selection in reconfigurable architectures, in *18th International Symposium on VLSI Design and Test* (IEEE, New York, 2014), pp. 1–6
5. K. Rahmani, S. Proch, P. Mishra, Efficient selection of trace and scan signals for post-silicon debug. IEEE Trans. Very Large Scale Integr. (VLSI) Syst. **24**(1), 313–323 (2016)

6. K. Basu, P. Mishra, P. Patra, Efficient combination of trace and scan signals for post silicon validation and debug, in *2011 IEEE International Test Conference (ITC)* (IEEE, New York, 2011), pp. 1–8
7. K. Rahmani, P. Mishra, Feature-based signal selection for post-silicon debug using machine learning, IEEE Trans. Emerg. Top. Comput. (2017)
8. K. Rahmani, S. Ray, P. Mishra, Postsilicon trace signal selection using machine learning techniques. IEEE Trans. Very Large Scale Integr. (VLSI) Syst. **25**(2), 570–580 (2017)
9. P. Thakyal, P. Mishra, Layout-aware selection of trace signals for post-silicon debug, in *2014 IEEE Computer Society Annual Symposium on VLSI (ISVLSI)* (IEEE, New York, 2014), pp. 326–331
10. K. Basu, P. Mishra, Efficient trace signal selection for post silicon validation and debug.,in *2011 24th International Conference on VLSI Design (VLSI Design)* (IEEE, New York, 2011), pp. 352–357
11. D. Chatterjee, C. McCarter, V. Bertacco, Simulation-based signal selection for state restoration in silicon debug, in *2011 IEEE/ACM International Conference on Computer-Aided Design (ICCAD)* (2011), pp. 595–601
12. M. Li, A. Davoodi, A hybrid approach for fast and accurate trace signal selection for post-silicon debug, in *Design, Automation, and Test (DATE)* (2013), pp. 485–490
13. X. Liu, Q. Xu, Trace signal selection for visibility enhancement in post-silicon validation, in *DATE '09. Design, Automation Test in Europe Conference Exhibition* (2009), pp. 1338 –1343
14. K. Basu, P. Mishra, Rats: restoration-aware trace signal selection for post-silicon validation. IEEE Trans. Very Large Scale Integr. (VLSI) Syst. **21**(4), 605–613 (2013)
15. H. Ko, N. Nicolici, Algorithms for state restoration and trace-signal selection for data acquisition in silicon debug. IEEE Trans. Comput.-Aided Des. Integr. Circuits Syst. **28**(2), 285–297 (2009)
16. K. Rahmani, P. Mishra, S. Ray, Scalable trace signal selection using machine learning, in *2013 IEEE 31st International Conference on Computer Design (ICCD)* (2013), pp. 384–389

Part III
Generation of Tests and Assertions

Chapter 7
Observability-Aware Post-Silicon Test Generation

Farimah Farahmandi and Prabhat Mishra

7.1 Introduction

Post-silicon validation of an integrated circuit (IC) entails running tests on a fabricated, preproduction silicon to ensure that the design functions as expected under actual operating conditions and identify errors that have been missed during pre-silicon validation. Post-silicon validation is a highly complex activity, requiring elaborate planning, architectural support, and test development [17]. It is also an expensive activity accounting for more than 50% of the validation cost. Furthermore, post-silicon validation must succeed before mass production can begin. Thus, effectiveness of post-silicon validation affects product launch, company revenues, profitability, and market positioning [24].

Test generation has been widely studied for functional validation of integrated circuits. Majority of the test generation approaches are designed for pre-silicon validation [1, 3, 4, 8, 9, 16, 20]. Pre-silicon test generation has been performed at different abstraction levels. A vast majority of test generation efforts are focused on validation of register-transfer level (RTL) implementation [9]. Recently, there have been some targeted approaches for transaction-level model (TLM) validation [21]. Recent approaches show how to reuse transaction-level tests for RTL validation [8]. With the growing importance of post-silicon validation, there have been significant interest in test generation for post-silicon debug. Sousa and Sen [21] generated TLM testbenches using mutation testing. HYBRO [16] generates tests to cover branches using dynamic simulation data as well as static analysis of RTL control flow graphs. Adir et al. [2] presented a unified methodology for both pre-silicon and post-silicon validation that is based on common verification plan and similar languages for test templates and coverage models. Reversi [23] generates random test programs that are beneficial for post-silicon validation. Quick error detection [11] transforms existing post-silicon tests such that the error detection latency is significantly reduced.

The content of this chapter has been developed based on [10] with permission from EDAA.

F. Farahmandi (✉) · P. Mishra
Department of Computer and Information Science and Engineering, University of Florida, Gainesville, FL, USA
e-mail: ffarahmandi@ufl.edu

P. Mishra and F. Farahmandi (eds.), *Post-Silicon Validation and Debug*,
https://doi.org/10.1007/978-3-319-98116-1_7

Lin et al. [14] present how to create post-silicon validation tests that quickly detect bugs in multi-core SoCs. Zokaee et al. [25] have proposed a SAT-based ATPG method for enhanced-scan transition delay faults. Transition delay faults can be detected using a pair of test patterns which are generated using a 8-value logic encoding system. The encoding facilitates mapping of the ATPG into a SAT problem and results in a higher fault coverage. A probability-based scoring technique has been proposed to reduce the search space for fault diagnosis of scan-based designs [19]. The approach utilizes a signal probability analysis to rank faulty candidates.

Assertion-based validation is widely used in pre-silicon validation to create potential behavioral scenarios in order to increase coverage criteria [6]. However, in post-silicon validation, it is difficult to determine whether a set of assertions has been covered. There has been many efforts to generate tests for activating assertions [13, 15, 22]. Chen et al. [8] use model checker to generate directed test using TLM designs to overcome the complexity of RTL designs. However, none of these approaches consider observability constraints.

A fundamental problem in post-silicon validation is the lack of observability and controllability—only a few hundred among the millions of internal signals of an IC can be directly observed or controlled during silicon execution. This makes it difficult to diagnose bugs from observed failures of post-silicon tests, or even identify whether a test has passed, e.g., if the result of a test affects a signal which is not observable, it is difficult to determine whether the test has executed as expected.

To address this problem, it is critical that post-silicon tests be *observability-aware*, i.e., produce results whose values can be reconstructed from the available observability [10]. Unfortunately, this is difficult to achieve for several reasons. First, in an industrial IC development environment, observability architecture and (post-silicon) directed tests are developed independently and concurrently by different teams at different points of the design life cycle. It is often impossible for test generation teams to account for silicon observability since the observability architecture may not have been fully developed at the time of test generation. Furthermore, it is difficult to employ automated tools for creating (additional) observability-friendly directed tests after the observability architecture has been defined. Creating the observability architecture entails analysis of the RTL models to identify traceable signals; these signals are then routed through appropriate hardware instrumentation to an observation point such as an output pin or memory [5, 12, 18]. On the other hand, analysis of RTL models directly to identify test generation is typically infeasible. RTL models tend to be large and complicated (typically millions of lines of code) making such analysis beyond the capacity of test analysis tools. RTL models may also contain functional or design errors. Indeed, a key reason for post-silicon directed testing is to identify such errors. Consequently, if one develops the directed tests through analysis of the RTL, then the fidelity of the tests as well as any inference made on their effects on observability may become questionable.

In this chapter, we present an approach to develop observability-aware post-silicon directed tests through analysis of pre-silicon design collaterals. The key approach to overcome the scalability and relevance challenges mentioned above is to exploit more abstract TLM designs to perform the analysis. TLM is a high-level approach to modeling digital systems by hiding the details of the implementation of functional units or of the communication architecture. In other words, TLM definitions are much more abstract, structured, and compact, compared to RTL, which permits effective application of exploration to identify high-quality directed tests. A key challenge, however, is to map design functionality and observability collaterals between TLM and RTL so that the tests generated at TLM can be translated to effective, observability-aware tests for RTL. We discuss how to develop this mapping in practice.

The basic idea is to transform an RTL assertion as well as observability constraints to create a TLM assertion with observability constraints. The TLM assertion/property would be used to construct a TLM test. Therefore, test properties in the form of computation tree logic or linear temporal logic formulas are generated based on the fault models. In the next step, the properties (negated version) are applied on the formal model using a model checker to generate the required test cases (counterexamples). In other words, the proposed approach has four steps: (i) mapping observability constraints as part of test targets, (ii) mapping test targets from RTL to TLM, (iii) test generation using TLM description, and (iv) translating TLM tests to RTL. Finally, the TLM test would be translated to an RTL test that is debug-friendly. To improve the scalability of model checking based test generation methods, both property and design decomposition procedures can be developed using SAT-based bounded model checking (BMC).

The remainder of the chapter is organized as follows. Section 7.2 discusses the overall framework, some of the challenges faced, and the approach to overcome them. Section 7.3 discusses the experimental results. Finally, Sect. 7.4 concludes the chapter.

7.2 Observability-Aware Test Generation

We pose the problem of observability-aware test generation as follows. First, we present problem formulation. Next, we outline the assumptions and overview of the framework. Finally, we describe each of the four major steps in the framework.

Problem Description. Suppose we have an RTL model M, a set of checkers and coverage conditions $\mathscr{A}$ to be exercised in post-silicon, and a set of traceable signals $\mathscr{S}$. The goal is to develop directed tests for exercising $\mathscr{A}$ such that the results of the test can be inferred by observing the signals in $\mathscr{S}$.

Assumption and Viability. The approach uses TLM models for post-silicon test generation. Some key constraints should be imposed on the underlying design for viability of the approach. The requirements are based on experience with current industrial methodologies in SoC design, albeit somewhat abstracted to provide a clean formulation. The key requirement is the existence of a TLM definition of the system (in addition to RTL), where the TLM is assumed to be the "golden" specification. Although this requirement is not directly satisfied in industrial flows, it is indeed viable for the following reason. Exploration and development of SoC designs starts from architectural models which generates architectural specification documents. The specifications can be easily harvested to generate TLM descriptions, and indeed, such approaches are often used to enable early system level architectural validation before RTL. The second requirement is that TLM and RTL models must have the same external (input–output) interfaces. (Of course the internal registers and control variables are expected to be different between TLM and RTL, and the functionality definition in TLM is more abstract.) Correspondingly, for SoC designs composed of a number of hardware or software intellectual property (IP) blocks, it is required for the IPs to have the same interface variable definitions in TLM and RTL models at each network-on-chip (NOC) interface. These constraints are also viable since a primary purpose of TLM and architecture models in current industrial practice is to provide a prototyping environment for software and firmware development. Consequently, the definition of hardware interfaces and definitions of registers that are visible to firmware and low-level configuration, control, and debug software are defined together with architectural specification and integrated into the TLM model and thereby fixed at architecture definition time, even though the internal functionality of the RTL may not have been implemented yet. Finally, only assertions from $\mathscr{A}$ that are stutter-insensitive, i.e., oblivious to a refinement of the RTL in which a single transition is replaced by a finite sequence, will be considered. The general class of stutter-insensitive properties is LTL\X properties (X denotes next operator). Tests for stutter-insensitive properties are natural targets for generation based on TLM since the TLM models are untimed.

Figure 7.1 provides an overview of the approach. The approach involves four important steps: (i) mapping observability constraints as part of test targets, (ii) mapping test targets from RTL to TLM, (iii) test generation using TLM description, and (iv) translating TLM tests to RTL. The basic idea is to transform a RTL assertion (ϕ) as well as observability constraints (ψ) to create a modified RTL assertion with observability constraints (π). The modified assertion needs to be mapped to TLM assertion (α). The TLM assertion/property would be used to construct a TLM test. Finally, the TLM test would be translated to an RTL test. In the remainder of this section, we describe each of the steps in detail.

7.2.1 Mapping Observability Constraints

Let M_R and M_T be the RTL and TLM models of a design with (common) primary inputs $I = \langle I_1, I_2, \ldots, I_n \rangle$ and primary $O = \langle O_1, O_2, \ldots, O_m \rangle$. Let $R = \langle R_1, R_2, \ldots, R_l \rangle$ be the set of observable RTL signals. Consider a stutter-insensitive RTL assertion ϕ over I, O and R from $\mathscr{A}$. For the purpose of the discussion below, it is convenient to think of ϕ as an LTL$\setminus$ X formula. The method for generating observability-aware tests for ϕ involves the following steps.

1. **Trace Cone-of-influence Calculation**: The control/data flow of the RTL is traversed backwards from the signals in R to the variables in ϕ. This cone-of-influence calculation is made under the constraint that ϕ holds. All signals in R whose cone of influence does not include any variable in ϕ are discarded.
2. **Assertion Propagation**: Symbolic simulation is utilized to forward-propagate variables in ϕ along the cone of influence found in Step 1. The result is a restate-

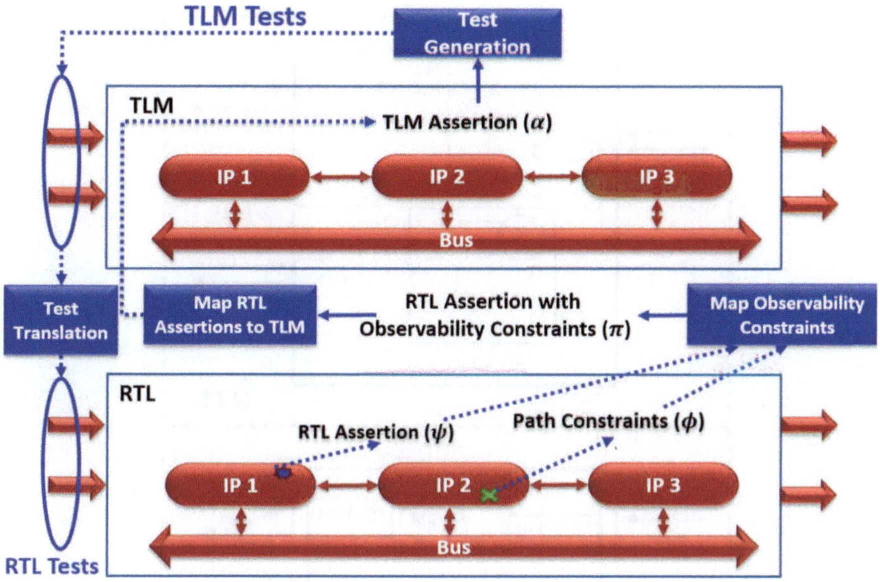

Fig. 7.1 The proposed methodology with four important steps

ment of ϕ into a new formula ψ stated in terms of traceable variables (including signals in R and O).

3. **Assertion Abstraction**: A formula π subsuming ϕ and ψ is constructed as follows. (i) If ψ is consequent of ϕ, $\pi : (\phi \rightarrow F\psi)$ and vice versa. (ii) If ϕ and ψ can be satisfied concurrently, $\pi : (\phi \wedge \psi)$.

To be able to accomplish our goal, RTL is analyzed to find the effect of the activation of RTL assertion ϕ on related trace signals first. In other words, control and data flow graph (CDFG) is traversed backward from trace signals to reach the points that conditions of ϕ are true. (Note that, if a trace signals is placed before one of the conditions of ϕ, forward traversing of CDFG is recommended, however, this case can be covered in next Section's operations).

Example 1 Figure 7.2 shows a router in RTL and TLM that receives a packet of data from its input channel. The router analyzes the received packet and sends it to one of the three channels based on the packet's address. F_1, F_2, and F_3 receive packets with address of 1, 2, and 3, respectively. Input data consists of three parts: (i) parity (*data_in*[0] in RTL and *pkt_in.parity* in TLM) (ii) payload (*data_in*[7..3] in RTL and *pkt_in.payload* in TLM) and (iii) address (*data_in*[2..1] in RTL and *pkt_in.addr* in TLM).

The RTL implementation consists of one FIFO connected to its input port (F_0) and three FIFOs (one for each of the output channels). The FIFOs are controlled by an FSM. The routing module reads the input packet and asserts the corre-

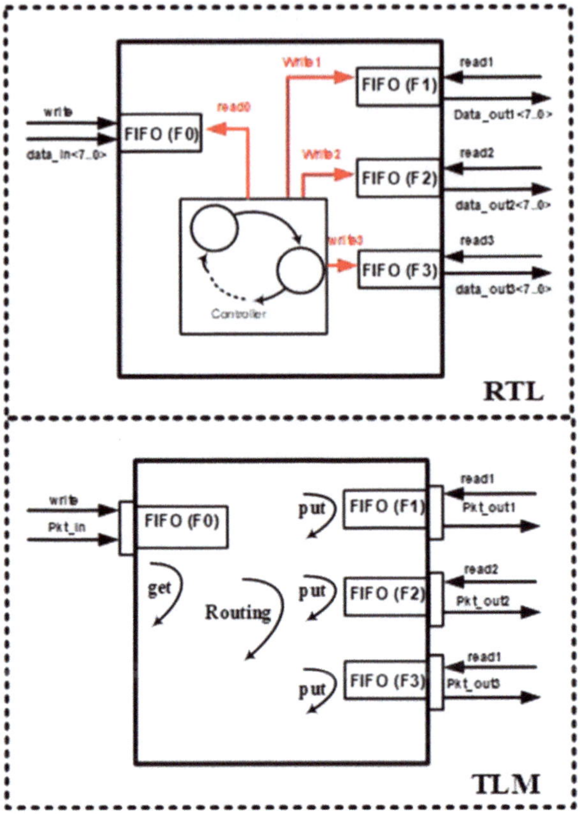

Fig. 7.2 Router design, RTL and TLM implementations

sponding target FIFO's write signal (write1, write2, and write3). Consider generating a test to check that signal $read0$ from F_0 (which is internal signal) is not stuck at zero. The corresponding assertion, written as an (LTL\ X) formula, is $(\phi : \mathsf{F}\ read0)$. Suppose that the address part of input data of F_1 ($F_1.data\_in[2..1]$) is selected as a trace signal. In order to observe activation's effect of ϕ through $F_1.data\_in[2..1]$, the following predicate must be true two cycles after $read0$ becomes true: $\psi : F_1.write1 \wedge F1.pkt\_in[2..1] = 1$. Thus, following the above steps we get

$$\pi : Fread0 \rightarrow XX(F_1.write1 \wedge F1.data\_in[2..1] = 1)$$

7.2.2 *Mapping Test Targets (Assertions) from RTL to TLM*

The key challenge in mapping assertions from RTL to TLM is to bridge the abstraction level between the two designs. This will be achieved by exploiting the commonality of interfaces. The goal is to find TLM property α that is *test equivalent* of RTL assertion π constructed in the previous section. Hereby *test equivalence*, we mean that they generate equivalent tests or counterexamples. The problem reduces to transforming π into a formula α such that (i) α is an LTL\X property over I and O, and (ii) If a test T is a counterexample of α in TLM then T is also a counterexample to π in RTL. If π contains internal RTL variables, it needs to be turned into a test equivalent RTL LTL\X property where it is only over the variables in the interface.

Assertion α can be defined through symbolic simulation of variables in π analogous to the previous section, but this time over the TLM model. Suppose that π is a temporal logic formula over $P_1, P_2, \ldots, P_n$ where each P_i shows one condition on interface or internal variables. For propagation to interface variables, CDFG is traversed backward from point/points that P_i is true to reach primary inputs.[1] As a result, each of P_i be restated as a temporal formula θ over $Q_i = q_1, \ldots, q_m$ where q_j denotes a condition on interface variables. Note that θ may not be LTL\X property since π (and hence θ). The next step is to remove exact timing notation from θ. The approach is based on the observation that the original assertion ϕ is stutter-insensitive. Thus, distributive property is applied such that their operands are atomic (e.g. $X(q_i \wedge q_j) \equiv X(q_i) \wedge X(q_j)$). In addition, the following rules can be applied.

- $F\ (Xp) = F\ p$. Thus, operator F can subsume X.
- A property $p \rightarrow XX...Xq$ can be replaced by $p \rightarrow Fq$.
- A property $p \wedge X\neg p$ can be replaced by $p \wedge F\neg p$.
- A property $p_1 \wedge X..Xp_2$ can be replaced by $p_1 \rightarrow F(p_2)$ when p_2 is a condition on variables from set O.

The modified assertion is an LTL\X that contains conditions on interface variables so it can be applied on TLM. In fact, assertion π is mapped as a sequence of *put* and *get* transactions. The next step is to perform name mapping when the interface signal names are not identical. The resultant assertion is the desired assertion α.

Example 2 Consider assertion π from Example 1, we want to turn it to TLM assertion α which is time insensitive and it is formulated over interface variables. From CDFG traversal, we know that signal *read*0 (shown in Fig. 7.2) is asserted when F_0 is not empty. Thus, $X(Xread0 = 1) \equiv (\neg F_0.empty)$. Having non-empty F_0 implies that *write* signal of F_0 has been asserted before. Thus, we can rewrite the formula as $XX(F0.read0 = 1) \equiv X(\neg F_0.empty) \equiv (F_0.write)$ Using the knowledge $(F1.data\_in[2..1] = 1 \wedge F_1.write1) \equiv (X(F1.read1) \wedge XX(F1.data\_out1$ $[2..1] = 1)$. The following formula is obtained:

$$\theta : write \rightarrow \mathbf{FXX}(read1 \rightarrow \mathbf{FX}data\_out1[2..1] = 1)$$

Finally, assertion α (after name mapping) can be written as

$$\alpha : Fwrite \rightarrow \mathbf{F}(read1 \rightarrow \mathbf{F}pkt\_out1.addr = 1)$$

[1] For some assertions like $P_i \rightarrow P_j$, backward traversal from P_i and forward traversal from P_j would be beneficial. However, in most of the cases performing one of them is enough.

Table 7.1 Directed TLM test cases for the router shown in Fig. 7.2

	Step 1	Step 2–3	Step 4
pkt_in.parity	1	...	...
pkt_in.addr	1	...	...
pkt_in.payload	11000	...	...
write	1	...	...
*read*1	0	0	1

7.2.3 Test Generation at TLM Level

An assertion α represents a functional property which holds in the design and violation of it exhibits a design fault. Assertion-based test generation methods take property $\neg\alpha$ and use model checkers to generate a counterexample for $\neg\alpha$. In other words, checking property $\neg\alpha$ leads to generate a test which can activate the scenario of the property α. Therefore, proper set of assertions results in higher fault coverage and guarantees the success of property based test generation.

SMV model checker [7] is utilized to find the counterexample of property $\neg\alpha$ over SMV model of TLM. The counterexample's assignments to primary inputs is the TLM test case.

Example 3 Consider TLM model of the router as shown in Fig. 7.2 and property of α from Example 2. The goal is to generate a TLM test which satisfies the property α. Thus, the model checker is run over assertion $\neg\alpha$. The generated test case can be seen in Table 7.1. The TLM test has assignments to various inputs (*parity*, *addr*, *payload* and *write*) at the first time step, and the corresponding *read1* is enabled in the fourth time step.

7.2.4 Translate TLM Tests to RTL Tests

The final step is to map TLM test vectors to the RTL tests. Since TLM test lacks the timing information in RTL implementation, they cannot be applied to RTL directly. The mapping process consists of two parts. First, the input/output variables are mapped. Next, templates are utilized to map TLM transactions to sequence of RTL computations. The template enables addition of timing relationship. This process is the inverse of the RTL to TLM assertion mapping. The timing relationship in templates can be provided by the designers or can be extracted by design analysis tools [8]. Example 4 shows how template can be utilized for TLM to RTL test translation.

Example 4 This examples translates the TLM test shown in Table 7.1 to an RTL test (Table 7.2). In the first step, all the names are mapped, e.g., the decimal value of

Timing Sequence

```
Template read (int fifoNum, bool val)
  #2 read%fifoNum = val;
end
Template forward (int fifoNum, data)
  F%fifoNum.data_in = data;
   write%fifoNum = 1' b 1 ;
   #1 -- wait to data places on FIFO
end
```

```
SPEC route(pkt_in, write, read1)
   Initialize();
   write = write;
   data_in = pkt_in;
    read (0, 1);
   # 1  -- wait to data places on output of F0
   forward(pkt_in.addr, pkt.payload);
   read (pkt_in.addr, read1);
end
```

Fig. 7.3 An example of timing sequence to translate TLM tests to RTL tests

Table 7.2 Directed RTL test case corresponding to Table 7.1

	Cycle 1	Cycle 2–6	Cycle 7
data_in[0]	1	...	...
data_in[2..1]	01	...	...
data_in[7..3]	11000	...	...
write	1	...	...
*read*1	0	0	1

pkt_in.addr is mapped to 2-bit value *data_in*[2..1]. Next, timing sequence is added to the RTL test by utilizing the template *SPEC route* in Fig. 7.3. In this case, all the RTL inputs (except *read1*) are initialized in the first clock cycle. The *read1* is enabled in clock cycle 7 based on the delay information in Fig. 7.3. This is due the fact that *SPEC route* template invokes *read(0,1)* function (consumes two cycles), followed by one cycle wait, *forward* function consumes another two cycles and the last read function needs two cycles. In other words, *read1* is enabled six cycles after the corresponding inputs are initialized. The translated test case is shown in Table 7.2.

The designers develop templates based on the relationship between TLM and RTL. For example, in Fig. 7.3, *read* function consumes two cycles due to the following events. When *write* signal is asserted in the first cycle, F_0 (shown in Fig. 7.2) receives the input packet in the second cycle and in the third cycle, the read signal of F_0 (*read*0) can be asserted as we are sure that the F_0 is not empty now.

It can be addressed in two phases. First, finding the corresponding values for control signals that have been abstracted for TLM level test generation. For instance, if there is a *data-enable* signal in RTL model which does not exist in TLM model, it should be asserted whenever the input data get a valid value. The second phase is to add time constraints to the test vectors. In this phase, each test vector contains all the valid values for RTL model's primary inputs. However, we have to define the exact clock cycles that their values should be assigned. This kind of information can be found from FSM and control diagram of the design.

7.3 Case Studies

We discuss the application of our approach on two case studies: a NoC switch protocol and a pipelined processor. In these experiments, we make use of Bounded SMV Model Checker [7] to optimize test generation time.

7.3.1 Wormhole Protocol on NoC Switches

Switches are used as the building block of a NoC. They receive packets as input and forward it to respective output ports. In this case study, the router uses wormhole routing protocol. We consider test generation for five interesting properties: property 1 is related to reservation of output port of channels, property 2 is about making two internal FIFO's full at the same time, property 3 is about receiving a packet with a specific value, property 4 is related to forcing the acknowledgment signal true, and property 5 is related to deadlock detection.

Table 7.3 shows the effectiveness of our method in generating observability-aware tests for these five properties. Since there are no prior efforts for observability-aware post-silicon test generation, we have tried to show the usefulness of our approach in two ways: (i) our approach (with observability constraints) takes reasonable test generation time compared to directed test generation (without observability), (ii) random test generation may be infeasible to activate buggy scenarios and propagate their effects to trace signals. We also tried to generate the test directly from RTL when the RTL is buggy; however, test generation failed because the counterexample cannot be produced. This observation emphasizes the importance of test generation in golden TLM. Table 7.3 provides statistics of test generation on four other selected properties. The column *DirectedTG* shows the needed time for generating directed test without considering observability constraints. The time consumption is comparable with the proposed approach but the effect is not observable on trace signals so these tests are not useful for post-silicon. Table 7.3 also shows that TLM random test generation is drastically worse than our approach and in most of the cases it cannot activate the scenario.

Table 7.3 Test generation's time for safety properties in network with four switches

Prop.	Random TG (min)	Our Proposed Method	
		Directed TG (min)	TG with Obs. Constraints (min)
Property 1	>600	4.83	6.11
Property 2	>600	3.45	7.72
Property 3	205	0.49	2.45
Property 4	502.6	1.85	4.73
Property 5	>600	8.54	13.49

Table 7.4 Test generation's time for safety properties in a pipelined processor

Prop.	Random TG (min)	Our Proposed Method	
		Directed TG (min)	TG with Obs. Constraints (min)
Property 1	>600	0.83	1.90
Property 2	92.10	1.12	1.17
Property 3	297.05	3.00	7.47
Property 4	416.02	1.12	1.36

7.3.2 Pipelined Processor

We have applied the method on a MIPS processor. The processor has five stages: Fetch (it fetches the new instruction from memory), Decode (it decodes the instruction and reads the possible operands), Execute (it executes the instruction), MEM (it is responsible for load and store operations) and WriteBack (stores the results in instruction's target register). These stages are implemented by one or more IP block in both RTL and golden TLM implementations. These IP blocks are connected by FIFOs together.

The result of test generation based on properties related to testing fetch, decode execution units of the processor are reported in Table 7.4. It is obvious that test generation with observability constraints is most beneficial for post-silicon validation.

7.4 Conclusion

This chapter presented a high-level directed test generation method based on a golden TLM model of the design. The proposed method generates directed tests at TLM level and maps them back to RTL level to measure the effectiveness of the generated tests. The tests are generated in such a way that not only activate buggy scenarios (especially the scenarios that are hard to activate), but also they guide the effect of buggy scenario to observable points in order to help the debugger to root-cause the source of faults. The approach has several merits. First, it enables test generation for buggy RTL designs since the tests are generated using golden TLM model. Second, the proposed method overcomes the size limitation of RTL models since TLM models are significantly less complex than RTL implementation. Thus, the complexity of applying model checkers at TLM level is lower. Finally, the test generation takes observability into consideration by forcing results of the buggy scenario activation to the trace signals. Case studies using NOC router, and processor designs demonstrated the effectiveness and feasibility of observability-aware test generation.

References

1. A. Adir, E. Almog, L. Fournier, E. Marcus, M. Vinov, A. Ziv, Genesys-pro: innovations in test program generation for functional processor verification. **2**, 84–93 (2004)
2. A. Adir, S. Copty, S. Landa, A. Nahir, G. Shurek, A. Ziv, C. Meissner, J. Schumann, A unified methodology for pre-silicon verification and post-silicon validation, in *Design Automation and Test in Europe (DATE)* (2011), pp. 1–6
3. A. Ahmed, P. Mishra, Quebs: qualifying event based search in concolic testing for validation of RTL models, in *2017 IEEE 35th International Conference on Computer Design (ICCD)* (IEEE, 2017), pp. 185–192
4. A. Ahmed, P. Mishra, Directed test generation using concolic testing of RTL models, in *IEEE/ACM Design Automation and Test in Europe (DATE)* (IEEE, 2018)
5. K. Basu, P. Mishra, Efficient trace signal selection for post silicon validation and debug, in *2011 24th International Conference on VLSI Design (VLSI Design)* (2011), pp. 352 –357
6. N. Bombieri, R. Filippozzi, G. Pravadelli, F. Stefanni, RTL property abstraction for TLM assertion-based verification, in *Design Automation and Test in Europe (DATE)* (2015), pp. 85–90
7. Cadence Berkeley Lab, http://www.kenmcmil.com. The Cadence SMV Model Checker
8. M. Chen, P. Mishra, D. Kalita, Automatic RTL test generation from SystemC TLM specifications. **11**(2012)
9. M. Chen, X. Qin, H. Koo, P. Mishra, *System-Level Validation - High-Level Modeling and Directed Test Generation Techniques* (Springer, Berlin, 2012)
10. F. Farahmandi, P. Mishra, S. Ray, Exploiting transaction level models for observability-aware post-silicon test generation, in *2016 Design, Automation and Test in Europe Conference and Exhibition (DATE)* (IEEE, 2016), pp. 1477–1480
11. T. Hong, Y. Li, S. Park, D. Mui, D. Lin, Z. Kaleq, N. Hakim, H. Naeimi, D. Gardner, S. Mitra, QED: quick error detection tests for effective post-silicon validation, in *International Test Conference (ITC)* (2010)
12. H.F. Ko, N. Nicolici, Algorithms for state restoration and trace-signal selection for data acquisition in silicon debug. IEEE Trans. Comput. Aided Des. Integr. Circuits Syst. **28**(2), 285–297 (2009)
13. T. Li, Y. Guo, S. Li, Assertion-based automated functional vectors generation using constraint logic programming, in *ACM Great Lakes Sympoisum on VLSI* (2004), pp. 288–291
14. D. Lin, T. Hong, F. Fallah, N. Hakim, S. Mitra, Quick detection of difficult bugs for effective post-silicon validation, in *Design Automation Conference (DAC)* (2012), pp. 561–566
15. L. Liu, D. Sheridan, W. Tuohy, S. Vasudevan, Towards coverage closure: using goldmine assertions for generating design validation stimulus, in *Design Automation and Test in Europe (DATE)* (2011), pp. 173–178
16. L. Liu, S. Vasudevan, Efficient validation input generation in RTL by hybridized source code analysis, in *Design Automation and Test in Europe (DATE)* (2011)
17. P. Patra, On the cusp of a validation wall. IEEE Des. Test Comput. **24**, 193–196 (2007)
18. K. Rahmani, P. Mishra, S. Ray, Scalable trace signal selection using machine learning, in *2013 IEEE 31st International Conference on Computer Design (ICCD)* (2013), pp. 384–389
19. H. Sabaghian-Bidgoli, P. Behnam, B. Alizadeh, Z. Navabi, Reducing search space for fault diagnosis: a probability-based scoring approach, in *2017 IEEE Computer Society Annual Symposium on VLSI (ISVLSI)* (IEEE, 2017), pp. 545–550
20. E. Sadredini, R. Rahimi, P. Foroutan, M. Fathy, Z. Navabi, An improved scheme for precomputed patterns in core-based soc architecture, in *2016 IEEE East-West Design and Test Symposium (EWDTS)* (IEEE, 2016), pp. 1–6
21. M. Sousa, A. Sen, Generation of TLM testbenches using mutation testing, in *International Symposium on Harwdare Software Codesign and System Synthesis (CODES+ISSS)* (2012)
22. J. Tong, M. Boule, Z. Zilic, Airwolf-tg: A test generator for assertion-based dynamic verification, in *International Workshop on High Level Design Validation and Test (HLDVT)* (2009), pp. 106–113

23. I. Wagner, V. Bertacco, Reversi: post-silicon validation system for modern microprocessors, in *International Conference on Computer Design (ICCD)* (2008), pp. 307–314
24. S. Yerramilli, Addressing post-silicon validation challenge: leverage validation and test synergy. Keynote Int. Test Conf. (2006)
25. F. Zokaee, H. Sabaghian-Bidgoli, V. Janfaza, P. Behnam, Z. Navabi, A novel sat-based ATPG approach for transition delay faults, in *2017 IEEE International High Level Design Validation and Test Workshop (HLDVT)* (IEEE, 2017), pp. 17–22

Chapter 8
On-Chip Constrained-Random Stimuli Generation

Xiaobing Shi and Nicola Nicolici

8.1 Post-silicon Constrained-Random Validation

The unique requirements for generating functionally compliant sequences differentiate the constrained-random methodology both for verification and validation from the traditional directed and random test methods. The generated stimuli are required to be constrained, so that it eliminates a large percent of useless stimuli that are generated randomly. The constrained-random verification methodology has become the mainstream to verify system-wide functionality. For pre-silicon verification, the Pseudo-Random Number Generator (PRNG) embedded in the simulator can be constrained to generated stimuli that satisfy user-defined constraints [11]. The state-of-the-art simulation tools, e.g., [12], rely on built-in algorithms that can guarantee both stimuli randomness, as well as constraint satisfaction, which generally comes at an acceptable overhead in simulation speed.

The process of post-silicon validation requires a large volume of *random, yet functionally compliant* stimuli extensive periods of time to expose the design errors that have escaped to the silicon prototypes [7]. However, this is a challenging task because the methods applied in pre-silicon verification cannot be reused in a straightforward manner due to the unique environment at the post-silicon phase. On the one hand, transmitting constrained-random stimuli from a simulation environment to a silicon prototype is impractical due to bandwidth limitations. Also, the volume of stimuli that can be generated in a given time is limited by the relatively slow speed of software tools running on the host, thus becoming impractical to meet the real-time constraints of post-silicon validation. On the other hand, the attempt of emulating the behavior of simulator algorithms in hardware is impractical because of both the

X. Shi (✉) · N. Nicolici
McMaster University, Hamilton, ON, Canada
e-mail: shix6@mcmaster.ca

N. Nicolici
e-mail: nicolici@mcmaster.ca

P. Mishra and F. Farahmandi (eds.), *Post-Silicon Validation and Debug*,
https://doi.org/10.1007/978-3-319-98116-1_8

complexity of the algorithms and excessive cost of hardware resources. Furthermore, the constraints formalized in hardware verification languages, such as SystemVerilog, are not supported natively for synthesis. Even if the expressions for constraints were to be converted into synthesized logic blocks, these blocks would be fixed and could not be updated/reprogrammed when the constraints need to be changed during the post-silicon validation process. An exception is for the microprocessor-centric designs, where the microprocessor itself is used for stimuli preparation and application; nevertheless, precautions need to be taken to circumvent the potentially long delay due to the rejection of invalid stimuli. Therefore, there is a need to deal in a systematic way with designs that have no programmable embedded microprocessors or have many blocks that are not directly accessible by microprocessors, such as high-speed peripherals or hardware accelerators (e.g., video or graphics). For such cases, on-chip constrained-random stimuli generator structures can be employed to generate at-speed functionally compliant stimuli for the silicon prototypes. The stimuli are subjected to constraints consistent with the specification and format of data packets fed to the design-under-validation in an application environment.

Considering the unique requirements for stimuli generation during post-silicon validation for generic logic blocks, this chapter presents programmable solutions for on-chip constrained-random stimuli generation. The main building block is a Constrained-Random Stimuli Generator (CRSG), which can be configured at design time and programmable by the user at validation time. The in-system programmability is achieved using binary cubes, which can be translated from user-defined constraints on the host, and are subsequently loaded on-chip to constrain the random sequences generated in hardware at- speed. Therefore, the CRSG can generate functionally compliant stimuli that can be modified and refined from one post-silicon validation experiment to another. The key benefit lies not only in generating randomized functionally compliant stimuli at-speed, but also in the ability to control the debug experiments and to bias the constrained-random sequences as the validation progresses.

In this chapter, Sects. 8.2–8.4 will elaborate on the solutions to provide support at the post-silicon stage for the functional constraints that are widely used during pre-silicon verification. Subsequently, Sects. 8.5–8.8 will discuss how to control the distribution of the stimuli. A summary of the chapter is provided in Sect. 8.9.

8.2 Representation of Functional Constraints

This section provides the technical background for stimuli constraints and how they are described in a hardware verification language. It also provides an overview of hardware-oriented representation of constraints, i.e., cubes.

```
class GreaterEqual;
  rand bit[1:0] x;
  rand bit[1:0] y;
  constraint good {
    x>=y;
  }
endclass
```

(a) A constraint for $x \geq y$.

```
typedef enum {ADD, SUB, SLL, SRL, SRA} OP_T;
class StimuliForALU;
  rand OP_T op;
  rand bit[7:0] opr1, opr2;
  constraint opr2_range {
    if (op==SLL || op==SRL || op==SRA)
      opr2 inside {[0:7]};
  }
endclass
```

(b) A class for generating ALU instructions.

Fig. 8.1 SystemVerilog examples for logic constraints

8.2.1 Functional Constraints in SystemVerilog

Functional constraints provide a way of specifying relationships between input signals of a hardware module. SystemVerilog [2], as a widely adopted hardware description and verification language, provides a variety of programming features to formalize functional constraints for input stimuli and requirements on their distribution. Although the SystemVerilog language is aimed at pre-silicon verification, the functional constraints described with it can be used also for post-silicon validation, as elaborated in this chapter.

SystemVerilog can formalize the requirements for both the logic relationship between signals, as well as the sequential relationship between clock cycles. On the one hand, logic constraints capture the static features and the format of a pattern, e.g., relational expressions, Boolean expressions, if-else implications, or set relations. For instance, Fig. 8.1a illustrates a simple SystemVerilog code for the constraint $x \geq y$, in which x and y are 2-bit unsigned integers. Another example in Fig. 8.1b combines different types of logical constraints to generate instructions randomly for the functional verification of a simple microprocessor, where different instructions have distinct constraints on the operand fields. Specifically, the second operand for left shift (**SLL**), arithmetic right shift (**SRA**) and logical right shift (**SRL**) instructions is constrained to be less than 8. This constraint captures the relationships between fields within the same clock cycle. Note that, for this particular example, there are no constraints in between two consecutive instructions.

Some designs require stimuli where the randomized patterns in each clock cycle follow a prescribed order and/or format for specific protocols or standards. Sequential constraints capture the dynamic behavior between consecutive stimuli. The sequential constraint in the **randsequence** block in Fig. 8.2 is used to generate one frame in multiple cycles. It employs 3 **class** blocks to define logic constraints for slices (**head**, **tail** and **body**).

```
logic [15:0] stimulus;
task GenerateRandomFrame;
  parameter body_size=16;
  begin
    FrameHead head=new; // The class for the frame head
    FrameBody body=new; // The class for the frame body
    FrameTail tail=new; // The class for the frame tail
    randsequence (frame)
      frame: fhead repeat(body_size) fbody ftail;
      fhead: {head.randomize(); stimulus=head.o; @(posedge clk);};
      fbody: {body.randomize(); stimulus=body.o; @(posedge clk);};
      ftail: {tail.randomize(); stimulus=tail.o; @(posedge clk);};
    endsequence
  end
endtask
```

Fig. 8.2 An example of a valid frame in multiple cycles using sequential constraints

Fig. 8.3 The K-map for valid elements compliant to $x \geq y$, where x and y are 2-bit unsigned integers

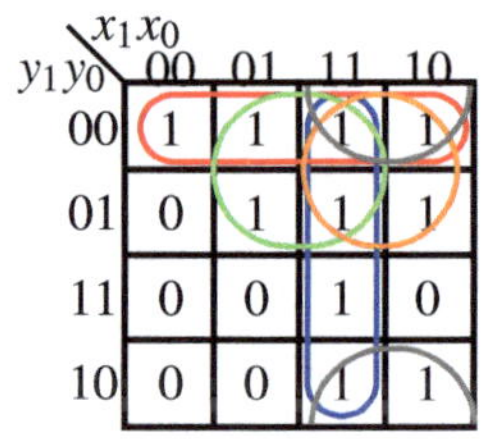

8.2.2 *Equivalent Cubes*

In order to reuse function constraints for on-chip validation, an equivalent type of representation, based on *cubes*, is used. A cube is an m-symbol vector comprised of "0", "1", and "X" (don't-care bits). It implies a set of vectors, (i.e., implied space) where "X"s are replaced with either "0"s or "1"s. For instance, the 4-symbol cube "1X0X" implies the set {1000, 1001, 1100, 1101}.

The equivalent set of cubes exactly covers all the *true* elements in the Karnaugh map (K-map) compliant to user-defined logic constraints. Figure 8.3 shows the K-map for the constraint in Fig. 8.1a. The equivalent set of cubes (formatted as $x_1x_0y_1y_0$) is {1X0X, 11XX, XX00, X10X, 1XX0}.

The left part of Fig. 8.4 sketches the cube content processing flow. The K-map exemplifies the step of converting simple constraints to cubes based on Boolean function minimization. The step identifies product terms (minterms) and lowers the total number of the minterms that can exactly cover all the valid stimuli. For complex constraints, a third-party two-level logic minimization tool, e.g., Espresso [6] can be used. Implementations for cube conversion have been detailed in [8], which rely on customized tools and hardware synthesis algorithms.

Then cubes are encoded into binary cubes. The symbols "0", "1", and "X" are encoded as 00, 01, and 10 respectively. For instance, the 4-symbol cube "1X0X" is encoded into an 8-bit binary cube 01100010. Optionally, they can be compacted to reduce the size of on-chip cube memory, yet at a cost of inserting a hardware module (decompression logic) to restore binary cubes on-chip. A run-length compaction for cubes featuring a low-area cost decompression logic is discussed in detail in [10].

8.3 On-Chip Generator for Functional Constraints

At design-time, the configuration of the on-chip generator (i.e., CRSG shown in the right of Fig. 8.4) is selected, e.g., whether to support only logic constraints or sequential constraints as well. At runtime, CRSG is programmed by loading equivalent cubes. The user is given the freedom to reconfigure cubes in CRSG, so as to apply functionally compliant sequences with different (user-programmable) constraints as the validation process evolves. During stimuli generation, the controller fetches a cube from the memory (and decompresses it if it is compacted); meanwhile, pseudo-random generator produces a primitive stimulus per clock cycle. The correction logic rectifies the primitive stimulus to comply with the fetched (or *active*) cube. Hence, CRSG can continuously generate compliant stimuli for the target design. In the following, the details and the variations of CRSGs are elaborated according to the types of constraints that are supported.

8.3.1 Supporting Logic Constraints

The top architecture of CRSG is shown on the right side of Fig. 8.4, in which the modules are customized to support logic constraints.

Cube Memory, which also has a **controller** for addressing, stores the binary cubes (or compact binary cubes). The controller can keep a cube activated for an arbitrary number of cycles to generate a user-controlled amount of valid stimuli. The cube memory can be implemented either as a first-in, first-out (FIFO) addressed by implicit increment, or a dual-port RAM. The throughput can be flexibly adapted to the validation environment. In the case the volume of cubes is very large, CRSG

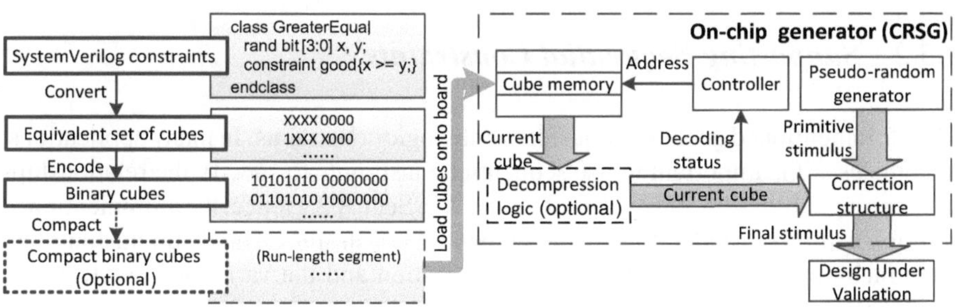

Fig. 8.4 The cube processing flow and the top level of on-chip generator

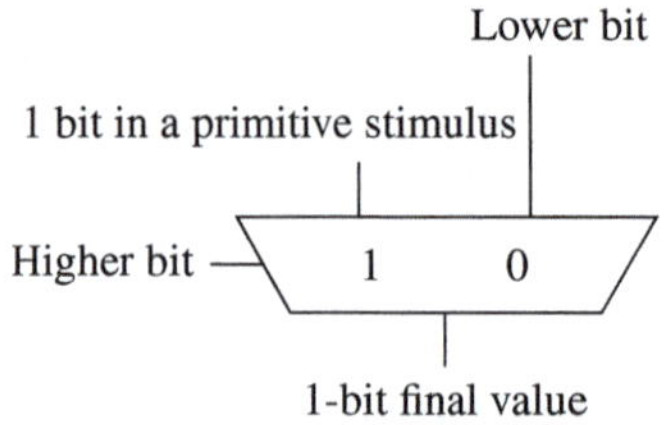

(a) Logic structure to rectify primitive stimulus from an LFSR.

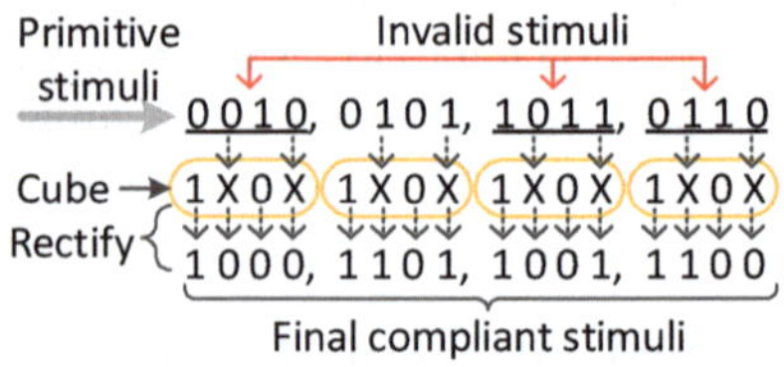

(b) Example for rectification of primitive stimuli.

Fig. 8.5 Rectification of primitive stimuli from an LFSR using 2-bit codes for each bit of the mask

only requires a small on-chip cube memory for buffering a few cubes which will be used during a limited time window for stimuli application. The subsequent subset of cubes can be uploaded via a low-bandwidth interface from a host concurrently with the application of the stimuli expanded from the current subset of cubes.

Decompression Logic is required only if compact binary cubes are loaded. This logic needs to be efficient in terms area cost, while guaranteeing a low decoding latency. For instance, the decompression logic for run-length compaction [10] can be flexibly configured with multiple byte-wise decoders for parallel decompression, so as to balance the compression efficiency, on-chip area, and decoding latency.

Pseudo-random Generator consists of an m-bit maximum-length Linear Feedback Shift Register (LFSR), which can be configured to generate $2^k - 1$ patterns [4]. This type of LFSR has been traditionally used as an on-chip random stimuli generator.

Correction Structure is the essential logic used to rectify each primitive stimulus from the LFSR to a valid stimulus based on the current cube as shown in Fig. 8.5b. It essentially forces the specified bits from a cube and it fills the don't-care bits with random values from the LFSR. It consists of m bit-wise multiplexers shown in Fig. 8.5a. Each multiplexer reads a 2-bit binary code (i.e., 00 is for the symbol "0", 01 for "1" and 10 for "X") and arbitrates whether to output the corresponding bit from the primitive stimulus or to correct it to a constant 0 or 1. The higher bit in the binary code can directly serve as the selection signal and the lower bit is the constant output when it is corrected.

8.3.2 *Supporting Sequential Constraints*

The aforementioned generator considers only logic constraints. In many applications, nonetheless, it is important to use sequential constraints to specify the relationships between signals in adjacent cycles, as outlined in Sect. 8.2. Hence, the aforementioned solution can be expanded to support sequential constraints. The distinguishing features including the cubes with timing information and the variations in CRSG are emphasized.

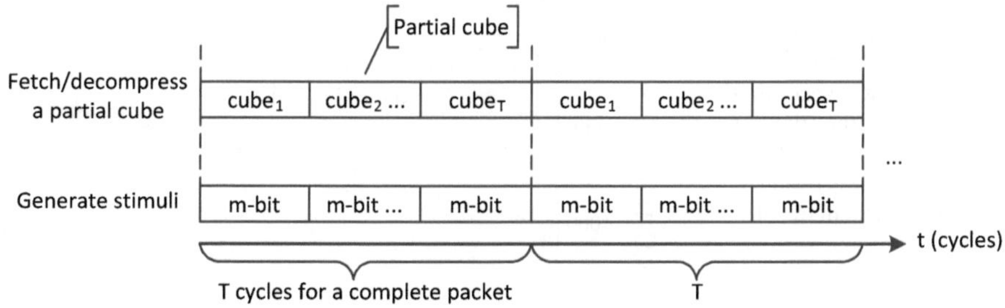

Fig. 8.6 The timeline for generating stimuli with partial binary cubes

Cubes with Timing Information express a sequential constraint which can be seen as a series of partial logic constraints in a specific order. The final cubes combine the partial cubes converted from the partial logic constraints in the order specified in the sequential constraint, which can be activated on-chip one by one in adjacent cycles during stimuli generation. Note that if an optional compaction method is used, it should be simple enough for decompressing process, in order to meet the timing requirement of sequential constraints. For instance, in Fig. 8.2, the stimulus for **FrameBody** is to be generated immediately after a **FrameHead** is generated. Consequently, the partial cubes for **FrameHead** have to be generated no later than this point. A feasible compaction method using loose-coupling run-length algorithm [10] facilitates on-chip logic to restore each cube within one clock cycle and hence meets the need (i.e., constantly switching cubes in each adjacent cycle).

Addressing Controller in CRSG distinguishes from the one discussed in Sect. 8.3.1 in order to support both cubes and partial cubes, which stems from using a single clock cycle for decompression and correction. Because the design supports decoding of a partial cube within one cycle, all the parts of a single cube are processed together within the same clock cycle. In each cycle, a binary cube is fetched from the memory (and decompressed if compacted) to feed other logic blocks as shown in Fig. 8.4. The timeline in Fig. 8.6 illustrates a typical flow of scheduling partial binary cubes in a cube. It generates a complete n-bit stimulus within T cycles, during which the partial cubes are switched in order and an m-bit ($n = mT$) stimulus is generated. After generating a complete n-bit packet, the controller can issue the address to the next cube or repeat the previous cube.

8.4 Experimental Assessment for Functional Constraints

In this section, the performance and the cost of the solutions (including CRSGs discussed in Sects. 8.3.1 and 8.3.2) are examined in applying extensively long constrained-random sequences during post-silicon validation.

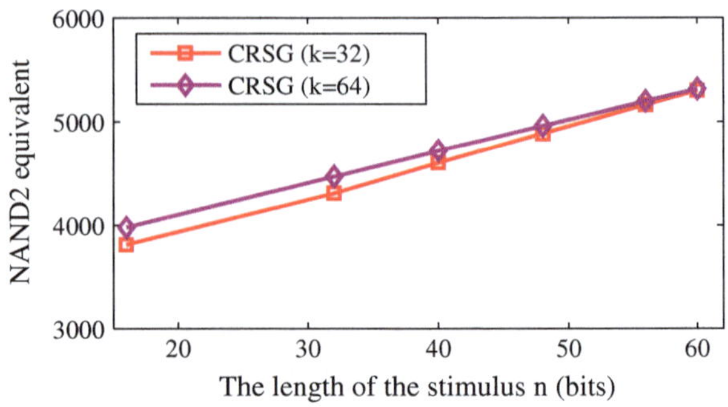

Fig. 8.7 The hardware cost (exclusive of RAM) of the proposed CRSGs according to the length of the stimulus n. The data is collected from the logic synthesis results with a 90 nm standard cell library

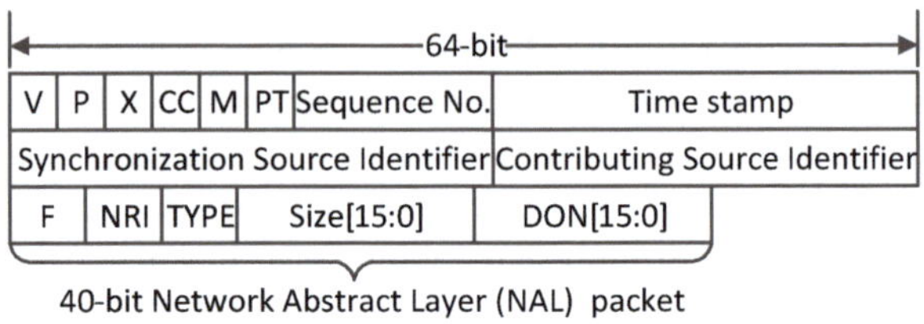

Fig. 8.8 A typical packet format of H.264 RTP

8.4.1 Area Cost

The synthesis results in Fig. 8.7 show that the area cost of CRSGs is dependent on the maximum bit length of the generated stimulus (denote as n-bit) that is applied to the design-under-validation. The adoption of decompression logic can impact the total area cost, which depends on the adopted compaction method, e.g., the decompression logic based on run-length coding method evaluated in [10].

8.4.2 On-Chip Storage

In order to evaluate the effectiveness in reducing the storage requirements for large validation sequences, the CRSG is configured to generate stimuli for resembling 168-bit packet heads for H.264 real-time transport protocol (RTP) [13] shown in Fig. 8.8, as well as 160-bit packet heads in the PCI-express (PCIe) 3.0 transaction layer packet (TLP) format [3] shown in Fig. 8.9.

To generate stimuli compliant to protocols, each field in the packet head must satisfy the requirements specified in the spec, including the format, defined/reserved values, and the coordination among fields. The fields that can be randomized are extracted for the design of constraints, thus leaving the nonrandom CRC field to be attached by the CRC computation logic. The results of converting the logic constraints to cubes according to the cube processing flow shown in the left part of Fig. 8.4 using the Espresso tool [6], are listed in Table 8.1.

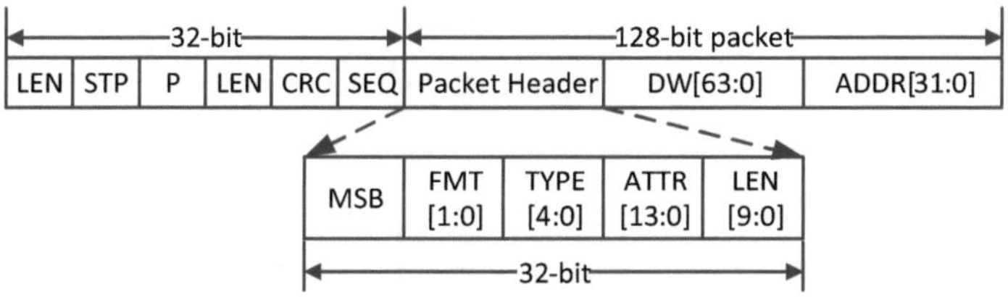

Fig. 8.9 A typical packet format of PCIe 3.0 TLP

Table 8.1 The encoding results of the cube sets for H.264 RTP and PCIe TLP

Packet head	Free bits	Cube count	Binary cube set size	No. of patterns
H.264 RTP	158	335	14.91 KB	1.22×10^{50}
PCIe TLP	127	5119	204.76 KB	8.71×10^{41}

Only the cubes in the binary format are required to be loaded to the cube memory. Considering that the total number of constrained-random patterns that can be applied to the design using only one cube is exponential to the number of free bits in that cube, it is evident that by loading a small amount of binary cubes, the number of stimuli that can be used for validation can easily meet the objectives of real-time execution that lasts for hours.

8.4.3 *The Distribution Quality of Generated Stimuli*

In principle, the distribution of the stimuli generated by CRSGs should rely in part on the uniform distribution associated with stimuli generated by LFSRs; however, it is also strongly dependent on the position of "0"s, "1"s, and "X"s in each cube; in addition, the distribution is also biased by the state of the LFSR when a particular cube is activated. Taking a 4-bit LFSR as an example, if the LFSR has two adjacent states "1100" and "1001", i.e., the LFSR shifts left by one position and it feeds "1' at the rightmost position, an unconstrained CRSG (i.e., the activated cube is "XXXX") will output the two binary values as two consecutive stimuli; hence the distribution of samples at the output of the LFSR will contain each of the samples "1100" and "1001" exactly once. If the activated cube is "XX10", the two generated stimuli are "1110" and "1010", thus each of them will again count once in the distribution. However, if the activated cube is "X1X0", both stimuli from the output of the LFSR will be corrected to "1100" and hence this particular sample will be accounted for twice in the distribution.

Concerning the quality of the randomized stimuli during on-chip generation, the distributions of the stimuli produced by CRSGs are evaluated. Figure 8.10a illustrates the relation between the number of stimuli generated and the number of unique stimuli

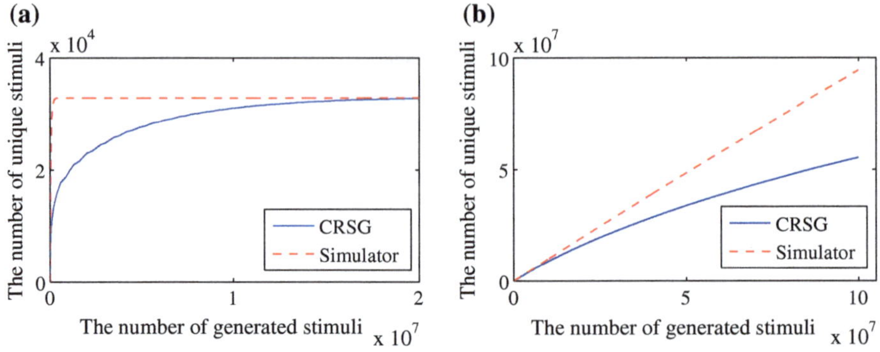

Fig. 8.10 The relation between the number of generated stimuli and the number of unique stimuli based on the constraint $x \geq y$. The unsigned variables x and y are set to 8 bits in **a** and 16 bits in **b** respectively

based on a simple constraint $x \geq y$, in which x and y are unsigned 8-bit variables. If the random values are drawn from the uniform distribution then, until the entire valid space (as defined by the constraint) is exhausted, the value on the vertical axis should almost match the value on the horizontal axis; thereafter, the value on the vertical axis saturates to the maximum number of unique stimuli. The maximum number of valid pairs that satisfy $x \geq y$ is 32,896 and the software-based random generator in a SystemVerilog compliant simulator [12] reaches this saturation point after 543,085 patterns; it should be noted that these results have been obtained using the **rand** type for the randomized variables for the software random generator. As the number of generated patterns increases, the number of unique patterns generated by the hardware method is approximately half of the ones generated in a software simulator (for the same number of total patterns). Figure 8.10b shows this trend more clearly, where x and y are set to be 16 bits each. It should be noted that during post-silicon validation it is reasonable to have experiments with a significantly larger number of clock cycles than during pre-silicon verification; therefore, though the valid randomized stimuli are repeated more often in the hardware implementation, as confirmed by this experiment, the extensive number of clock cycles exercised on silicon prototypes, which is at least four orders of magnitude more than during pre-silicon simulation [5], are expected to compensate for this repetition of constrained-random stimuli.

8.5 Stimuli Distribution During On-Chip Validation

The previous sections have focused on functional constraints. For the post-silicon validation cases, where the distribution of stimuli is important, the observations from this section can be leveraged.

```
class FPNumber;
  rand bit sign;
  rand bit [7:0] exponent;
  rand bit [22:0] fraction;
  constraint number_range {
    exponent == 127;
    fraction[22:20] dist { [6:7]:/3, [1:5]:/2 };
  }
endclass
```

Fig. 8.11 A constraint with **dist** for producing single precision floating-point numbers

8.5.1 Motivation for Distribution Control at Validation Time

The results in Fig. 8.10 show that the unique patterns applied in-system are approximately half of the ones generated by a software simulator. Considering that the redundant cycles due to the repeated stimuli can be reduced, the saved cycles can be allocated to previously unused compliant stimuli that can stress the target design into valid states which otherwise could not have been explored. Correspondingly, SystemVerilog [2] provides the features of random modifiers for pre-silicon verification, **rand** and **randc**, to ensure that the random variables are sampled from a uniform distribution, while only **randc** (referred to as *random-cyclic*)guarantees that a random variable does not get assigned the same value twice until the entire constrained space has been exhaustively enumerated.

Another type of **dist** constraints in SystemVerilog is for weighted distributions, as exemplified in Fig. 8.11 for verifying floating-point unit (FPU) in a microprocessor. It produces real numbers compliant to the IEEE 754 format [1]. The **dist** statement guides some bits in **fraction** to be evaluated in the ranges of [1:5] and [6:7] with the ratio of 2:3 in order to, for example, identify a potentially erroneous division calculation for which the **FPNumber** falls in a suspicious range.

The following question motivates the discussions in the following sections: what would be the cost, in terms of content (e.g., cubes) preparation and on-chip hardware, to support the application of stimuli compliant to specified distributions for in-system validation, while constraints can be dynamically reconfigurable? To answer this question, the enhancement to CRSG design will be presented. First, it is critical to develop the insights into the causes that lead to stimuli repetition.

8.5.2 Explore the Causes of Stimuli Repetition

The repetition of stimuli can be caused by either the overlaps between cubes or sampling the same pattern multiple times from the same cube as elaborated next.

Overlapped Cubes will imply repetitive stimuli. For example, consider two cubes "1X0X" and "11XX" derived from the constraint $x \geq y$ shown in Fig. 8.1a. As illus-

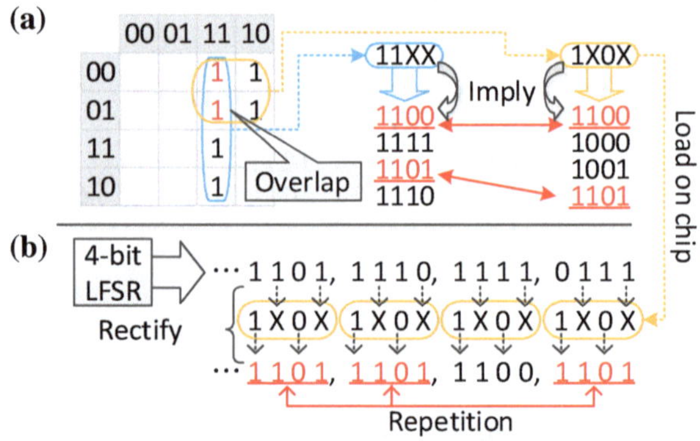

Fig. 8.12 Stimuli repetition due to **a** overlapped cubes and **b** LFSR and correction strategy

trated in Fig. 8.12a, each of these cubes will imply both "1100" and "1101" patterns; therefore, the CRSG will generate these two patterns pattern twice if both cubes are loaded. To address this problem, Sect. 8.6 presents an algorithm to rectify the overlap between cubes, thus guaranteeing that the intersection between any two cubes is void, without any loss of the valid pattern space.

Duplication from Sampling Patterns from the Same Cube is the other cause for stimuli repetition. This problem stems from the interaction between the autonomous sequence generated by the LFSR and the on-chip stimuli correction strategy. As shown in Fig. 8.12b, given a cube "1X0X" and a 4-bit unconstrained sequences {..., 1101, 1110, 1111, 0111}, the correction logic would mask the left first bit with "1" and the third bit with "0". So the stimuli sequence will be {..., 1101, 1100, 1101, 1101}. As a result, although all the masked (rectified) stimuli are made compliant to the constraint (based on cube "1X0X"), pattern "1101" is generated multiple times because the LFSR values for the 'X' positions will happen to be the same for multiple clock cycles. To address this problem, the hardware generator including a dynamic LFSR and a flexible correction component (vector assembler) that can avoid repetition for the *same* cube will be presented. The on-chip correction procedure ensures that, as constrained patterns are generated on-the-fly from any given cube, they are never repeated.

8.6 Generation of Nonoverlapped Cubes

The overlapped cubes are introduced in the cube processing flow as shown in the left part of Fig. 8.4. In particular, the employed logic minimizer tool Espresso is designed for minimizing the number of cubes [6]; hence, it is not concerned with recognizing and removing overlaps between cubes. For instance, it minimizes a given set {110X, 100X, 111X} and deduces the result set as {11XX, 1X0X}, which contains overlapped cubes. Therefore, an add-on algorithm (shown in Fig. 8.13) is used to process the result produced by a logic minimizer.

1: **function** RECTIFYSET(s_{in}) ▷ Rectify all cubes in the given cube set
2: $s_{out} \leftarrow \emptyset$
3: **while** $s_{in} \neq \emptyset$ **do**
4: $a \leftarrow$ FINDREFERENCECUBE(s_{in})
5: Replace each overlap cube b in s_{in} with the result of RECTIFYCUBE(a,b)
6: $s_{out} \leftarrow s_{out} \cup \{a\}$
7: **end while**
8: **return** s_{out}
9: **end function**

Fig. 8.13 The algorithm to rectify overlapped cubes in s_{in} and produce the equivalent set s_{out}

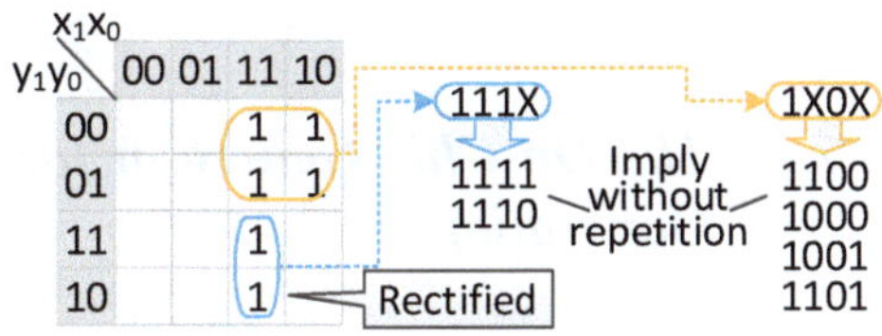

Fig. 8.14 Eliminate repetition by rectifying overlapped cubes

Given a cube set s_{in}, the algorithm produces the rectified set s_{out} *without any loss of the patterns that can be implied*, in which any two cubes are mutually exclusive (i.e., nonoverlapped cubes). In each iteration, FINDREFERENCECUBE chooses a reference cube $\boldsymbol{a}$ from s_{in} (Line 4) and RECTIFYCUBE rectifies any other cube $\boldsymbol{b}$ that overlaps $\boldsymbol{a}$. Figure 8.14 rectifies s_{in} ={1X0X,11XX} from Fig. 8.12a into s_{out} ={1X0X,111X}, which implies the same sample space as s_{in}.

Conceptually FINDREFERENCECUBE can choose an arbitrary cube in s_{in}. The discussion in [9] introduced cube weight to choose a cube in order to achieve the goal of lowering the total number of the final cubes and specified bits.

RECTIFYCUBE produces the relative of complement of $\boldsymbol{a}$ in $\boldsymbol{b}$, i.e., $\boldsymbol{b} - \boldsymbol{a}$, by deducing a series of mutually exclusive cubes to reconstruct the vector space of $\boldsymbol{b} - \boldsymbol{a}$. The function iteratively applies Shannon's expansion shown as (8.1) on $\boldsymbol{b}$ at overlap bit positions i between $\boldsymbol{a}$ and $\boldsymbol{b}$ (i.e., the bit in one cube is specified bit "0" or "1", while the bit in the other cube at the same position is don't-care bit 'X').

$$f(b_i, b_0, b_1, \cdots, b_{m-1}) = b_i f(1, b_0, b_1, \cdots, b_{m-1}) + \overline{b_i} f(0, b_0, b_1, \cdots, b_{m-1}) \tag{8.1}$$

8.7 Variations for Supporting On-Chip Stimuli Distribution Control

This section discusses the implementation of on-chip generators that supports the features of distributions control (i.e., random-cyclic and weighted distributions) introduced in Sect. 8.5. In particular, the hardware modules that eliminate duplication in sampling the same cube are emphasized. In addition, a refined on-chip cube scheduling strategy is presented concerning the issue of unevenly cube sampling.

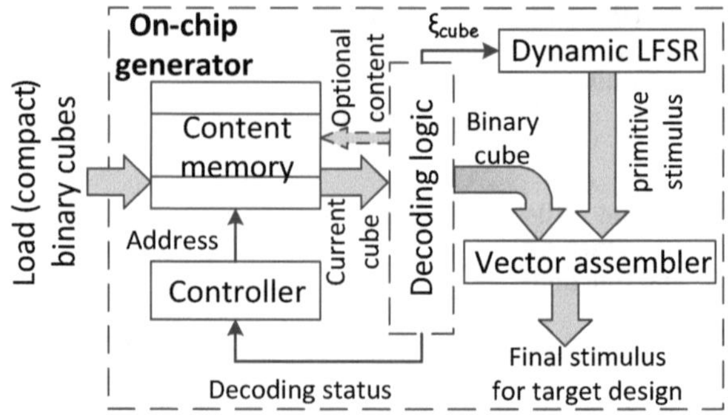

Fig. 8.15 The top level of the on-chip CRSG for random-cyclic stimuli generation

8.7.1 *The On-Chip Generator for Random-Cyclic Stimuli Generation*

Compared with CRSGs for functional constraints, the design for supporting distribution control shown in Fig. 8.15 consists of content memory with addressing controller, decoding logic (including optional decompressing logic), dynamic LFSR as pseudo-random generator, and vector assembler instead of correction logic. The content memory stores binary cubes that have been loaded on chip; it also keeps additional information when needed by the process. The decoding logic counts the number of free bits in the current cube, denoted as ξ_{cube}, which is used by the dynamic LFSR. Optionally, the decompression logic can be integrated if the loaded cubes are in the compact format. In the following, the process of stimuli rectification with the dynamic LFSR and vector assembler is discussed.

The m-bit LFSR produces m-bit vectors. It consists of m 1-bit registers and XOR-gates in its feedback. The behavior can be modeled by a characteristic polynomial $f_m(x) = 1 + c_1x + c_2x^2 + \cdots + c_mx^m$, in which switching function $c_i \neq 0$ $(1 \leq i \leq m)$ if the ith register is used in the feedback. Note if $f(x)$ is a primitive polynomial, the LFSR enumerates nonrepetitive $2^m - 1$ vectors except the all-0 vector in a period of $T = 2^m - 1$ clock cycles [4]. However, such fixed-length LFSR cannot vary its output size according to the number of don't cares in different cubes, which causes stimuli repetition as elaborated in Sect. 8.5.2.

The dynamic LFSR, therefore, changes c_i from a constant to a variable dependent on the number of X's in the cube (denoted as ξ). Figure 8.16 depicts an 4-bit dynamic LFSR. If and only if $c_i(\xi) \neq 0$, the ith register is connected to the feedback. In particular, if $c_m = c_{m-1} = \cdots = c_{m-k+1} = 0$ and $c_{m-k} \neq 0$, the m-bit LFSR is degenerated to $(m-k)$-bit LFSR. The sets of c_i are designed based on the binary primitive polynomials, so that the LFSR can dynamically degenerate to ξ-bit LFSR, while at the same time keeping $f_\xi(x)$ as a primitive polynomial. For example, $\xi \in \{4, 9, 13\} \rightarrow c_4 = 1$ for 16-bit dynamic LFSR. A listing of c_i for up to 64-bit dynamic LFSR is given in [9]. By adding the extra logic to generate all-0 vector, the dynamic LFSR, therefore, can enumerate nonrepetitive 2^ξ vectors in the first ξ bits of the output in consecutive 2^ξ cycles. For example, compared with the fixed-length

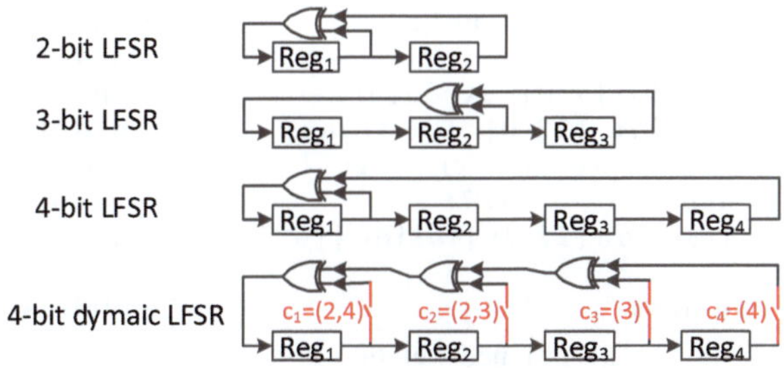

Fig. 8.16 The sketch of 4-bit dynamic LFSR evolving from fixed-length LFSRs, which makes c_i as a switching function based on the on-set

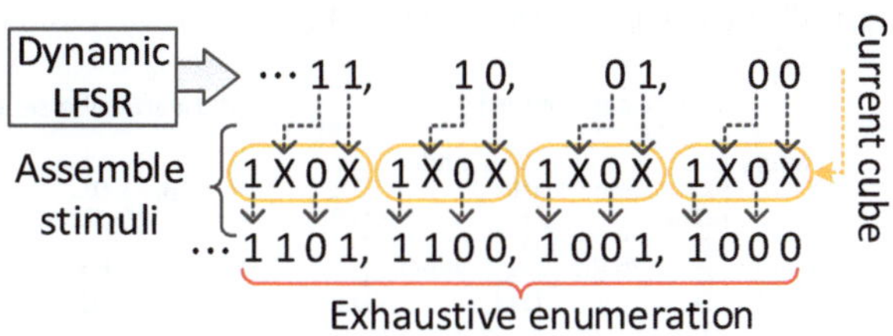

Fig. 8.17 Exhaustive enumeration by dynamic LFSR and the correction strategy, compared with Fig. 8.12b

LFSR in Fig. 8.12b, the dynamic LFSR shown in Fig. 8.17 degenerates to 2-bit LFSR for the cube "1X0X" and it exhaustively enumerates four 2-bit primitive stimuli. The step of assembling a final compliant stimulus is elaborated in the following.

Vector assembler uses the first ξ_a bits from the dynamic LFSR and the specified bits in the cube a to assemble a final compliant stimulus. As shown in Fig. 8.17, each bit of the primitive stimulus replaces an "X" in the cube one by one. Thus, the vector assembler puts together a complete 4-bit stimulus with the 2-bit primitive stimulus and the 2 specified bits in the cube. The above features can enable the generation of uniformly distributed (more specifically, *random-cyclic*) stimuli in real time.

8.7.2 Design with Support for Weighted Distributions

On-chip stimuli generation with weighted distributions introduces unique circumstances that need to be supported, including converting **dist** constraints to equivalent cubes and the correspondent on-chip stimuli generation by CRSG.

Rewriting dist Constraints is integrated in the conversion step of the cube processing flow shown in Fig. 8.4. It rewrites the constraint block into several logic constraint blocks with **inside** operator to specify a set of ranges in which the constrained variable should be sampled in. The total number of newly formed blocks equals the number of ranges specified in the **dist** operator to be rewritten. For instance, Fig. 8.18 shows two constraint blocks rewritten from the **dist** constraint in Fig. 8.11.

Consequently, each produced constraint block can be converted to an equivalent set of cubes by the existing flow. Meanwhile each set derives its weight from the original **dist** statement, which is loaded on-chip along with the corresponding cube

```
class FPNumber_sub1;
 rand bit sign;
 rand bit[7:0] exponent;
 rand bit [22:0] fraction;
 constraint number_range {
  exponent == 127;
  fraction[22:20]inside{[6:7]};
 }
endclass
```

(a) Assigned weight= 3

```
class FPNumber_sub2;
 rand bit sign;
 rand bit[7:0] exponent;
 rand bit [22:0] fraction;
 constraint number_range {
  exponent == 127;
  fraction[22:20]inside{[1:5]};
 }
endclass
```

(b) Assigned weight= 2

Fig. 8.18 Rewriting the constraint from Fig. 8.11 to two constraint blocks using the **inside** operator

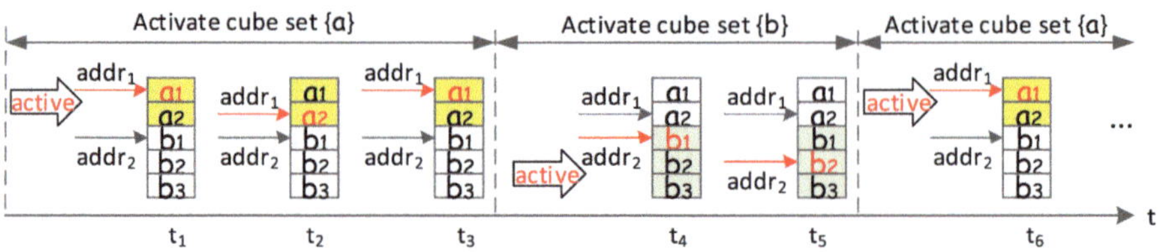

Fig. 8.19 The timeline for scheduling two sets $\{a_i\}$ (with weight = 3) and $\{b_j\}$ (with weight = 2)

set. The on-chip generator uses both *the multiple sets of cubes* and *their weight values* to produce compliant stimuli.

Scheduler with support of Multiple Sets of Cubes is embedded in the controller of CRSG shown in Fig. 8.4. It arranges the cube sets by a round-robin strategy at the level of cube sets. It complies with the constraint expressed in SystemVerilog in the sense that the total cycles for each set should be proportional to its weights. For instance, in a running period 1000 cycles, the two cube sets for the blocks in Fig. 8.18 are interleaved, and the total cycles for the two sets should converge to 600 and 400, respectively, so as to comply with the weight ratio of 3:2.

It schedules at both cube sets level (which set should be selected for the current cycle) and cubes level (which cube in the current cube set should be fetched) by using sets of registers. Figure 8.19 exemplifies the timeline for scheduling two cube sets $\{a_i\}$ (weight = 3) and $\{b_j\}$ (weight = 2). At the cube set level, the scheduler periodically selects each set for a number of cycles equal to its weight. When selected, the cube set can arrange a cube at the cube level independently. Consequently, the strategy implements the weighted distributions as 3:2 in runtime.

8.8 Experimentation Evaluation of the Solutions for Supporting Distribution Control

The solutions of supporting features of on-chip control on stimuli distributions are evaluated, including random-cyclic and weighted distributions discussed in Sect. 8.5.

Table 8.2 The test cases for assessing the runtime of Fig. 8.13

Case	Constraints	Stimulus length	Space size	Run time
(1)	$1000 \leq x + 2y \leq 8000$	24 bits	1.2×10^7	1.3 s
(2)	$y \geq 5x - 6000$, and (1)	24 bits	5.2×10^6	0.9 s
(3)	$x \geq 600$, and (1), (2)	24 bits	3.1×10^6	0.5 s
(4)	$y \geq 2000$, and (1), (2), (3)	24 bits	1.6×10^6	0.2 s
(5)	PCIe TLP	160 bits	8.7×10^{41}	3.5 s
(6)	H.264 RTP	168 bits	1.2×10^{50}	0.2 s

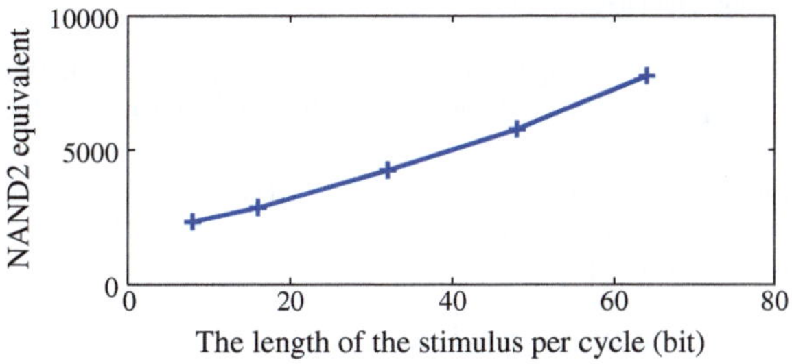

Fig. 8.20 The hardware cost (exclusive of RAMs) of the CRSG for supporting random-cyclic distribution according to the length of stimulus per cycle

8.8.1 *Runtimes for Cube Rectification*

The runtimes of the algorithm to rectify overlapped cubes in Sect. 8.6 are evaluated based on test cases targeting different types of constraints, as shown in Table 8.2. Cases (1)–(4) target solving inequalities in integer linear programming (shown in the second column), which typically specify numerical relationship between variables by a series of arithmetic and relational constraints. Cases (5) and (6) generate packet heads of PCIe 3.0 TLP and H.264 RTP introduced in Sect. 8.4 respectively. The third column gives the stimulus length and the fourth column shows the dimension of the constrained space from which valid patterns are sampled. The last column gives the runtimes, which indicate the practical feasibility of the algorithm.

8.8.2 *Evaluating the Solution for Random-Cyclic Distributions*

The area cost of the CRSG to support **randc** distributions presented in Sect. 8.7.1 is shown in Fig. 8.20. It requires more hardware as the stimulus per cycle increases, in which case the vector assembler dominates that area.

Figure 8.21a, b illustrate the ability to generate random-cyclic sequences, i.e., the on-chip generator can produce t unique stimuli in t cycles.

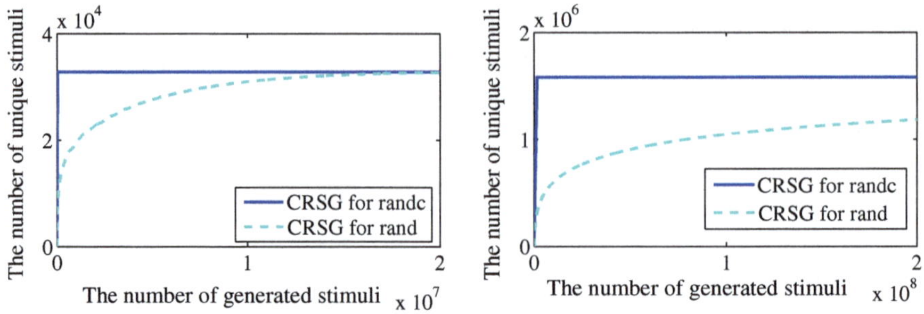

(a) The result for the constraint $x \geq y$ where x and y are 8-bit unsigned integers

(b) The result for the constraint #4 in Table 2.

Fig. 8.21 Assessing the repetition in the generated stimuli by the CRSG with support for random-cyclic patterns (i.e., specified with **randc**) and the CRSG without it (specified with **rand**)

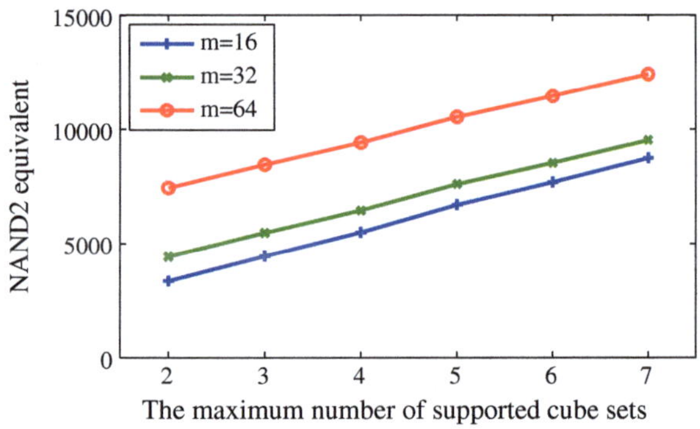

Fig. 8.22 The area cost of the generator (exclusive of RAM) according to the number of supported cube sets. The length of stimuli is m bits

8.8.3 Evaluating the Solution for Supporting Weighted Distribution

The area cost of the CRSG with support of weighted distributions depends on both the maximum length of a stimulus (denoted as m-bit) and the maximum number of supported sets of cubes, as shown in Fig. 8.22. For each case with the same m, the area cost of CRSG increases linearly, because the addressing scheduler logic needs one more set of addressing registers in order to schedule an additional cube set.

The distribution quality is assessed as follows. CRSG is configured to generate stimuli sequences for resembling packet heads in PCIe TLP as introduced in Sect. 8.4. If some specific types of packets are deemed to be unlikely to be fallible (or were previously validated in more detail), the constraints for the two fields can be composed to guide the on-chip generator to prioritize the packets with the higher weight. Table 8.3 exemplifies a case that can utilize weighted distributions to tune the frequencies of specific types of packets during a validation session.

Table 8.3 The specification for generating specific types of TLP packets with different frequencies

FMT	TYPE	Type of the packet	Frequency	Comments
001	00000	Memory read request	50%	Most concerned
011	00000	Memory write request	30%	Less concerned
000	01010	Completion	20%	Supported packet
(Other values)		Other types	N/A	Not for this session

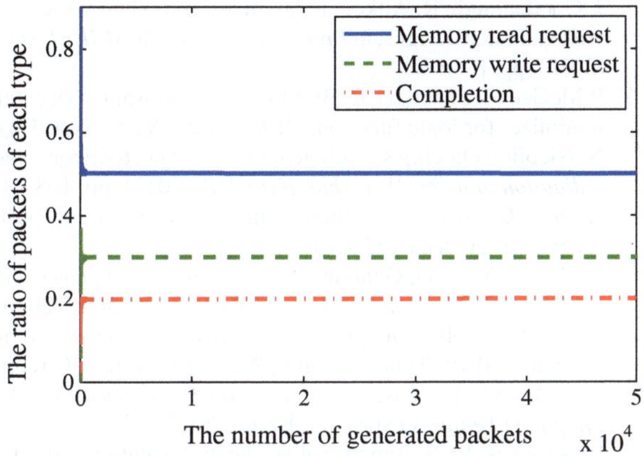

Fig. 8.23 The ratio for the packets of each type among the total number of packets that have been produced during the on-chip generation process according to the specification from Table 8.3

The distribution of packets when using the constraints with weights from Table 8.3 to generate TLP packets is assessed. Figure 8.23 shows the ratio of packets. It can be observed that as the validation time increases the measured ratio of packets belonging to each type converges to the expected weight (0.5, 0.3, and 0.2, respectively).

8.9 Summary

This chapter has presented several solutions for on-chip constrained-random stimuli generation. These solutions support constraints on both functionality and stimuli distribution. With the usage of a cube-based representation for stimuli constraints and on-chip hardware structures for stimuli generation, one can reuse a subset of the SystemVerilog constraints during the post-silicon validation phase. The main application is to generate functionally compliant randomized stimuli that can be biased from one validation experiment to another.

References

1. IEEE standard for floating-point arithmetic. IEEE Std 754-2008 (2008), pp. 1–70
2. IEEE standard for SystemVerilog–unified hardware design, specification, and verification language. IEEE Std 1800-2012 (2013), pp. 1–1315
3. J. Ajanovic, PCI express 3.0 overview, in *Proceedings of Hot Chip: A Symposium on High Performance Chips* (2009)
4. P. Bardell, W. McAnney, J. Savir, *Built-in Test for VLSI: Pseudorandom Techniques* (Wiley-Interscience Publication, Wiley, 1987)
5. J. Goodenough, R. Aitken, Post-silicon is too late avoiding the $50 million paperweight starts with validated designs, in *Proceedings of the ACM/IEEE Design Automation Conference (DAC)* (2010), pp. 8–11
6. P. McGeer, J. Sanghavi, R. Brayton, A. Sangiovanni-Vicentelli, Espresso-signature: a new exact minimizer for logic functions. IEEE Trans. VLSI Syst. **1**(4), 432–440 (1993)
7. N. Nicolici, On-chip stimuli generation for post-silicon validation, in *IEEE High Level Design Validation and Test Workshop (HLDVT)* (2012), pp. 108–109
8. X. Shi, Constrained-random stimuli generation for post-silicon validation, Ph.D. thesis, McMaster University (2016)
9. X. Shi, N. Nicolici, Generating cyclic-random sequences in a constrained space for in-system validation. IEEE Trans. Comput. **65**(12), 3676–3686 (2016)
10. X. Shi, N. Nicolici, On-chip cube-based constrained-random stimuli generation for post-silicon validation. IEEE Trans. Comput. Aided Des. Integr. Circuits Syst. **35**(6), 1012–1025 (2016)
11. C. Spear, G. Tumbush, *SystemVerilog for Verification: A Guide to Learning the Testbench Language Features* (Springer, Berlin, 2012)
12. I. Synopsys, VCS - functional verification solution (2015), http://www.synopsys.com/VCS
13. Y.K. Wang, R. Even, T. Kristensen, Tandberg, R. Jesup, RTP payload format for H.264 video. RFC 6184 (2011)

Chapter 9
Test Generation and Lightweight Checking for Multi-core Memory Consistency

Doowon Lee and Valeria Bertacco

9.1 Introduction

Modern microprocessor chips integrate multiple cores to provide high computing capability under a limited power budget. Multi-core processors are indeed widely deployed in consumer electronics and data centers, and the number of cores integrated in a single chip is ever growing since the inception of multi-core architecture. For example, in the server domain, a recent Intel Xeon Skylake-SP processor integrates 28 powerful cores, each of which supports 2 threads of execution. In the mobile system-on-chip domain, a recent Qualcomm's Snapdragon consists of 8 single-threaded cores, as well as a variety of heterogeneous application-specific accelerators. For these multi-core processors, the software can be designed to utilize the multiple cores available in two different ways: either multiple independent tasks run separately on each core, or a single task (application) spawns multiple threads of execution, each assigned to a different core. In this latter case, the threads must share the same memory space (multithreading). In recent years, multithreading has become increasingly popular, as it collaboratively advances a single application with the potential of greatly reducing the application's completion time.

Multi-core systems deploy a complex memory subsystem to enable efficient multiprocessing and multithreading computation, by processing tens or hundreds of memory operations in parallel. The memory subsystem typically includes multiple

---

This work is based on an earlier work: "MTraceCheck: Validating Non-Deterministic Behavior of Memory Consistency Models in Post-Silicon Validation" by Doowon Lee and Valeria Bertacco in the 44th Annual International Symposium on Computer Architecture (ISCA '17) ©ACM 2017. https://doi.org/10.1145/3079856.3080235.

D. Lee · V. Bertacco (✉)
University of Michigan, 2260 Hayward Street, Ann Arbor, MI 48109-2121, USA
e-mail: valeria@umich.edu

D. Lee
e-mail: doowon@umich.edu

P. Mishra and F. Farahmandi (eds.), *Post-Silicon Validation and Debug*,
https://doi.org/10.1007/978-3-319-98116-1_9

caches and memory channels. These memory components interact with the processor cores via on-chip communication networks such as buses, crossbars, and mesh networks. In this complex memory subsystem architecture, many memory-access requests can be executed in an arbitrary order that is different from the sequential order defined in the program code. This memory-level parallelism, on one hand, is an essential element of a high-performing multi-core system. On the other hand, it comes with great verification and validation challenges. There are various memory-access interleaving patterns that the system can exhibit—many of which are difficult to foresee during design time.

Various microprocessor vendors define their own specifications on correct memory ordering behaviors, called memory consistency models. There are a wide range of memory consistency models, ranging from weak models, as used by ARM and IBM, to stronger ones, like the one adopted by Intel, to the strongest model of MIPS systems. A weak memory consistency model allows memory operation re-orderings that are disallowed by a stronger memory model. From the viewpoint of hardware verification, every memory ordering pattern that an implementation may exhibit must be checked to determine whether it abides by the memory consistency model. Note that, however, modern processor designs are usually too complex to be formally verified within the relatively short amount of time allowed for verification and validation. As a result, functional bugs related to memory consistency have slipped through mainstream multi-core processors, as reported in many specification update documents [8, 28].

9.1.1 Brief Introduction to Memory Consistency Models

Memory consistency models (MCMs) are the specifications regarding memory ordering behaviors in multi-core systems, providing a protocol between hardware designers and software programmers [2, 46]. A multi-core microprocessor can arbitrarily re-order its memory operations under execution, as long as such re-ordering obeys the memory ordering rules specified by the MCM. At the same time, programmers must understand the microprocessor's re-ordering behaviors to write correct multithreaded programs.

In addition, every execution of a multithreaded program can exhibit different memory-access interleaving patterns, depending on micro-architecture state at the time the program is executed. In order to tame such nondeterministic memory interleavings, programmers must insert fence operations whenever necessary to ensure correct program execution. A fence operation is a special operation that restricts memory re-ordering behaviors—the order among memory operations before and after a fence operation is strictly enforced. In other words, memory operations issued before the fence operation must be completed before those following the fence operation can be issued.

There are several memory consistency models that are worth noting. Here, we briefly introduce only three groups of models; please refer to [2, 46] for a more

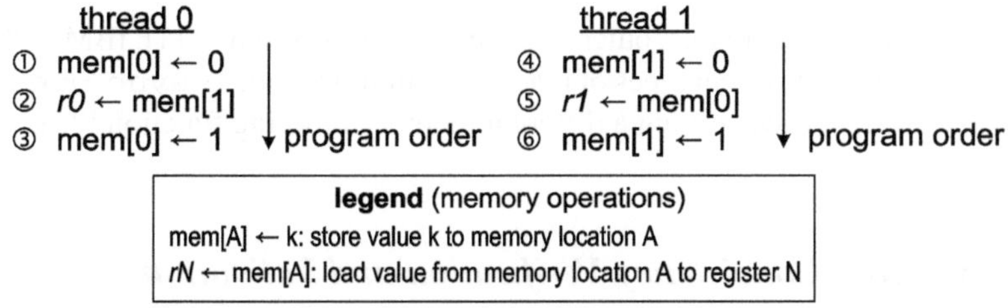

Fig. 9.1 A 2-threaded test sharing 2 memory locations. Under the total store order memory model, load operations ② and ⑤ are allowed to perform in advance of preceding store operations ① and ④, respectively

in-depth discussion. *Sequential consistency* (SC) [31] is the most intuitive, strongest memory model among most models in the literature. No memory re-ordering within a thread is allowed by this model, so memory operations are performed in the order specified by the program. However, this model still allows an arbitrary interleaving of memory operations across different threads—the inter-thread interleaving is the source of nondeterministic results of memory operations in this model.

Total store order (TSO) [27, 50] allows load operations to be performed before preceding store operations, relaxing the store→load ordering requirement. Figure 9.1 illustrates a two-threaded program sharing two memory locations. In each thread of this program, the load operation (second operation) can be performed before its preceding store operation (first operation). This re-ordering allows the load operation to retrieve data from caches or main memory components while the store operation still resides in the core. Relaxing the store→load ordering enables some micro-architectural optimizations, such as store buffering, to greatly enhance memory-level parallelism. Note that, however, the last store operation (third operation) cannot be performed before any of its preceding operations, meaning that the store→store and load→store orderings are still preserved in this model. Many multi-core processors, such as Oracle SPARC and Intel Core and Xeon processors, deploy the TSO memory model or a variation of the same.

Weakly ordered memory models [10, 11, 25, 49] are also widely adopted by many processors such as ARM Cortex and IBM POWER processors. In this group of memory models, any memory operation can be performed in advance of its preceding memory operation in the absence of a fence. Multi-core processors with this kind of memory models can deploy a variety of micro-architectural optimizations to fully utilize the memory bandwidth and reduce the memory-access latency without being subject to memory ordering rules.

There are slight differences among memory models within the same MCM category. For example, Intel processors allow the values written by store operations to be observed as occurring in different orders by different threads [27], while SPARC processors do not allow such different orderings [50]. For weakly ordered memory models, the ARMv8 architecture provides many variants of the memory barrier operation (`dmb`) for different operation types (load and store) and memory domains (non-shareable, inner shareable, outer shareable, and full system) [11]. IBM POWER v2.07 supports a few different types of fence operations (`sync`, `lwsync`, `ptesync`, `eieio` and `mbar`), each serving a different memory ordering function [25].

9.1.2 Memory Consistency Verification and Validation

Ensuring that the memory-access patterns observed by the memory consistency model stipulated for the architecture under consideration can be achieved by various verification and validation efforts. In the early design stages, the memory consistency model of a multi-core processor is determined by computer architects. They decide how the processor's micro-architecture should enforce memory ordering rules to follow the memory consistency model. Therefore, the goal of memory consistency verification and validation is to check that the processor implementations (e.g., functional simulators, RTL models, and silicon chips) respect the memory ordering rules.

Formal Verification of Memory Consistency

Formal methods have been applied to detect memory consistency violations in the literature. With formal methods, the memory subsystem under verification is modeled as mathematical formulas that capture memory ordering behaviors of the memory subsystem. For example, Alglave [3, 6] uses the Coq theorem prover [47] to verify several weak MCMs, showing that it can expose a subtle memory consistency bug in the ARM Cortex-A9 processor [8]. Expanding this Coq-based framework, PipeCheck [34] specifies micro-architectural behaviors in detailed happens-before graphs that capture interactions among pipeline stages. RTLCheck [39] automatically generates SystemVerilog Assertions from micro-architectural axioms and performs formal property verification using JasperGold.

Litmus tests [4, 5, 7, 36, 37] are often paired with the formal methods mentioned earlier. These tests are a small set of multithreaded programs that are tailored to trigger specific memory consistency bugs when they are present. A typical litmus test includes several instructions with a few threads of execution, and targets a very specific consistency violation scenario. For a given litmus test and mathematical model of a processor design, all possible memory ordering patterns are explored and checked for the targeted violation scenario. Figure 9.2 illustrates a memory consistency violation that can be detected by a 2-threaded litmus test, as the one shown on the left. While litmus tests are very powerful in uncovering many consistency violations, they are often insufficient to validate full multi-core systems [20]. This is partly because litmus tests often miss very subtle corner cases that have not been captured by the mathematical model. In other words, an inaccurate model often leads to false positives (i.e., falsely proven to be correct) [44].

Dynamic Validation of Memory Consistency

Dynamic validation methods check memory ordering patterns observed in processor implementations while *running* multithreaded programs. This validation approach uses various types of test programs, such as litmus tests, constrained-random tests, and real applications. The major difference with formal verification is that dynamic validation analyzes the results of *concrete execution* in *real design implementations*. In other words, dynamic validation does not involve mathematical models, exhaustive enumeration, nor symbolic execution.

Hangal et al. [24] present an industrial memory consistency validation tool, called TSOtool, that is used to validate SPARC chip-multiprocessor systems [50]. Roy et al. [43] try to speed up memory consistency validation by approximating the result-checking algorithm. Meixner and Sorin [40] include small additional hardware modules within a memory subsystem so as to track memory ordering behaviors observed at runtime. Similar hardware modifications on the memory subsystem have been proposed by other research works [16, 38]. In addition, Axe [41] provides a memory trace generator that can be applied to the Berkeley's RISC-V and Cambridge's BERI memory subsystems. It also provides an open-source memory consistency checker supporting various memory models.

Constraint Graph

In order to check the correctness of memory ordering, the order among memory operations is analyzed using their happens-before relationships. Each happens-before relationship indicates a partial order of two memory operations. By collecting all happens-before relationships, we construct a constraint graph, also known as a happens-before graph, where each vertex corresponds to a memory operation, and each edge represents the happens-before relationship between two vertices. If there is a cycle in a constraint graph, it implies a memory consistency violation. How to construct a constraint graph has been extensively investigated in the literature [3, 15–17, 34, 43].

In Fig. 9.2, the constraint graph on the right represents the result of the 2-threaded litmus test on the left. This constraint graph is built under the assumption that the system adopts a TSO memory model. As the graph shows, there is a cyclic dependency along the four memory operations, thus the execution violates the TSO model. In general, to check for cyclic dependencies in a constraint graph, we can use a topological sorting or depth-first search, which has high computational complexity, $\Theta(V + E)$, where V and E are the set of vertices and edges, respectively [18]. Therefore, the

multi-threaded test

initial memory values: mem[0]=mem[1]=0	
thread 0	thread 1
①: *r0* ← mem[0]	③: *r1* ← mem[1]
②: mem[1] ← 1	④: mem[0] ← 1
register values after a test run: *r0*=*r1*=1	

constraint graph

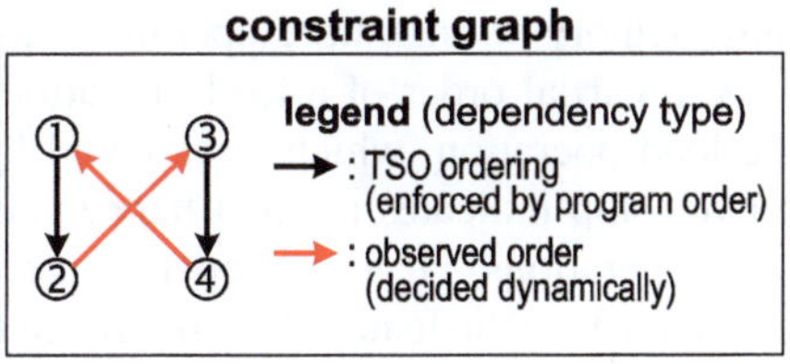

Fig. 9.2 A 2-threaded test execution (left) and its corresponding constraint graph (right). The TSO model allows various memory operation results, except for the *r0*=*r1*=1 shown on the bottom left. The constraint graph on the right identifies this violation with a cyclic dependency

result-checking can quickly become the critical bottleneck of the entire memory consistency validation process.

9.1.3 Post-silicon Microprocessor Validation

Post-silicon microprocessor validation strives to detect subtle errors that slip through silicon chips. In this validation stage, constrained-random tests are widely used along with directed tests. Constrained-random tests are essential to verify unexpected use cases that have not been identified by verification and validation engineers, while directed tests target corner cases that are expected by the engineers. These two types of tests are often designed in a way that overcomes the challenge of limited observability in post-silicon validation. Specifically, these tests can be created in a way such that their test results can be easily checked on the same platform that is under validation (self-checking). Self-checking solutions [1, 21, 33, 48] need to observe only minimal test results to check the correctness of test runs, and thus they mitigate the overhead of transferring test results to a host machine. For example, QED [33] duplicates instructions and compares the results of the original and the duplicate instructions. Reversi [48] generates a sequence of reversible instructions where the result of the sequence becomes a simple value to be checked easily—for instance, it uses a pair of addition and subtraction operations with the addend and the subtrahend (i.e., the number to be subtracted) being equal. Foutris et al. [21] leverage the fact that an instruction set consists of various instructions, most of which can be replaced by a sequence of other instructions—for instance, a series of additions can replace a multiplication instruction.

Memory Consistency Post-silicon Validation

Both directed tests and constrained-random tests are also commonly used in memory consistency validation. Litmus tests are popular directed tests for memory consistency validation. These are often hand-crafted by engineers [7, 26] or can be automatically generated from a formal model of a memory subsystem [5, 36]. Constrained-random tests are tailored to exhibit rare memory ordering behaviors, by intensively accessing a small number of memory locations shared across multiple threads. These tests are often generated in a way that every store operation writes a unique value into memory so that store operations can be easily disambiguated. Also, in these tests, a partial order of a load operation is determined based on the value read by the load operation, which is captured by a *reads-from* relationship. A reads-from relationship indicates a direct happens-before relationship between some load and store operations: "if a load operation reads from a store operation, the store must happen before the load." With reference to Fig. 9.2, the load operation ③ reads from the store operation ②, which implies that ② happens before ③.

Both litmus tests and constrained-random tests are run repeatedly to strive to exhibit various distinct memory-access interleaving behaviors. There are many different valid memory-access interleaving outcomes, especially for constrained-random tests where many more memory operations are included. In other words, every run of a test may lead to a unique memory-access interleaving behavior that has not been seen by prior test runs. Consequently, the activity of validating the test outcomes requires much more analysis and computation than other post-silicon validation activities, where test outcomes are deterministic and can be self-checked. Thus, it is important to minimize the result-checking computation to improve the overall efficiency of the validation process.

In the rest of this book chapter, we present MTraceCheck, an efficient memory consistency validation framework specialized for post-silicon validation [32].

9.2 MTraceCheck: Efficient Memory Consistency Validation

MTraceCheck is a memory consistency validation framework that efficiently checks many distinct memory ordering patterns. It is tailored to post-silicon validation environments with very limited observability—however, it is also applicable to pre-silicon dynamic validation environments such as instruction set simulation, RTL design simulation, or FPGA prototyping.

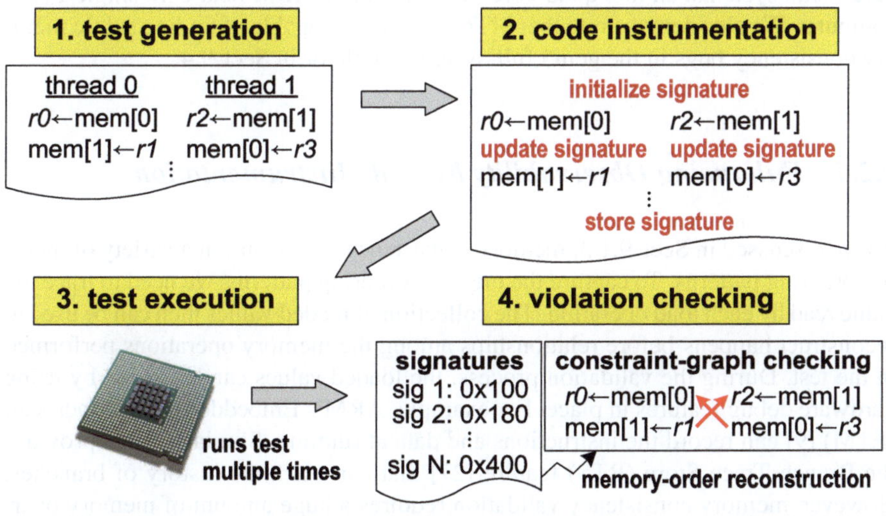

Fig. 9.3 MTraceCheck overview. MTraceCheck first generates multithreaded constrained-random tests, which are then instrumented with our observability-enhancing code. The instrumented code computes a memory-access interleaving signature at runtime. Each instrumented test is run multiple times to obtain a collection of signature values. These signature values are then decoded to reconstruct constraint graphs capturing happens-before relationships among the memory accesses. The constraint graphs are then collectively validated by our checking algorithm that leverages the similarities among the graphs

Figure 9.3 outlines the validation workflow using the MTraceCheck framework in four steps. In the first step, it generates constrained-random tests using several test-generation parameters. The generated tests include multiple threads of execution where each thread performs many memory operations on a small memory region shared across all threads, as in [20, 24]. In the next step, the generated tests are augmented with signature computation code that calculates a signature value corresponding to the memory-access interleaving behaviors observed at runtime. We explain this code instrumentation process in Sect. 9.2.1. The augmented tests are then loaded onto the platform under validation and repeatedly executed on the platform. From the repeated test runs, we collect signatures and store them in a thread-local storage. Lastly, we examine the signatures in a collective manner that minimizes repetitive checking computation. This checking process is detailed in Sect. 9.2.2. Compared to prior related work on post-silicon memory consistency validation, MTraceCheck provides the following contributions:

- MTraceCheck's *code instrumentation process* computes a signature representing the order among memory operations. The signature minimizes the amount of data transfers for the purpose of result-checking.
- MTraceCheck's *lightweight graph-checking algorithm* leverages the similarities of constraint graphs across repeated executions of a memory consistency test.

In Sect. 9.3, we quantify the benefits and drawbacks of MTraceCheck in two bare-metal systems: an x86 quad-core desktop and an ARM octa-core single-board computer. We also demonstrate that MTraceCheck is capable of finding subtle memory consistency bugs in the gem5 full-system simulator in Sect. 9.4.

9.2.1 Enhancing Observability by Code Instrumentation

As we discussed in Sect. 9.1.3, memory consistency tests result in a variety of memory ordering patterns. To capture the memory ordering patterns, we need to trace the value read by each load operation. The collection of loaded values then can be used to reconstruct happens-before relationships among the memory operations performed in the test. During the validation process, the loaded values can be traced by using hardware debug features in place. For example, ARM's Embedded Trace Macrocell (ETM) [9] can record the instructions and data at runtime. Similarly, Intel provides the Branch Trace Store (BTS) feature [27] that can track the history of branches. However, memory consistency validation requires a huge amount of memory operations to be traced on the fly, and using these features is often insufficient to do so. Specifically, ETM's on-chip trace buffers are usually small—tests must be halted whenever the trace buffers become full. Also, BTS shares the memory subsystem to store the results of branches, thus potentially perturbing the memory ordering behaviors of test execution. Therefore, in this work, we strive to trace the memory operations efficiently for the purpose of memory consistency validation. In addition,

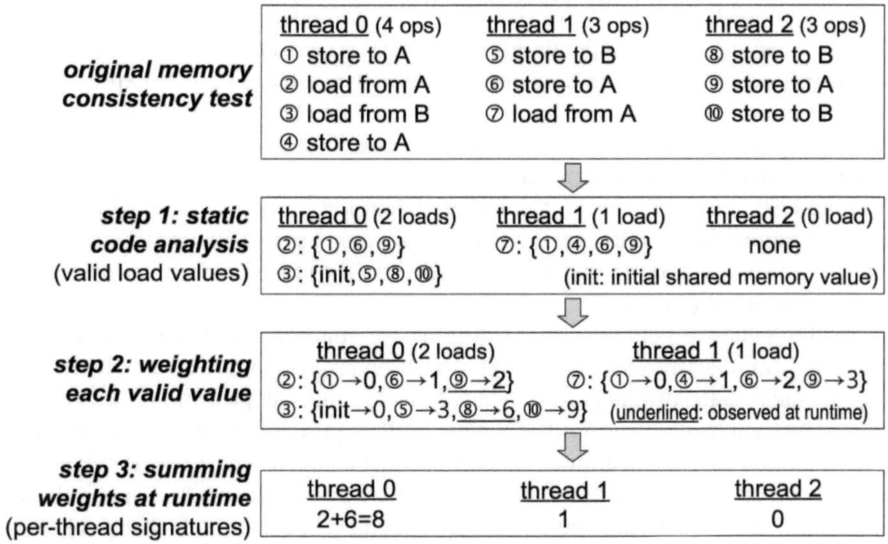

Fig. 9.4 Memory-access interleaving signature. Each signature value represents a unique memory ordering pattern observed during the test run. To compute the signature, we first profile all values that can be read by each load operation. We then assign an integer weight for each of the values. These two steps are performed statically. At runtime, the weights of the observed values are accumulated for each thread independently, creating a per-thread signature. Finally, all per-thread signatures are concatenated to create an execution signature

we advocate a software-based technique in tracing the memory operations, which can be also paired with these existing hardware debug features.

A conventional software-based memory-tracing method has been shown in TSOtool [24]. In TSOtool, the values read by load operations are temporarily stored in registers, and then flushed back to memory. Note that, due to the limited amount of architectural registers available, this register-flushing method frequently interrupts the test execution. In addition, it is bound to perform an excessive amount of store operations to record the history of loaded values. These store operations are subject to perturb the original test's memory-access behaviors, altering the intention of the original test. In summary, we need to minimize the amount of store operations required to trace the results of the original test's memory operations.

9.2.1.1 Computing Memory-Access Interleaving Signature

Here, we describe our signature computation method to encapsulate the results of memory operations in a small signature value. Our method is inspired by a prior control-flow profiling method proposed by Ball and Larus [13]. This method presents a systematic way to compute a small sum of values accumulated for each basic block executed. We take a similar approach to compute a signature representing the results of memory operations.

As mentioned earlier, the values read by load operations need to be traced to decide the order among memory operations, or memory-access interleaving behaviors. In this work, we propose a *memory-access interleaving signature* that expresses the loaded values in a compact format. Each distinct value in the memory-access interleaving signature indicates a unique memory ordering pattern of a multithreaded program. This small-sized signature is stored for each test run; unlike the register-flushing technique [24], our signature-based solution introduces only a handful of store operations for the purpose of validation. By doing so, we aim at minimizing the interference of memory tracing in order to observe the results of memory operations while running the test.

To this end, we compute a signature as we run a memory consistency test. Figure 9.4 summarizes our signature computation process in three steps. In the first step, we analyze the original test statically so as to collect all possible values that can be read by each load operation. The top box of the figure illustrates a 3-threaded test program created by a constrained-random test generator. The test program includes 10 memory operations, which share 2 memory addresses (`A` and `B`). Among the 10 operations, there are 3 load operations (②, ③, and ⑦), each of which can take any of the values listed in the set in the second box of the figure.

Analyzing all possible loaded values can be a demanding task depending on the test under validation. For constrained-random tests, test generators usually have all the information on the generated operations including the address of each memory operation. Thus, memory addresses can be extracted from the test generators and our static analysis can be easily done. However, in real-world multithreaded programs (e.g., graph analytics), our static analysis can be limited and some memory addresses cannot be disambiguated. In this case, we can use a binary instrumentation tool in order to disambiguate memory addresses as much as possible via profiling. We can also exclude memory operations with unknown addresses from the signature computation, and record the results of these operations as the conventional register-flushing method does.

Algorithm 1 Weight assignment

```
Input: all possible reads_from relationships
Output: weights, multipliers, reads_from_maps
multiplier ← 1
for each load operation L from first to last in test program do
    multipliers[L] ← multiplier
    index_reads_from ← 0
    for each reads_from for the load operation L do
        weights[L][index_reads_from] ← index_reads_from × multiplier
        reads_from_maps[L][index_reads_from] ← reads_from
        index_reads_from ← index_reads_from + 1
    end for
    multiplier ← multiplier × index_reads_from
end for
return weights, multipliers, reads_from_maps
```

```
thread 0
    initialize: sig = 0
① store to A
② load from A
    // update signature for ②
    if (②'s value==①)        sig += 0
    else if (②'s value==⑥) sig += 1
    else if (②'s value==⑨) sig += 2
    else assert error
③ load from B
    // update signature for ③
    if (③'s value==init)     sig += 0
    else if (③'s value==⑤) sig += 3
    else if (③'s value==⑧) sig += 6
    else if (③'s value==⑩) sig += 9
    else assert error
④ store to A
    finish: store sig to memory
```

```
thread 1
    initialize: sig = 0
⑤ store to B
⑥ store to A
⑦ load from A
    // update signature for ⑦
    if (⑦'s value==①)        sig += 0
    else if (⑦'s value==④) sig += 1
    else if (⑦'s value==⑥) sig += 2
    else if (⑦'s value==⑨) sig += 3
    else assert error
    finish: store sig to memory

Thread 2 always stores sig=0 to
memory, because it has no load
operation.
```

Fig. 9.5 Code instrumentation. Branch and arithmetic instructions are instrumented after each load operation, and these instructions update the signature variable (`sig`) as the test runs. When the test finishes, the computed signature is stored in a thread-local memory region

In step 2, we decide an integer weight for each possible value read by load operations. This step is illustrated in the third box of the figure and also summarized in Algorithm 1. We perform the weight assignment for each thread individually, while the previous step is performed globally considering all threads. With reference to thread 0 in the figure, the first load operation ② has three possible loaded values (or three possible reads-from relationships), stored by ①, ⑥, and ⑨. We assign weights 0, 1, and 2 for these values, respectively. For the next load operation ③, we assign multiples of 3 since there were three options in the previous load operation. The way we allocate the weights guarantees a unique correspondence between signatures and the orders of memory operations. Our weight assignment resembles the way that [13] assigns an integer value for each edge in a control-flow graph.

In the last step, we accumulate the weights that correspond to the loaded values at runtime. Suppose in the third box in the figure, the value written by ⑨ (weight 2) is read by the first load operation and the value written by ⑧ (weight 6) is read by the second load operation, as illustrated by the two underlined values in thread 0 of the box. As a signature for this thread, we obtain the sum of the two weights, which is 8, at the end of the execution. Similarly, suppose that the load operation ⑦ in thread 1 reads the value underlined, and its weight (1) becomes a signature of this thread. Thread 2 has no load operation, so it always returns 0 as its signature. These per-thread signatures are illustrated in the bottom box of the figure. Note that the signature computation is done for each thread individually at runtime. At the end of the test run, we form the *execution signature* by concatenating all the per-thread signatures obtained for each test run.

Code Instrumentation

Figure 9.5 shows the test program after we conduct our code instrumentation to the original test in Fig. 9.4. The instrumented code includes three parts. First, the `sig` variable is initialized to 0 at the beginning of the test. The `sig` variable is then increased after each load operation. This increase is carried out by the second instrumented section: a chain of branch and addition operations conditioned on the loaded value. Note that we append an assertion statement at the tail of each branch chain. This assertion statement is intended to catch obvious errors that need no constraint-graph checking. For instance, it can catch a program-order violation that the load operation ② takes the value from the younger store operation ④. In the last segment of instrumented code, the accumulated signature `sig` is stored in a thread-local storage when the test finishes.

Depending on the memory-interleaving possibilities, the signature can exceed the size of a single register (typically 32 bits or 64 bits), in which case we save the `sig` variable before it overflows, reinitialize it (i.e., `sig`=0), and start accumulating it from there. In other words, we split a per-thread signature into multiple words. To this end, we statically identify any signature overflow when we compute weights during our code instrumentation process. In this case, we store the current signature (i.e., the signature that has been computed until the point we detect the overflow) to a thread-local memory region, then start over the signature computation by resetting the *multiplier* (i.e., *multiplier*=1). By doing so, the previous signature represents the previous code segment, while the new signature represents the next code segment.

While our instrumentation increases the code size to some degree, its impact on test runtime is much smaller compared to the increase of the code size. We provide experimental evaluations on the code size and runtime overheads later in Sect. 9.3. Note that the instrumented routines after load operations minimally perturb the memory-access patterns intended by the original test, because the routines are composed of mostly branch and arithmetic operations, but not memory operations. There are two exceptional cases where memory-access patterns can be affected: (1) when the signature `sig` needs to be reset due to an overflow possibility as mentioned earlier—in this case, the previous `sig` is flushed to memory before it is cleared, and (2) when the program execution hits any of the assertion statements (`assert error`). Note that modern microprocessors are equipped with out-of-order execution engines with highly accurate branch predictors, so our instrumented routines can be effectively executed in parallel with memory operations.

9.2.1.2 Reconstructing the Order Among Memory Operations

We decode each signature following the pseudocode described in Algorithm 2 so as to reconstruct the order among memory operations captured by a signature. The decoding process is basically a reverse process of the signature computation process in Algorithm 1. The goal of the decoding process is to obtain a set of reads-from relationships represented by a signature. Before we apply the decoding procedure in Algorithm 2, we first split an execution signature into a set of per-thread signatures. For each thread, we then reconstruct the reads-from relationships, one at a time,

Algorithm 2 Signature decoding procedure

```
1: Input: signature, multipliers, reads_from_maps
2: Output: reads_from relationships observed at runtime
3: for each load L from last to first in test program do
4:     multiplier ← multipliers[L]
5:     index_reads_from ← signature / multiplier
6:     signature ← signature % multiplier
7:     reads_from[L] ← reads_from_maps[L][index_reads_from]
8: end for
9: return reads_from
```

starting from the last load operation in the thread. The reconstruction process walks backward to the first load operation in the thread (line 3 in the algorithm). The *multipliers* array has the multiplier used to calculate each load operation's weights. For each load operation, the *signature* is divided by the multiplier specified in the *multipliers* array (lines 4 and 5). The quotient indicates the index of a reads-from relationship observed for the load operation (line 5). The remainder represents a signature value before being accumulated for this load operation, and thus it replaces the *signature* for the next load operation under reconstruction (line 6). The index of the reads-from relationship is then used to look up the *reads_from_maps* array which provides the mapping relationships between indices and reads-from relationships (line 7).

At the same time we assign weights, we gather additional information necessary to reconstruct the constraint graph. Besides the *multipliers* and *reads_from_maps* arrays, we also collect the intra-thread memory ordering required by the MCM (e.g., ①→② shown in Fig. 9.2), and the write-serialization order (e.g., store operations to the same address within a thread must be observed in the program order, such as ① and ④ in Fig. 9.4). All the aforementioned information is necessary to reconstruct constraint graphs as well as reads-from relationships.

9.2.1.3 Signature Size

The size of a signature is proportional to the number of reads-from relationships across all memory operations. Specifically, our memory-access interleaving signature is designed to capture all allowed memory-access interleaving patterns under a very weak MCM where no memory ordering is enforced. In a stronger MCM where fewer memory-access interleavings are allowed, however, our signature can express a value corresponding to a memory ordering *disallowed* by the MCM, making the signature unnecessarily large for the MCM. These invalid signature values will be eventually caught as consistency violations during our graph checking in Sect. 9.2.2. In addition, we try to keep our signature small by considering intra-thread program-order requirements. For example, by a program-order requirement, a load operation cannot take a stale value from a store operation older than the most recent store operation within the same thread. This kind of violations is caught by the assertion statements in Fig. 9.5.

Here we estimate the signature size of a test generated under a constrained-random testing environment used in our experimental evaluations. In our testing environment, we have several test-generation parameters specifying the characteristics of a test: the number of threads (T), the number of stores per thread (S), the number of loads per thread (L), and the number of shared memory locations (A). We created memory addresses in a random fashion. With these parameters, the size of a per-thread signature can be expressed by the following equation:

$$\texttt{per-thread signature cardinality} = \left\{1 + \frac{S}{A}(T-1)\right\}^{L} \quad (9.1)$$

The curly brackets in the equation include two components for each load operation. The first component (1 in the equation) captures the case where the load reads from the same thread (i.e., reading the value written by the most recent store to the same address in the same thread), or the case where the load reads the initial memory value if there is no prior store operation to the same address. The second component ($\frac{S}{A}(T-1)$) represents the cases where the load reads from any of the other threads. $\frac{S}{A}$ indicates the number of store operations with the same address as the load. The expression enclosed by the curly brackets is then multiplied L times to compute all possible combinations across all load operations in the thread. The size of an execution signature is T times the equation above, because an execution signature is the collection of all per-thread signatures.

As an example, the smallest test configuration we used (Table 9.2) has the following test-generation parameters: $T = 2$, $S = L = 25$, $A = 32$. A per-thread signature in this configuration can represent $\{1 + \frac{25}{32}(2-1)\}^{25} \approx 1.9 \times 10^6 \approx 2^{21}$ sets of loaded values, requiring approximately a 21-bit storage. In Sect. 9.3.4, we will show the average size of the execution signature across various constrained-random tests, and discuss how to reduce the signature size in Sect. 9.5.

9.2.2 *Collective Graph Checking*

To check the correctness of memory ordering observed at runtime, we analyze the happens-before relationships among all memory operations performed during test execution. The happens-before relationships, as studied in [3, 34], can be classified into three types of observed happens-before relationships as follows: *reads-from*, *from-read*, and *write-serialization*. Besides these observed relationships, we take into account intra-thread ordering rules as defined by the MCM. With reference to Fig. 9.6, which assumes the TSO memory model, the store operation ① should happen before ③ as illustrated with the *TSO ordering* edge. But ① does not necessarily happen before ②, so there is no edge between these two operations. Also, suppose that the load operation ② takes the value of the store operation ❶; this happens-before

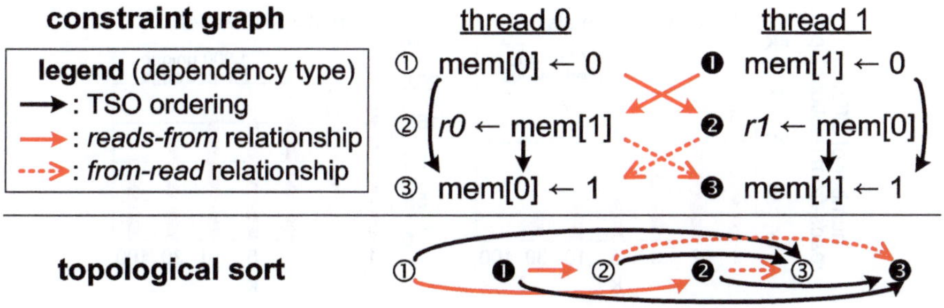

Fig. 9.6 A constraint graph (top) and its topological sort (bottom). No consistency violation can be found from this constraint graph because there exists a topological sort for the graph, following all happens-before constraints

relationship is captured by the red reads-from edge between the two operations in the figure. From this reads-from relationship, we can derive that a *from-read* relationship to the next store operation ❸ in thread 1, because store operations to the same address must be ordered (*write-serialization*).

Once we construct a constraint graph with happens-before relationships, we check for a cyclic dependency along the relationships. For instance, suppose three happens-before relationships among three memory operations (A, B, and C) in the following: (1) A happens before B, (2) B happens before C, and (3) C happens before A. These three relationships can be summarized as A→B→C→A, which is a cyclic dependency. This cyclic dependency is an indication of memory consistency violation.

Topological sorting, as introduced in Sect. 9.1.2, is a conventional method to detect a cycle in a graph, by trying to find an order among all the vertices in a way that there is no backward edge. In a formal definition, *a topological sort of a directed graph is a linear ordering of all its vertices such that if the graph contains an edge* (u, v), *then* u *appears before* v *in the ordering* [18]. The constraint graph in Fig. 9.6 can be topologically sorted as shown on the bottom of the figure, so we find no memory consistency violation in this example. However, the constraint graph in Fig. 9.2 has no topological sort, indicating a consistency violation.

Prior works apply topological sorting to each individual constraint graph obtained from post-silicon memory consistency tests. However, we notice that *many test runs actually exhibit memory-access interleaving patterns that are similar to each other*. This observation leads us to our novel *collective* constraint-graph checking solution that exploits the similarities among test runs. With our collecting checking solution, we strive to reduce the amount of computation required to validate the results of memory consistency tests. In the following subsections, we discuss how to estimate the similarities among constraint graphs (Sect. 9.2.2.1), then present our topological re-sorting technique (Sect. 9.2.2.2).

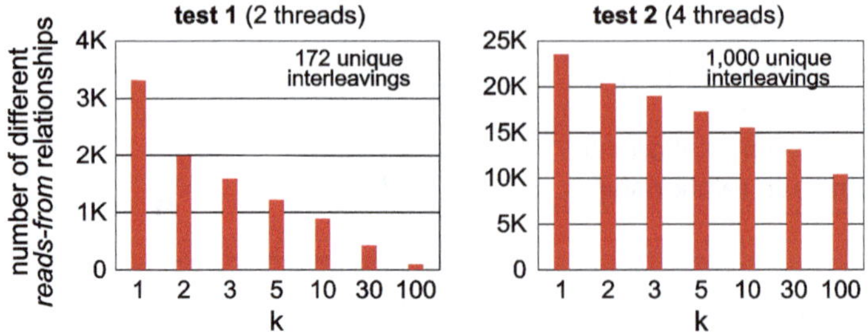

Fig. 9.7 Measuring similarities among constraint graphs using k-medoids clustering for two tests (one with 2 threads and the other with 4 threads). The medoid graphs are a set of "representative" constraint graphs. In order for the medoids to be truly representative, the number of differing reads-from relationships should be very small. We found that as k increases, the number of differing reads-from relationships decreases as shown in both tests. However, many reads-from relationships in test 2 remain different from the medoid graphs, even in a large k value. In other words, the cluster tightness is very low because of many memory-access interleaving opportunities in this test

9.2.2.1 Estimating the Similarities Among Constraint Graphs

As mentioned earlier, we would like to reduce the computation required to detect cyclic dependencies in constraint graphs. To this end, we propose an incremental verification scheme—we first compare a graph under validation with another graph that was validated previously. We then isolate the portion of the new graph that differs from the previous graph, and check only the differing portion by applying our modified topological sorting.

Limit Study—k-Medoids Clustering to Evaluate Graph Similarities

In the process of designing our incremental checking scheme, we first evaluated the similarities among graphs obtained from the same test. To this end, we conducted a preliminary study to identify a small set of representative graphs from the entire set of constraint graphs generated from repeated executions of two constrained-random tests. The first test includes 2 threads, 50 memory operations per thread, and 32 distinct shared memory addresses. The second test has more threads (4 threads), but the rest of the parameters are identical to the first test. For each of the two tests, we applied a k-medoids clustering analysis [29] to the set of constraint graphs obtained from 1,000 test runs so as to choose k representative graphs. In this analysis, our distance metric was the number of different reads-from relationships between the two graphs. To obtain reads-from relationships in this preliminary experiment, we used an in-house architectural simulator that mimics the behavior of sequential consistency (SC) [31]. This simulator randomly chooses memory operations without violating SC.

Figure 9.7 shows the total number of different reads-from relationships, where each graph is compared to its closest medoid graph. In the first test, where only 172 executions out of 1,000 executions are unique, the number decreases rapidly as k increases. On the contrary, in the second test, where every execution results in a unique interleaving behavior, the number still remains high even in the highest k value ($k = 100$). In other words, the selected medoid graphs are still vastly different from the individual graphs.

From these experiments, we conclude that the k-medoids clustering is unfit for two reasons in the following. (1) There are a large amount of discrepant reads-from relationships after applying this clustering algorithm. (2) The computational complexity of k-medoids clustering is very high [29], exceeding the complexity of topological sorting. Therefore, using this clustering negates the potential time saving of a fast topological sorting that exploits graph similarities.

Our Solution—Sorting Signatures and Diffing Corresponding Graphs

Instead of the k-medoids analysis, we strive to leverage a lightweight computation to find a previous graph sufficiently similar to a graph under validation. To this end, we repurpose our memory-access interleaving signature (Sect. 9.2.1) as a graph similarity metric.

To be specific, we first collect all execution signatures from multiple executions of a test. We then *sort* these execution signatures in ascending order. Note that signatures adjacent in this order have reads-from relationships similar to each other. The difference between two adjacent signatures is likely to appear on low bits of the signatures, which translate to differing reads-from relationships at the beginning of the test. We then reconstruct a constraint graph for each of the signatures in the sorted order. These graphs are finally checked by our novel re-sorting procedure (Sect. 9.2.2.2) that compares the two adjacent graphs and uses the previous one as a baseline graph when validating the next one. Note that our sorting also provides another benefit—all duplicated executions (i.e., executions whose memory-access interleaving patterns have been observed by previous executions) can be easily removed during the sorting process.

We treat each execution signature as a multi-word integer value during the sorting process. Thus, our method is sensitive to how we place signature words within the execution signature. From our preliminary experiment, we settled on the following integer layout. We place the first thread's signature in the most significant position, and the last thread's signature in the least significant position. If a per-thread signature consists of multiple words, we place the first word of the per-thread signature in the most significant position within the thread's position mentioned earlier, and the last word in the least significant position. We compared this layout with an alternative layout that places signature words from relevant code sections side by side. We found that this alternative layout created more diverse reads-from relationships between adjacent signatures.

9.2.2.2 Re-sorting Topological Order

MTraceCheck's graph checking examines constraint graphs in a collective manner. Specifically, MTraceCheck verifies each of the constraint graphs, one at a time, in ascending order of their corresponding signatures. The first graph is checked as a

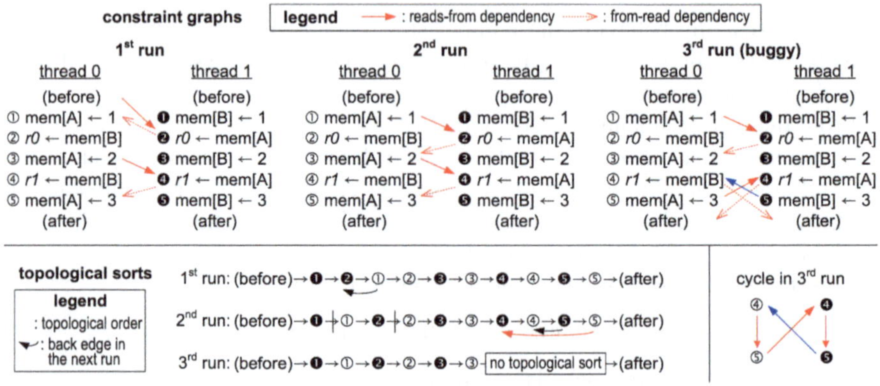

Fig. 9.8 Re-sorting a topologically sorted graph for three test runs. The topological sort from the previous run is partially rearranged for the next run. The sorting boundaries (leading and trailing) are calculated based on backward edges that are introduced by the next run. The absence of topological sort in the segment between the two boundaries indicates a consistency violation

conventional graph checking that topologically sorts all vertices of the graph. Starting from the second graph, MTraceCheck performs a partial re-sorting that applies to only the portion that differs from the prior graph in the sorted order.

MTraceCheck's re-sorting technique uses the result of topological sorting on the previous graph. From the previous topological sort, it re-sorts the region determined by two boundaries: leading boundary and trailing boundary. We decide the two boundaries in a way that the topological sort outside the boundaries can be untouched. To achieve this goal, we decide the leading boundary as the first vertex (in the previous sort) that is connected to a backward edge in the current graph under validation. The trailing boundary is decided in a similar manner; it is the last vertex that is adjacent to a backward edge introduced by the current graph. When deciding leading and trailing boundaries, we consider only new backward edges, but neither forward nor removed edges. In the best case, our re-sorting can skip the entire graph if there is no backward edge.

Figure 9.8 illustrates our re-sorting procedure on three constraint graphs obtained from a 2-threaded program. Suppose that we sorted three execution signatures and reconstructed the three graphs in ascending order of the signatures. MTraceCheck's graph-checking scheme starts with the first graph—this first-time topological sorting is identical to a conventional complete graph checking. MTraceCheck then uses the result of this first-time sorting for the second graph. It examines each of the edges newly introduced in the second graph, ①→❷ and ❷→③, and finds that only the former is backward. This backward edge is illustrated with the backward arrow below the first run's topological sort. From this backward edge, we decide ❷ as the leading boundary and ① as the trailing boundary. The order of the two nodes is then swapped in the topological sort as shown in the second run's diagram. The third run demonstrates a buggy scenario; there is no topological sort of the four vertices enclosed by the backward edge ⑤→❹. This consistency violation is also illustrated by the cyclic happens-before relationships highlighted in the bottom right of the figure.

9.3 Experimental Evaluation

We evaluated MTraceCheck in two real systems in various test configurations. We first explain our experimental setup in Sect. 9.3.1, then discuss the characteristics of our constrained-random test programs in Sect. 9.3.2. We quantify the benefits and drawbacks of our signature-based memory-tracking method in Sect. 9.3.3 and our collective graph checking in Sect. 9.3.4, respectively.

9.3.1 Experimental Setup

We evaluated MTraceCheck in two different real systems: an x86-based desktop and an ARM-based single-board computer, as summarized in Table 9.1. For each of the systems, we built a bare-metal operating environment that uses no operating system. This bare-metal environment is specialized for our validation tests and performs minimal initialization tasks for caches, MMUs, and page tables. This environment enables an uninterrupted execution of our validation tests with no context switching.

Our bare-metal operating environment initializes each of the systems slightly differently. In our x86 bare-metal environment, the boot-strap processor awakens the secondary cores using interprocessor interrupt (IPI) messages, followed by initializations for caches, page tables, and MMUs. After the initializations, we first allocate test threads in the secondary cores, and then in the primary boot-strap core, but only when no secondary core is available (i.e., 4-threaded tests in our experiments).

Table 9.1 Specifications of the systems under validation

System	x86-64 desktop	ARMv7 single-board computer
Processor	Intel Core 2 Quad Q6600	Samsung Exynos 5422 (big.LITTLE)
MCM	x86-TSO	weakly-ordered memory model
Operating frequency	2.4 GHz	800 MHz (scaled down)
Number of cores	4 (no hyper-threading)	4 (Cortex-A7) + 4 (Cortex-A15)
Cache organization	32+32 KB (L1), 8 MB (L2)	A7: 32+32 KB (L1), 512 KB (L2) A15: 32+32 KB (L1), 2 MB (L2)
Cache configuration	Write back (both L1 and L2)	Write back (L1), write through (L2)

Table 9.2 Test-generation parameters

Test-generation parameter	Values
Number of test threads	2, 4, 7
Number of static memory operations per thread	50, 100, 200
Number of distinct shared memory locations	32, 64, 128

In our ARM bare-metal environment, a primary core in the Cortex-A7 cluster runs the Das U-Boot boot loader [19], which in turn launches our test programs. At the beginning of each test program, the primary core powers up other secondary cores. The secondary cores are then switched to supervisor mode, while the primary core remains in hypervisor mode in order to keep running the boot loader. All caches, page tables, and MMUs are then initialized before the test starts. At the beginning of each test, we allocate test threads to the big cores in the Cortex-A15 cluster then to the little cores in the Cortex-A7 cluster. For example, 4-threaded tests utilize all the four big cores in the Cortex-A15 cluster.

Table 9.2 shows three key parameters used when we generate our constrained-random test programs. We chose these parameter values considering the prior work [20, 24]. We then created 21 representative *test configurations* by combining these parameter values, as shown on the x-axes of the two graphs in Fig. 9.9. The naming convention for the test configuration is as follows: [ISA]-[test threads]-[static memory operations per thread]-[distinct shared memory locations]. For example, ARM-2-50-32 represents a test configuration for the ARM ISA with 2 threads, each issuing 50 memory operations to 32 distinct shared memory locations. For each test configuration, we generated 10 distinct tests with different random seeds and ran each random test 5 times. To remove unintended dependencies across the 5 test runs, we applied a hard reset before starting each test run. Each memory operation is either load or store operation generated with equal probability (i.e., load 50% and store 50%), and transfers a 4-byte data.

Each test run iterates a test-routine loop that encloses the generated memory operations. The repeated execution of the loop is intended to exhibit various memory-access interleaving behaviors. Unless otherwise noted, we iterated this loop 65,536 times. In addition, the beginning of the loop includes a synchronization routine waiting until the previous iteration is completed for all threads, followed by a shared memory initialization and a memory barrier instruction (`mfence` for x86 and `dmb` for ARM). This custom synchronization routine is implemented with a conventional sense-reversal centralized barrier.

9.3.2 Nondeterminism in Memory Ordering

In this section, we quantify the diversity of memory ordering patterns observed from test programs. Figure 9.9 presents the number of unique memory-access interleaving patterns across our 21 test configurations (averaged over 10 tests for each test configuration), measured by counting unique execution signatures. As explained, each multithreaded test program runs 65,536 times in a loop, except for ARM-2-200-32* where its test routine is iterated 1 million times. To summarize the figure, we observe almost no duplicates in several configurations on the top linear-scale graph, while we observe very few distinct interleavings in several test configurations on the bottom logarithmic-scale graph. The dark-red bars in the figure report the results for our baseline address-generation scheme where there is no false sharing among shared memory locations. In those bars, for example, ARM-7-200-64 presents 65,536 unique memory-access interleaving patterns (100%), while ARM-2-50-32 only reveals 11 unique patterns on average (0.02%).

Among the three test-generation parameters of Table 9.2, the number of threads is the most dominant parameter diversifying memory-access interleaving behaviors. For example, we observed about 7 distinct patterns in ARM-2-50-64, 22,124 patterns in ARM-4-50-64, and 65,374 patterns in ARM-7-50-64. Note that the total number of memory operations differs in these three configurations. Under the constant total number of operations, we again observe a significant increase of interleaving patterns: ARM-2-100-64 with 123 patterns versus ARM-4-50-64 with 22,124 patterns.

The number of memory operations is the second influential parameter affecting nondeterministic interleaving behaviors, but this parameter is much less significant than the number of threads. For example, we observed 11 patterns in ARM-2-50-32, 508 patterns in ARM-2-100-32, and 35,679 patterns in ARM-2-200-32. In addition, increasing the number of shared memory locations leads to fewer memory accesses per memory location, thus reducing the number of unique interleaving patterns. ARM-2-200-64 exhibits only 9,638 patterns, much fewer than the 35,679 patterns in ARM-2-200-32.

In most 7-threaded configurations, almost all iterations exhibit very different memory-access patterns. This is partly because of the relatively low loop-iteration count that we used (65,536). As a sensitivity study, we evaluate the impact of iteration count for ARM-2-200-32, where we compared the results from two iteration counts: 35,679 unique patterns out of 65,536 (54%) versus 311,512 unique patterns out of 1,048,576 (30%), as shown in the two left-most bars in the linear-scale graph.

Moreover, we notice that our two systems show different levels of memory-access interleaving. Specifically, our x86-based system exhibits fewer distinct interleaving behaviors, compared to our ARM-based system. This is partly because of the stricter memory ordering rules enforced by the system's TSO model. To follow the TSO model, the x86-based system needs to restrict some memory re-ordering behaviors, while such re-ordering behaviors are allowed by the ARM-based system. We believe that the memory model is the major contributor to the difference between the two systems, among other factors such as load-store queue (LSQ) size, cache organiza-

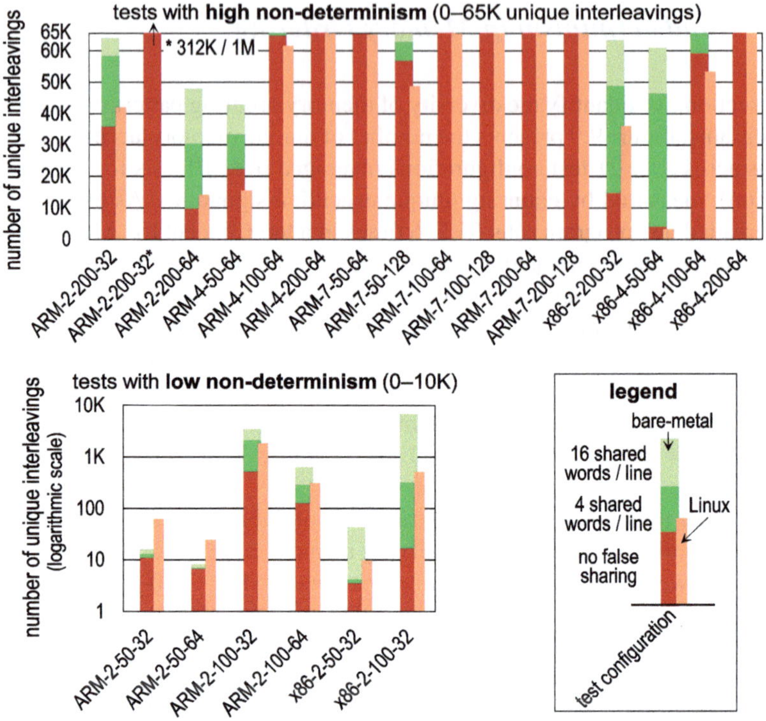

Fig. 9.9 The number of unique memory-access interleaving patterns. For each multithreaded test, we measured the number of unique memory-access interleaving signatures from 65,536 repeated runs. Most of the 2-threaded tests exhibit only a handful of distinct interleavings as shown in the logarithmic-scale graph on the bottom, while long 7-threaded tests show unique interleavings in almost all iterations as shown in the linear-scale graph on the top. False sharing of cache lines contributes to diversifying interleavings, as shown in the green and light-green bars. Compared to our bare-metal operating environment, the Linux environment creates more unique interleavings in 2-threaded tests, while it restricts interleavings in 4-threaded and 7-threaded tests

tion, interconnect, etc. For a fair comparison, each x86 test is generated in a way that it has the identical memory-access patterns as in the ARM counterpart.

Impact of False Sharing of Cache Lines

Modern multi-core systems provide coherent caches among cores. The memory consistency in these systems is implemented in consideration of the underlying cache coherence mechanism. In cache-coherent systems, a cache line is the minimal granularity of a data transfer among caches. Our prior explanation refers to the dark-red bars in Fig. 9.9 that are obtained from tests with no false sharing. In other words, only one shared word is placed in each cache line, while the rest of the cache line is not accessed at all.

Placing multiple shared words in a cache line creates additional contention among threads, and diversifies memory-access interleaving behaviors. The green and light-green bars in Fig. 9.9 present the numbers of unique memory-access interleaving patterns for two different data layouts; 4 and 16 shared words per cache line, respectively. By the false sharing, x86-4-50-64 shows the most dramatic increase among others: from 3,964 (no false sharing) to 46,266 (4 shared words) to 60,868 (16 shared words) unique patterns. We observe a larger increase in the x86-based system than in the ARM-based system.

Impact of the Operating System

In our previous discussion, we exclusively used bare-metal operating environments explained earlier. Besides the bare-metal environments, our test programs can be targeted to the Linux operating system, which allows context switching and interferences with other concurrent tasks sharing the memory subsystem. The operating system also introduces another source of test perturbation: the OS can shuffle the memory addresses by paging, while our bare-metal environment has a direct mapping between virtual and physical addresses. To quantify the perturbation of an OS, we re-targeted the same set of tests to the following Linux systems: Ubuntu MATE 16.04 for the ARM-based system and Ubuntu 10.04 LTS for the x86-based system. Under these Ubuntu systems, test threads are launched via the m5threads pthreads library [14], although, after the launch, synchronizations among test threads are carried out by our own synchronization primitives as in our bare-metal environments.

The light-red bars in Fig. 9.9 show our experimental results from the Linux environment with no false sharing. We observe two noticeable trends. In 2-threaded tests, the number of unique interleavings increases compared to the bare-metal counterparts. We believe that fine-grained (i.e., instruction-level) interferences dominate in these tests, thus diversifying the interleavings. For example, OS system tasks (which are unrelated to test threads) can perform their memory operations that can perturb the timing behavior of test threads' memory operations. In both 4-threaded and 7-threaded tests, on the contrary, the opposite trend holds. We believe that this tendency comes from coarse-grained (i.e., thread-level) interferences. For instance, some of the test threads can be preempted by OS, and then resume running after all other threads are already done, in which case this stranded thread is subject to experience much fewer memory-access interleavings.

9.3.3 Validation Performance

We measured two major components of validation time: the time spent on result-checking and the time spent on test execution. In summary, MTraceCheck achieves a significant speedup in result-checking at the expense of a small runtime increase in test execution. We provide detailed experimental results and insights below.

Result-Checking Speedup

We performed a constraint-graph checking on a host machine equipped with an Intel Core i7 860 2.8 GHz and 8 GB of main memory, running Ubuntu 16.04 LTS. This host machine is more powerful than our systems under validation (Table 9.1) because

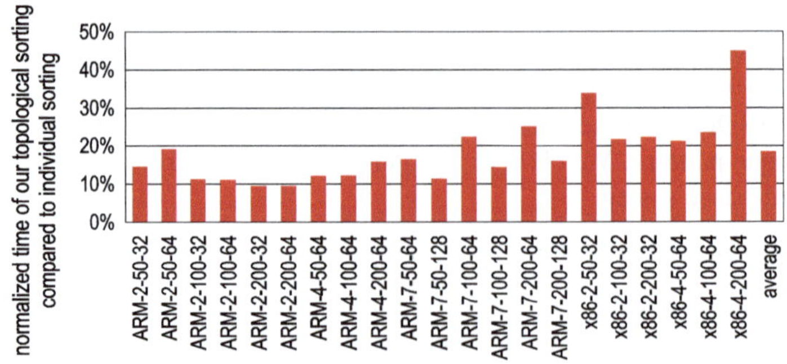

Fig. 9.10 Normalized violation-checking computation. The collective graph checking reduces topological sorting time by 81% on average, compared to a conventional individual-graph checking

the result checking takes significantly more time than the test execution. For reproducibility of our results, we adopted a well-known topological sort program, `tsort` that is part of GNU core utilities [23]. We modified the original `tsort` program so as to efficiently check multiple constraint graphs generated from a test. Specifically, for both the baseline individual checking and MTraceCheck's collective checking, vertex data structures are recycled for all constraint graphs, while edge data structures are not. We excluded the time spent in reading input graph files, assuming that all graphs are loaded into memory beforehand. We ran the `tsort` program 5 times for each evaluation to mitigate random interferences from the host machine's Linux environment. We used the signatures obtained in the Linux environment (the light-red bars in Fig. 9.9) instead of the signatures obtained in the bare-metal environment. For a fair comparison, we considered only unique constraint graphs for both checking techniques.

Figure 9.10 plots the time spent on topological sorting in our collective graph-checking method, normalized to a conventional baseline approach that examines each constraint graph separately. We observe that our collective checking consistently reduces the overall computation by 81% on average. In addition, we observe a noticeable difference between our ARM and x86 platforms: the benefit of our technique is smaller in the x86 platform. This difference is also shown in the wall-clock time spent on our topological sorting, which is not shown on the graph. For the graphs obtained from the ARM platform, the wall-clock time ranges from 2.4 μs for ARM-2-50-64 to 1.2 s for ARM-7-200-64. The wall-clock time for the x86 platform's graphs ranges from 8.6 μs for x86-2-50-32 to 4.6 s for x86-4-200-64.

We conducted an in-depth analysis to investigate the difference between the ARM and x86 platforms. From our analysis, we found that two major factors that contribute to the speedup of our collective graph checking. The first speedup factor we notice is that many constraint graphs can be validated immediately without modifying the previous topological sort. This situation is highlighted especially in the left-most bar (ARM) of Fig. 9.11, where most constraint graphs except for the first graph do not require any re-sorting. This is because the `tsort` program unwittingly places store

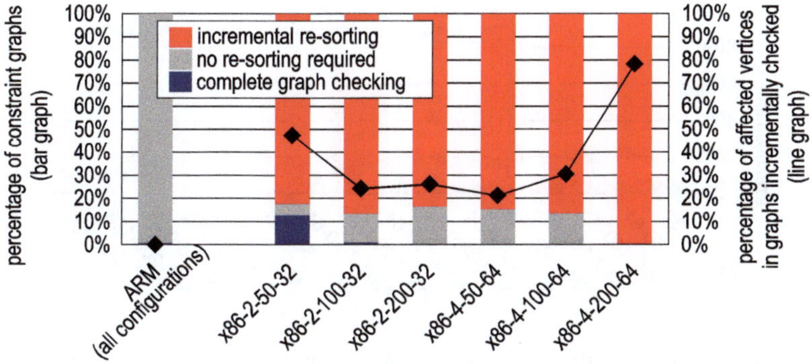

Fig. 9.11 Sources of speedup in our collective graph checking. Most graphs from ARM tests are skipped and do not require any topological re-sorting. A majority of graphs from x86 tests needs to be re-sorted partially; the re-sorting affects up to 78% of the vertices

operations prior to load operations since stores do not depend on any load operations in the absence of fence operations, which is the case for our ARM tests. Even in this case, however, our collective checking still needs to check every new edge whether the edge is backward or not, because the program is MCM-agnostic and not aware of the fact that no backward edge exists before the program examines all of the new edges. Consequently, this edge checking is the major computation performed while checking the ARM test results.

Second, for our x86 tests, we also observe that a small portion of constraint graphs, up to 16% of the graphs, can be validated without re-sorting. But a majority of the x86 graphs needs incremental re-sorting as shown in the red segments in the figure, which ranges from 82% in x86-2-50-32 to almost 100% in x86-4-200-64. For these graphs, we compute the percentage of the vertices affected by re-sorting, as shown in the line plotted in the figure. This percentage ranges from 21% in x86-4-50-64 to 78% in x86-4-200-64. This x86-4-200-64 test configuration has the largest re-sorting portion with respect to both of the metrics: the percentage of re-sorted graphs and the percentage of affected vertices. Thus this test configuration gains the least benefit from our collective checking as shown in Fig. 9.10.

Test Execution Overhead

We report MTraceCheck's runtime overheads, measured in our ARM bare-metal system. We used built-in hardware performance monitors available in the ARM system [10] to accurately measure the runtime. Figure 9.12 summarizes two types of runtime overheads introduced by MTraceCheck: (1) signature computation (i.e., the time spent in running chains of branch and arithmetic operations and storing signatures) and (2) execution signature sorting. We implemented our signature-sorting routine using a balanced binary tree written in the C programming language, and ran

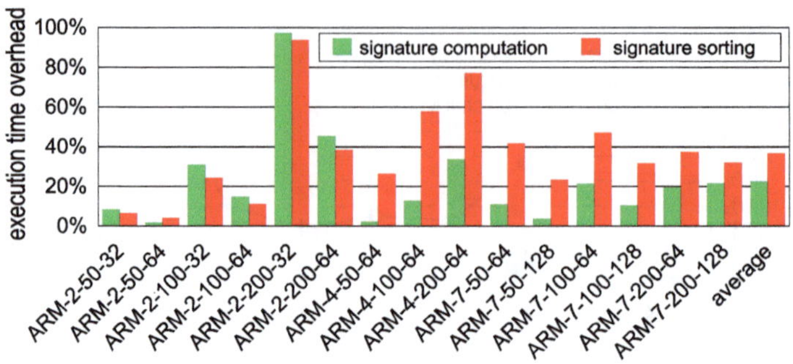

Fig. 9.12 Test execution overhead. The test runtime is increased by 22% on average due to our instrumented code for signature computation. In addition, signature sorting also increases the runtime by 38% on average

this sorting routine on the primary core in the Cortex-A7 cluster[1] after all repeated test executions completed.

The original test takes 0.09–1.1 s to run 65,536 iterations. On top of this baseline, MTraceCheck's signature computation incurs 22% additional time and its signature sorting adds 38% more time on average. Specifically, the signature computation overhead ranges from as low as 1.5% in ARM-2-50-64 to up to 97.8% in ARM-2-200-32, which is an exceptional case. The lowest overhead in ARM-2-50-64 can be explained from the fact that there are only about 7 unique interleaving patterns among 65,536 iterations, thus the branch predictor can almost perfectly predict the directions of the branch operations in our instrumented code. On the contrary, ARM-2-200-32 exhibits a unique interleaving pattern for almost every iteration, thus the branch misprediction penalty becomes much noticeable. The signature-sorting overhead ranges from 3.9% in ARM-2-50-64 to 93.5% in ARM-2-200-32. For the 1 million-iteration version of ARM-2-200-32, we found a 140% signature-sorting overhead, which is not shown in the graph.

9.3.4 *Intrusiveness of Code Instrumentation*

To quantify the intrusiveness of MTraceCheck's code instrumentation, we first measure the additional memory accesses introduced to track the order among memory operations. These additional memory accesses are not part of the original test and only needed for the validation purpose. As explained in Sect. 9.2.1, reducing the additional memory accesses is a key challenge while tracking the results of memory operations in post-silicon memory consistency tests.

[1]The signature-sorting time can be significantly reduced by two optimizations: (1) the sorting routine can utilize multiple cores available, and (2) the sorting routine can be run on the more powerful Cortex-A15 cluster. These two are not implemented in our experimental evaluations.

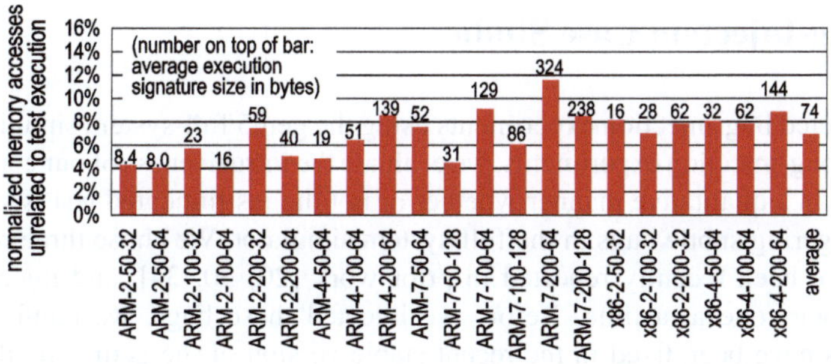

Fig. 9.13 Intrusiveness of verification. Our signature-based tracking greatly reduces memory accesses unrelated to test execution compared to a prior register-flushing approach

We report the amount of these additional accesses in Fig. 9.13, normalized to a conventional register-flushing technique [24] where all loaded values must be flushed back to memory. Compared to the conventional technique, our signature-based technique requires only 7% additional memory accesses on average. The amount of additional memory accesses ranges from 3.9% in ARM-2-100-64 to 11.5% in ARM-7-200-64. Tests with more contention (i.e., more threads, more memory operations, and fewer shared memory locations) have a larger signature footprint, which in turn require more data to be transferred.

The average size of an execution signature is shown on the top of each bar in the figure. Under low-contention tests, the signature size is bounded by the register bit width. In the tests with 2 threads, 50 operations and 32 locations (ARM-2-50-32 and x86-2-50-32), the per-thread signature barely exceeds 32 bits, as explained earlier in Sect. 9.2.1.3. However, we use all the bits of a register to store a per-thread signature: 64 bits for our x86 system and 32 bits for our ARM system. The register-width difference results in a significant difference (almost twice) in the average size of an execution signature between the two systems: 16 bytes for the x86 system and 8.4 bytes for the ARM system. Under high-contention tests, the gap between 32-bit and 64-bit registers becomes narrow, as the per-thread signature indeed requires more than one words. Among the test configurations we tested, ARM-7-200-64 has the largest signature; its size is 324 bytes on average.

In addition, our code instrumentation increases the code size, which is another source of perturbation introduced for the purpose of verification. To quantify this aspect, we measured the size of the instrumented code compared to the original test routine that excludes initialization and signature sorting routines, illustrated in Fig. 9.14. The ratio of the code size ranges from 1.95 in ARM-2-50-64 to 8.16 in ARM-7-200-64. On the top of each bar in the figure, we show the aggregated instrumented code size for all threads. While this increase is fairly significant, the code is still small enough to be accommodated in a processor's L1 instruction cache, which is usually tens of KB in modern processors.

9.4 Bug-Injection Case Studies

We conducted bug-injection experiments using the gem5 full-system simulator [14]. From our bug-injection experiments, we evaluate the intrusiveness of our code instrumentation in a qualitative manner whether or not the instrumented test is still capable of triggering subtle bugs in the full-system simulator. We chose three real bugs, which have been recently reported in prior work [20, 30, 34], and injected each of these bugs one at a time. Before we injected these bugs, we confirmed that these bugs have been fixed in the recent stable version of the gem5 simulator (tag `stable_2015_09_03`). To recreate each of the bugs, we searched the relevant bug fix from the gem5 repository [22] and reverted the fix.

Our gem5 is configured to simulate eight out-of-order x86 cores connected with a 4×2 mesh network. The cache coherence is maintained with a directory-based MESI cache coherence protocol where the directories are located on the four mesh corners. For bugs 1 and 3, we purposefully calibrated the size and the associativity of the L1 data cache (1 KB with 2-way associativity) to intensify the effect of cache evictions under a small working set used by our constrained-random tests, while the rest of the cache configuration is modeled after a recent version of the Intel Core i7 processor. We compiled our tests with the m5threads library as in our Linux environment explained in Sect. 9.3.2, and used the gem5's syscall emulation mode to run our tests.

9.4.1 Bug Descriptions

Bug 1 – An improper cache-invalidation handling leads to load→load violation: This bug is modeled after "MESI,LQ+SM,Inv" presented in [20]. It was fixed in June 2015. It is a variant of the Peekaboo problem [46], where a load operation is speculatively performed earlier than its preceding load operation in the TSO memory model. It is triggered very rarely when a cache receives an invalidation message for

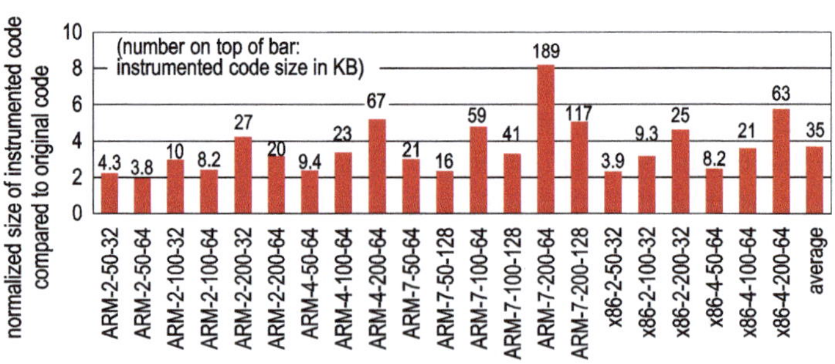

Fig. 9.14 Normalized code size. Our code instrumentation increases the code size by 3.7 times on average

a cache line that is currently transitioning from shared state to modified state (i.e., the cache line is in shared-to-modified transient state). When the cache receives the invalidation message, subsequent load operations to this cache line must be replayed after handling the invalidation message. However, due to this bug, the cache does not squash the subsequent load operations, resulting in a load→load violation.

Bug 2 – Lack of LSQ replay leads to load→load violation: This bug is reported by two independent works: [34] and [20]. It was fixed in March 2014. It manifests in a similar way as bug 1 but is caused by a buggy LSQ that does not invalidate subsequent loads when it receives an invalidation message.

Bug 3 – A race condition in cache coherence protocol: This bug is modeled after the "MESI bug 1" detailed in [30], which is evaluated also in [20]. It was fixed in January 2011. It is triggered when a race condition happens between an L1 writeback message (PUTX) and a write-request message (GETX) from another L1 cache.

9.4.2 Bug-Detection Results

For each of the three bugs, we deliberately chose the test configuration in the second column of Table 9.3 after analyzing the bug description, and generated 101 different constrained-random tests under the same test configuration. The iteration count was set to 1,024, which is greatly reduced from 65,536 in Sect. 9.3, because the gem5 simulation is much slower than the native execution of the systems in Table 9.1.

With MTraceCheck, we were able to successfully detect all the injected bugs. The third column of Table 9.3 summarizes our bug-detection results: how many tests and signatures reveal the bugs. Among the three bugs, bug 1 was the most difficult to find. It was detected by only one test, which generated 29 invalid signatures. We also tried other constrained-random tests and hand-crafted litmus tests for a few days, but these tests failed to trigger the bug. Bug 2 manifested itself in 11 tests. Among the 11 tests, one test generated two incorrect signatures, while each of the other tests generated one incorrect signature. While these two bugs are subtle and appeared only in some tests and iterations, bug 3 caused much more significant failures. For all the 101 tests, the gem5 simulations ended prematurely with internal errors (a protocol deadlock and an invalid transition) from the gem5's ruby memory subsystem.

Table 9.3 Bug-detection results

Bug	Test configuration	Detection results
1	x86-4-50-8 with 4 shared words per cache line	1 test, 29 signatures
2	x86-7-200-32 with 16 shared words per cache line	11 tests, 12 signatures
3	x86-7-200-64 with 4 shared words per cache line	all 101 tests, crash

Figure 9.15 shows a code snippet of the test that exposed bug 1. The first 20 memory operations of thread 0 are omitted for simplicity. In thread 0, there is no store operation to memory address `0x1` until the starred instruction ★, although the omitted code includes store operations to the other addresses in the same cache line (i.e., memory addresses `0x0`, `0x2` and `0x3`). Each of threads 1 and 2 performs a store operation to `0x1` as shown in the figure (gray bold operations). Thread 3 executes three consecutive load operations from memory address `0x1`. Among these three load operations, the first and the third load operations read the initial value of the memory location, while the second reads the value from the store operation ★. From the second and third loaded values, we identify a cyclic happens-before relationship illustrated with blue arrows in the figure: ★ happens before ② (a reads-from relationship), which should happen before ③ (a load→load ordering), which in turn happens before ★ (a from-read relationship).

9.5 Discussions

Reducing the signature size by pruning invalid memory-access interleaving patterns We make a conservative assumption in our code instrumentation (Sect. 9.2.1) to support a wide range of MCMs in a unified framework. In our assumption, each memory operation can be independently interleaved, regardless of memory ordering rules required by MCMs. This conservative assumption leads to two major overheads discussed earlier in Sect. 9.3.4: the increased signature size (Fig. 9.13) and code size (Fig. 9.14). To address these two overheads, we can leverage micro-architectural information during the code instrumentation (*static pruning*). For instance, we can compute a window of outstanding memory operations (as in [45]) by taking into account the number of LSQ entries, etc. By doing so, we can prune infeasible options for loaded values. Unfortunately, we could not gather sufficient micro-architectural details to apply this static pruning optimization in our real-system evaluations.

Another possible optimization relies on a runtime technique (*dynamic pruning*) computing a frontier of memory operations in strong MCMs (e.g., TSO). In these MCMs, reading from memory operations older than the frontier is often considered

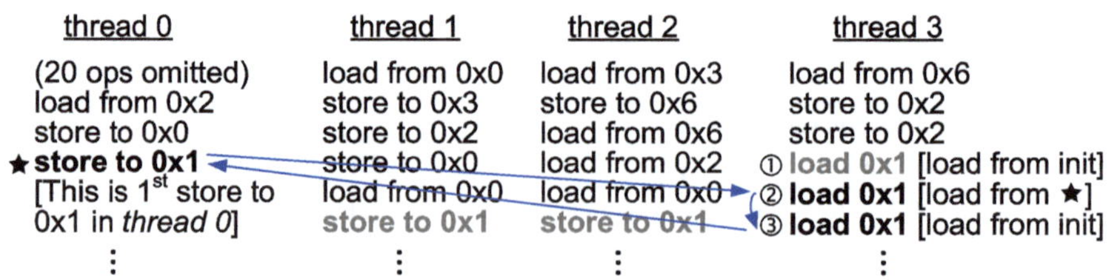

Fig. 9.15 A load→load ordering violation detected by a 4-threaded test. In thread 3, loads ② and ③ are incorrectly re-ordered due to bug 1

to be invalid. To implement this optimization, each thread needs to track the set of recent store operations performed. This dynamic pruning technique, however, can complicate the signature computation and decoding processes.

Improving scalability As the size of the test program grows, both the signature size and the instrumented code size also increase. While we did not experience scalability issues for the test configurations we tested, it is possible that our code instrumentation can be overly intrusive for a large test program with many memory-access interleaving opportunities. To improve the scalability, we can merge multiple independent code segments to create a single test program (as in [42]). The independent code segments share cache lines but do not share memory addresses within the cache lines, allowing only false sharing across the code segments.

Store atomicity We make no assumption on store atomicity (single-copy atomic, multiple-copy atomic, or non-multiple-copy atomic) in Sect. 9.2. However, we did not evaluate MTraceCheck in a single-copy atomic real system except for the limit study presented in Sect. 9.2.2.1. Constraint graphs for a single-copy atomic system contain additional dependency edges [12, 35] that are not necessarily included in a multiple-copy or non-multiple-copy atomic system, which may decrease the benefit of our collective checking due to larger re-sorting windows.

9.6 Conclusions

Multi-core processors experience a variety of different memory-access interleaving behaviors while they run multithreaded programs. Verifying such nondeterministic behaviors involves analyzing many different memory orders, if these orders abide by the memory consistency model. Post-silicon memory consistency validation aims at checking very subtle memory ordering behaviors that are rarely observed. To improve the efficiency of memory consistency validation, we propose MTraceCheck that tackles two major obstacles: limited observability and heavy result-checking computation. The first obstacle is mitigated by our novel signature-based memory-tracking method, which instruments observability-enhancing code. The instrumented code computes a compact signature value representing memory-access interleaving patterns observed at runtime. It also minimally perturbs the memory accesses in the original test, thus it is still capable of triggering subtle memory consistency bugs. The second obstacle is alleviated by our collective graph-checking algorithm that leverages the structural similarities among constraint graphs. Our comprehensive evaluations show that MTraceCheck can quickly detect subtle memory consistency bugs.

The MTraceCheck source code is available on our GitHub repository: https://github.com/leedoowon/MTraceCheck. The repository includes the constrained-random test generator for memory consistency (Sect. 9.3.1), the code instrumentation script (Sect. 9.2.1), the architectural simulator (Sect. 9.2.2.1), and the collective graph checker (Sect. 9.2.2.2). The current version supports three ISAs: x86-64, ARMv7, and RISC-V.

Acknowledgements We would like to thank Prof. Todd Austin, Biruk Mammo, and Cao Gao for their advice and counseling throughout the development of this project. The work was supported in part by C-FAR, one of six centers of STARnet, a Semiconductor Research Corporation program sponsored by MARCO and DARPA. Doowon Lee was also supported by a Rackham Predoctoral Fellowship at the University of Michigan.

References

1. A. Adir, M. Golubev, S. Landa, A. Nahir, G. Shurek, V. Sokhin, A. Ziv, Threadmill: a post-silicon exerciser for multi-threaded processors, in *Proceedings of the 48th Design Automation Conference* (2011), pp. 860–865. https://doi.org/10.1145/2024724.2024916
2. S.V. Adve, K. Gharachorloo, Shared memory consistency models: a tutorial. Computer **29**(12), 66–76 (1996). https://doi.org/10.1109/2.546611
3. J. Alglave, A formal hierarchy of weak memory models. Form. Methods Syst. Design **41**(2), 178–210 (2012). https://doi.org/10.1007/s10703-012-0161-5
4. J. Alglave, L. Maranget, S. Sarkar, P. Sewell, Fences in weak memory models, in *Computer Aided Verification: 22nd International Conference, CAV 2010* (2010), pp. 258–272. https://doi.org/10.1007/978-3-642-14295-6_25
5. J. Alglave, L. Maranget, S. Sarkar, P. Sewell, Litmus: running tests against hardware, in *Tools and Algorithms for the Construction and Analysis of Systems: 17th International Conference, TACAS 2011* (2011), pp. 41–44. https://doi.org/10.1007/978-3-642-19835-9_5
6. J. Alglave, L. Maranget, M. Tautschnig, Herding cats: modelling, simulation, testing, and data mining for weak memory. ACM Trans. Program. Lang. Syst. **36**(2), 7:1–7:74 (2014). https://doi.org/10.1145/2627752
7. ARM: Barrier Litmus Tests and Cookbook (2009)
8. ARM: Cortex-A9 MPCore Programmer Advice Notice Read-after-Read Hazards, ARM Reference 761319 (2011)
9. ARM: Embedded Trace Macrocell Architecture Specification (2011)
10. ARM: ARM Architecture Reference Manual, ARMv7-A and ARMv7-R edition (2012)
11. ARM: ARM Architecture Reference Manual, ARMv8, for ARMv8-A architecture profile (2017)
12. Arvind, J.W. Maessen, Memory model = instruction reordering + store atomicity, in *Proceedings of the 33rd Annual International Symposium on Computer Architecture* (2006), pp. 29–40. https://doi.org/10.1109/ISCA.2006.26
13. T. Ball, J.R. Larus, Efficient path profiling, in *Proceedings of the 29th Annual ACM/IEEE International Symposium on Microarchitecture* (1996), pp. 46–57. https://doi.org/10.1109/MICRO.1996.566449
14. N. Binkert, B. Beckmann, G. Black, S.K. Reinhardt, A. Saidi, A. Basu, J. Hestness, D.R. Hower, T. Krishna, S. Sardashti, R. Sen, K. Sewell, M. Shoaib, N. Vaish, M.D. Hill, D.A. Wood, The gem5 simulator. ACM SIGARCH Comput. Archit. News **39**(2), 1–7 (2011). https://doi.org/10.1145/2024716.2024718
15. H.W. Cain, M.H. Lipasti, R. Nair, Constraint graph analysis of multithreaded programs, in *Proceedings of the 12th International Conference on Parallel Architectures and Compilation Techniques* (2003), pp. 4–14. https://doi.org/10.1109/PACT.2003.1237997
16. K. Chen, S. Malik, P. Patra, Runtime validation of memory ordering using constraint graph checking, in *2008 IEEE 14th International Symposium on High Performance Computer Architecture* (2008), pp. 415–426. https://doi.org/10.1109/HPCA.2008.4658657

17. Y. Chen, Y. Lv, W. Hu, T. Chen, H. Shen, P. Wang, H. Pan, Fast complete memory consistency verification, in *2009 IEEE 15th International Symposium on High Performance Computer Architecture* (2009), pp. 381–392. https://doi.org/10.1109/HPCA.2009.4798276
18. T.H. Cormen, C.E. Leiserson, R.L. Rivest, C. Stein, *Introduction to Algorithms*, 3rd edn. (The MIT Press, Cambridge, 2009)
19. Das U-Boot – the universal boot loader (2016), http://www.denx.de/wiki/U-Boot
20. M. Elver, V. Nagarajan, McVerSi: a test generation framework for fast memory consistency verification in simulation, in *2016 IEEE International Symposium on High Performance Computer Architecture* (2016), pp. 618–630. https://doi.org/10.1109/HPCA.2016.7446099
21. N. Foutris, D. Gizopoulos, M. Psarakis, X. Vera, A. Gonzalez, Accelerating microprocessor silicon validation by exposing ISA diversity, in *Proceedings of the 44th Annual IEEE/ACM International Symposium on Microarchitecture* (2011), pp. 386–397. https://doi.org/10.1145/2155620.2155666
22. gem5 mercurial repository host (2016), http://repo.gem5.org
23. GNU coreutils version 8.25 (2016), http://ftp.gnu.org/gnu/coreutils
24. S. Hangal, D. Vahia, C. Manovit, J.Y.J. Lu, S. Narayanan, TSOtool: a program for verifying memory systems using the memory consistency model, in *Proceedings of the 31st Annual International Symposium on Computer Architecture* (2004), pp. 114–123. https://doi.org/10.1109/ISCA.2004.1310768
25. IBM: Power ISA Version 2.07B (2015)
26. Intel: Intel 64 Architecture Memory Ordering White Paper (2007)
27. Intel: Intel 64 and IA-32 Architectures Software Developer's Manual (2015)
28. Intel: 6th Generation Intel Processor Family Specification Update (2016)
29. k-medoids algorithm (2016), https://en.wikipedia.org/wiki/K-medoids
30. R. Komuravelli, S.V. Adve, C.T. Chou, Revisiting the complexity of hardware cache coherence and some implications. ACM Trans. Archit. Code Optim. **11**(4), 37:1–37:22 (2014). https://doi.org/10.1145/2663345
31. L. Lamport, How to make a multiprocessor computer that correctly executes multiprocess programs. IEEE Trans. Comput. **28**(9), 690–691 (1979). https://doi.org/10.1109/TC.1979.1675439
32. D. Lee, V. Bertacco, MTraceCheck: validating non-deterministic behavior of memory consistency models in post-silicon validation, in *Proceedings of the 44th Annual International Symposium on Computer Architecture* (2017), pp. 201–213. https://doi.org/10.1145/3079856.3080235
33. D. Lin, T. Hong, Y. Li, E. S, S. Kumar, F. Fallah, N. Hakim, D.S. Gardner, S. Mitra, Effective post-silicon validation of system-on-chips using quick error detection. IEEE Trans. Comput. Aided Design Integr. Circuits Syst. **33**(10), 1573–1590 (2014). https://doi.org/10.1109/TCAD.2014.2334301
34. D. Lustig, M. Pellauer, M. Martonosi, PipeCheck: specifying and verifying microarchitectural enforcement of memory consistency models, in *Proceedings of the 47th Annual IEEE/ACM International Symposium on Microarchitecture* (2014), pp. 635–646. https://doi.org/10.1109/MICRO.2014.38
35. D. Lustig, C. Trippel, M. Pellauer, M. Martonosi, ArMOR: defending against memory consistency model mismatches in heterogeneous architectures, in *Proceedings of the 42nd Annual International Symposium on Computer Architecture* (2015), pp. 388–400. https://doi.org/10.1145/2749469.2750378
36. D. Lustig, A. Wright, A. Papakonstantinou, O. Giroux, Automated synthesis of comprehensive memory model litmus test suites, in *Proceedings of the Twenty-Second International Conference on Architectural Support for Programming Languages and Operating Systems* (2017), pp. 661–675. https://doi.org/10.1145/3037697.3037723
37. S. Mador-Haim, R. Alur, M.M. Martin, Generating litmus tests for contrasting memory consistency models, in *Computer Aided Verification: 22nd International Conference, CAV 2010* (2010), pp. 273–287. https://doi.org/10.1007/978-3-642-14295-6_26

38. B.W. Mammo, V. Bertacco, A. DeOrio, I. Wagner, Post-silicon validation of multiprocessor memory consistency. IEEE Trans. Comput. Aided Design Integr. Circuits Syst. **34**(6), 1027–1037 (2015). https://doi.org/10.1109/TCAD.2015.2402171
39. Y.A. Manerkar, D. Lustig, M. Martonosi, M. Pellauer, RTLCheck: verifying the memory consistency of RTL designs, in *Proceedings of the 50th Annual IEEE/ACM International Symposium on Microarchitecture* (2017), pp. 463–476. https://doi.org/10.1145/3123939.3124536
40. A. Meixner, D.J. Sorin, Dynamic verification of memory consistency in cache-coherent multithreaded computer architectures. IEEE Trans. Dependable Secur. Comput. **6**(1), 18–31 (2009). https://doi.org/10.1109/TDSC.2007.70243
41. M. Naylor, S.W. Moore, A. Mujumdar, A consistency checker for memory subsystem traces, in *2016 Formal Methods in Computer-Aided Design (FMCAD)* (2016), pp. 133–140. https://doi.org/10.1109/FMCAD.2016.7886671
42. T. Rabetti, R. Morad, A. Goryachev, W. Kadry, R.D. Peterson, SLAM: SLice And Merge - effective test generation for large systems, in *Hardware and Software: Verification and Testing* (2013), pp. 151–165
43. A. Roy, S. Zeisset, C.J. Fleckenstein, J.C. Huang, Fast and generalized polynomial time memory consistency verification, in *Computer Aided Verification: 18th International Conference, CAV 2006* (2006), pp. 503–516. https://doi.org/10.1007/11817963_46
44. E. Seligman, T. Schubert, M.V.A.K. Kumar, *Formal Verification* (Morgan Kaufmann, San Francisco, 2015)
45. O. Shacham, M. Wachs, A. Solomatnikov, A. Firoozshahian, S. Richardson, M. Horowitz, Verification of chip multiprocessor memory systems using a relaxed scoreboard, in *Proceedings of the 41st Annual IEEE/ACM International Symposium on Microarchitecture* (2008), pp. 294–305. https://doi.org/10.1109/MICRO.2008.4771799
46. D.J. Sorin, M.D. Hill, D.A. Wood, *A Primer on Memory Consistency and Cache Coherence*, 1st edn. (Morgan & Claypool Publishers, 2011)
47. The Coq proof assistant (2016), https://coq.inria.fr
48. I. Wagner, V. Bertacco, Reversi: Post-silicon validation system for modern microprocessors, in *IEEE International Conference on Computer Design* (2008), pp. 307–314. https://doi.org/10.1109/ICCD.2008.4751878
49. A. Waterman, K. Asanovic, SiFive Inc., The RISC-V Instruction Set Manual, Volume I: User-Level ISA, Version 2.2 (2017)
50. D. Weaver, T. Germond, *The SPARC Architectural Manual (Version 9)* (Prentice-Hall Inc., Englewood Cliffs, 1994)

Chapter 10
Selection of Post-Silicon Hardware Assertions

Pouya Taatizadeh and Nicola Nicolici

10.1 Hardware Assertions

Before elaborating on the main steps in our methodology, we first motivate the idea of using hardware assertions for bit-flip detection. There exists two error scenarios: *silent errors* and *masked errors* [1]. A silent error occurs when an error propagates to an observable point but is missed due to insufficient checking. On the other hand, a masked error is an error that is produced but masked out before reaching an observable point. Due to the inherent lack of real-time observability in circuit blocks that are deeply embedded into the design under validation, depending on the workload, most bit-flips will not manifest themselves at an observable output, despite the fact that their presence proves an underlying problem with the design [2]. Moreover, even if these intermittent errors do propagate to primary outputs, checking them against a precomputed golden response is infeasible in post-silicon validation. This is because of the huge volume of clock cycles that are applied to silicon prototypes, which makes pre-computation of golden responses impractical in simulation environments. Another concern that arises when bit-flips are not detected soon after they occur is because failing experiments, which are caused by bit-flips, are not easily reproducible due to the electrical phenomena that cause them (e.g., unique temperature and power supply noise). Hence, one must ensure that the trace signals that are collected on embedded trace buffers (which are also constrained in size) are most relevant to the error that has occurred [3]. This recorded information is critical during root-causing

P. Taatizadeh (✉) · N. Nicolici
McMaster University, Hamilton, ON, Canada
e-mail: taatizp@mcmaster.ca

N. Nicolici
e-mail: nicolici@mcmaster.ca

P. Mishra and F. Farahmandi (eds.), *Post-Silicon Validation and Debug*,
https://doi.org/10.1007/978-3-319-98116-1_10

[4] and, to ensure that meaningful information is analyzed, the error detection latency must not exceed the depth of the trace buffers. Therefore, considering the goal of low error detection latency, employing assertions during post-silicon validation has been motivated by the following points:

- Assertions can perform property checking without requiring a golden response;
- Techniques such as [5], have been proposed for mapping assertions to hardware, thus making them suitable candidates for post-silicon validation;
- Traditionally, assertions have been carefully crafted by verification engineers before being deployed. The drawback of this approach was the omission of nonobvious assertions that were beyond the understanding of verification engineers. However, recent researches, have explored automatic assertion generation using different methods ranging from static or dynamic analysis as in [6] or by employing data mining as in [7].

Although the original goal of automated assertion generation is to aid pre-silicon verification, e.g., when the design implementation changes iteratively or design blocks are reused in different environments, we recognize the potential of these discovered assertions for post-silicon validation. Therefore, we build our methodology (illustrated in Fig. 10.1) by leveraging the recent advancements in this area [5, 8, 9]. A key observation is that bit-flips, unlike functional errors, are related to the design netlist, which facilitates both a concise definition of the error space to be covered, as well as the development of a method that does not rely on design functionality but rather on its structure. This, in turn, facilitates automation in a manner that resembles common tasks in the electronic design flow such as, for example, logic synthesis, place and route, or automatic test pattern generation. Nonetheless, to make the methodology practically feasible, one has to account for unique hardware constraints. Therefore, the huge number of assertions that are discovered during the pre-silicon phase need to be consciously selected before being mapped to hardware.

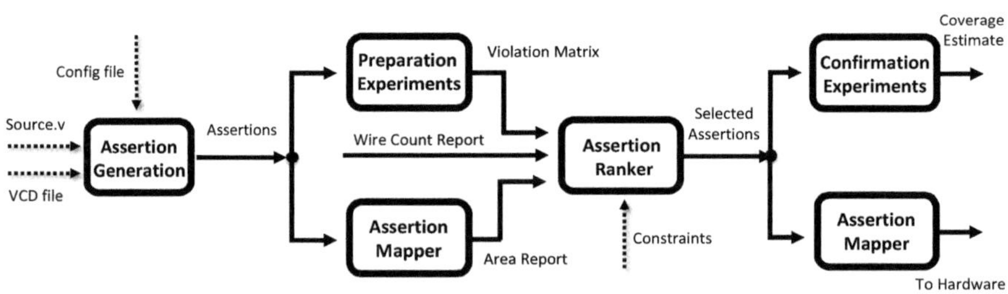

Fig. 10.1 Tool flow for selecting the most suitable assertions to embed on-chip under wire constraints [2]

10.2 Methodology

In this section, different steps of the methodology shown in Fig. 10.1 are elaborated.

10.2.1 Assertion Generation

As illustrated in Fig. 10.1, the first step of our methodology is to find assertions for a given design. Although one can develop assertions manually, in order to enable an automated methodology, it is necessary to rely on tools that can generate nonobvious assertions that cut across time cycles automatically which are also intermodular. In other words, assertions must be extracted from the given design (either netlist or RTL code) irrespective of the circuit's functionality.

There are two commonly used languages for writing assertions: Property Specification Language (PSL) and System Verilog Assertion (SVA). An example of an SVA assertion is shown below:

```
assr1: ((x == 1) && (y == 0) |=> ##2 (a == 0))
```

where *a* is the destination signal and *x* and *y* can be flip-flops, primary inputs or internal nets.

Most of the available tools for automatic assertion generation are customizable [6, 7], thus meaning that the user can choose the destination flip-flops. Since the objective is to detect bit-flips that occur in flip-flops, assertions that target flip-flops as destination signals are deemed to be the superior ones to be added to the discovered assertion pool. For instance, in the assertion statement above, if *a* is a flip-flop and a bit-flip occurs in the circuit changing its value from 0 to 1, and all other conditions hold, then this assertion would be violated. If the status of the assertions is monitored during circuit's operations, then this bit-flip can be detected as soon as the assertion is fired. We call a flip-flop as a *potentially covered* flip-flop if at least one assertion fires for at least one bit-flip that affects it. We will come back to the topic of coverage in Sect. 10.2.5.

10.2.2 Preparation Experiments

Our exploratory experiments (on the ISCAS89 circuits [10]) indicate that for a design block with 1–2000 flip-flops an assertion discovery tool can produce more than 20,000 assertions. Mapping all these assertions to hardware is obviously impractical due to both *area* and *wiring* constraints. Consequently, assertions need to be weighted and a subset of them must be selected as *candidate assertions* for hardware mapping. Since our objective is to improve the number of flip-flops that are potentially covered when bit-flips occur, assertions that are more likely to violate within a predefined time window in response to bit-flips are preferred over the other ones. For example, if, during a bit-flip injection experiment, *assr i* detects 12 different bit-flips and *assr j* detects 5 different bit-flips, but 4 of these bit-flips that are detected by *assr j* are also detected by *assr i*, then it would be logical to select *assr i* as the candidate assertion and dismiss *assr j*. There are many other factors, other than the violation count, that needs to be taken into account and they will be all elaborated in Sect. 10.3.

Since the potential of each assertion to detect specific bit-flips needs to be worked out such that ranking will be done based on this potential, following to assertions discovery, *preparation experiments* are carried out in order to determine the violation count for each assertion when bit-flips are randomly, but uniformly, injected in all the flip-flops. As shown in Fig. 10.2, the following steps are carried out to perform the preparation experiments:

1. Initially, each flip-flop is instrumented with a 2-to-1 mux and an inverter to facilitate error injections (bit-flips). The select signal of the mux determines when and where the bit-flip should occur. In addition to instrumenting flip-flops, all the discovered assertions are added to the design.
2. For each simulation, we load the circuit in a random state. We wait for a user-defined time (e.g., 10 clock cycles), during which we monitor assertions to make sure no violation happens due to the unlikely possibility that the initial state is an unreachable state.
3. If there is no violation after that user-defined period, we will target one flip-flop at a time, inject bit-flips and simulate the design using random input stimuli. In each simulation, bit-flips are injected at k random times and the circuit is simulated for a user-defined time (e.g., 256 clock cycles) after error injection; during which assertions are monitored and their violation is recorded, as illustrated in Fig. 10.2. As a result, if the circuit has m flip-flops, the total number of simulations that have to be performed will be $m \times k$.
4. Finally, by combining the violation reports for each simulation, a $m \times n$ matrix is created where m is the total number of flip-flops and n is the total number of assertions. Each entry in this *Violation Matrix* represents the total violation count of an assertion for a specific flip-flop in all simulations. For instance, entry (1,2) in Fig. 10.2 means that *assr1* has been violated 12 times for all the errors that were injected in *flip-flop 1*.

It is important to note that the violation matrix captures the error space of bit-flips that can be detected by the assertions that were assessed during the preparation experiments. Due to the randomness in preparation experiments, which is needed to account for the random occurrence of bit-flips on silicon prototypes, an assertion that fired for one bit-flip injection might not be violated for another bit-flip injection in the same flip-flop; this can happen due to multiple reasons. For instance, it is likely that the circuit is in a different state for which the signals that imply the assertion have different values. Also, it is extremely probable that the effect of the bit-flip might not propagate to the assertion checker due to a different input sequence. This is the reason for referring to the flip-flops from the violation matrix, for which at least one assertion has fired during the preparation experiments, as *potentially covered*. It is up to the user of the methodology to exclude or include flip-flops of interest for the

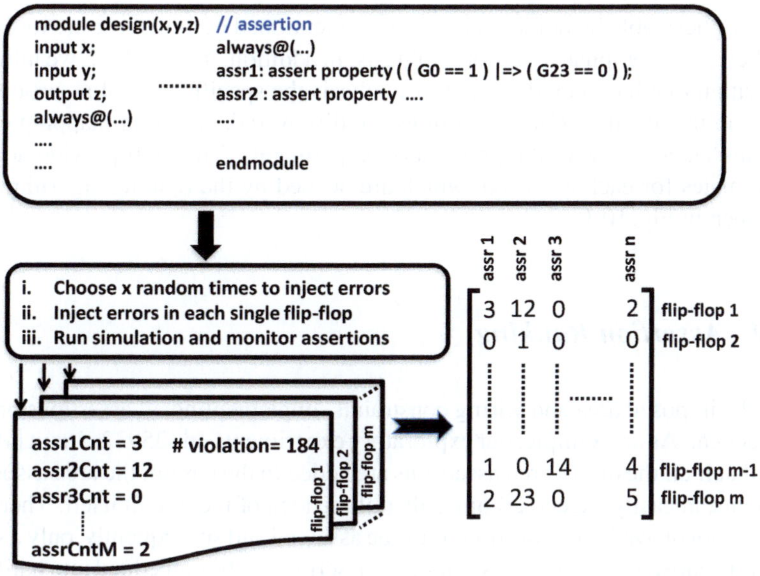

Fig. 10.2 Preparation experiments illustrating steps towards creation of the Violation Matrix. Each entry in this violation matrix shows the total number of violations of the assertion (from the corresponding column) when the bit-flip was injected in the respective flip-flop (identified by the row)

bit-flip injection experiments. Hence, the number of rows in the violation matrix is upper-bounded by the number of flip-flops in the design and the number of columns, i.e., the number of assertions to be considered by the preparation experiments, can be bounded based on, for example, the available computational resources. It should also be noted that the quality of the violation matrix is central to the accuracy of the assertion ranking algorithm. The more simulations we run during the preparation experiments, the more revealing is the information in the violation matrix. As it is clear, simulation speeds cannot keep up with that of actual silicon. Hence, to speed-up this step in our methodology, we will propose an FPGA-based emulation platform in Sect. 10.5 that significantly reduces the run-time of preparation experiments.

10.2.3 Mapping Assertions to Hardware

Traditionally, assertions have been developed for verification and are composed of logical and temporal operators and regular expressions. These statements can be added to the source code in pre-silicon verification to monitor errors using functional simulators. However, for using them in post-silicon validation, they must be mapped into hardware in order to do online property checking. Both PSL and SVA assertions are not synthesizable by default. However, as mentioned before, there are tools such as [5] that can accomplish assertion synthesis. In addition, in Sect. 10.4, we introduce an algorithm that has been designed to synthesis the specific type of assertions that are used in this chapter. Once assertions are discovered, assertion mapping can be done simultaneously with the preparation experiments. This will provide accurate area estimates for each assertion, which are needed by the ranking algorithm, as it can be seen in Fig. 10.1.

10.2.4 Assertion Ranking

Due to the imposed area and wiring constraints, implementing all assertions on-chip is impractical. As an example, our exploratory experiments on s35932 circuit [10] has shown that if all the discovered assertions are added to the circuit, the area associated with them can easily exceed 20 times that of the area of the circuit itself. Therefore, the large pool of available assertions must be assessed and subsequently, only a subset of them are chosen and marked as candidate assertions to be embedded into hardware. In this work, by having established a relationship between assertions and bit-flips that they might detect in Sect. 10.2.2, we tune the assertion ranker to focus on maximizing the number of flip-flops that are potentially covered under user-defined constraints. As it will be detailed in Sect. 10.3, the algorithm uses the violation matrix, area estimates, the wire count report and user-specified constraints. The required steps to prepare the violation matrix were thoroughly explained in Sect. 10.2.2; likewise, area estimates for assertions are obtained as explained in Sect. 10.2.3. The wire count report can be directly extracted from the assertions pool by counting the distinct number of wires that comprise each assertion statement. Finally, the constraints are provided by the user. In our current implementation, we provide the wire count as the constraint.

10.2.5 Confirmation Experiments

The last step in our methodology is to run confirmation experiments. During confirmation experiments, the circuit is instrumented by the assertions that have been selected by the specific algorithm in assertion ranker. The circuit is then simulated using random input stimuli during which one error is injected at a time (injections will be uniformly distributed across all the flip-flops throughout confirmation experiments) and the assertions violations are recorded. In the ideal case, whenever a bit-flip is injected into an arbitrary flip-flop, a violation will be detected if at least one of the selected assertions that are embedded into the design fires in response to the bit-flip. However, considering the random occurrence of bit-flips, it can happen that some bit-flip injections will not cause a violation despite the fact that based on the violation matrix, more than one of the selected assertions were expected to fire. Furthermore, though it is less likely, it might also happen that some of the selected assertions, that were not identified during the preparation experiments to be related to a particular flip-flop, will fire during the confirmation experiments when a bit-flip is injected in the respective flip-flop. In any case, if at least one bit-flip in a flip-flop causes a violation in at least one of the embedded assertions, that flip-flop is potentially covered by that assertion. At this point, we define *flip-flop coverage estimate* as a metric to evaluate the effectiveness of our algorithm, which can also be used as an internal feedback to the entire methodology. Similar to the preparation experiments, this step can also be accelerated using the architecture that is detailed in Sect. 10.5.

Definition 1 *Flip-flop coverage estimate* is defined as the ratio between the total number of flip-flops for which there exists at least one assertion that has detected at least one of the bit-flips that have been injected in that flip-flop over the total number of flip-flops.

The reason that the potential coverage of a specific flip-flop is evaluated only after all the error injections has occurred is that, as elaborated in Sect. 10.2.2, it does happen that an injection in a flip-flop causes all the associated assertions for that flip-flop (or other assertions if the error propagates to other flip-flops) to fire while another error in the same flip-flop might end up undetected. Since post-silicon validation sessions commonly run for extensively long durations, we believe that *flip-flop coverage estimate* is a remarkable metric. This is because if even a single-bit-flip is detected in a flip-flop over a long validation experiment (hours of real-time execution), this will highlight the presence of a subtle electrically induced error within the close proximity of the detection point that needs to be corrected before committing to high-volume manufacturing.

It is important to note why the coverage metrics is labeled as *estimate*. Bit-flips, unlike hard defects (modeled as, for example, stuck-at faults in the logic domain) occur only in some unanticipated clock cycles, depending on the logic state of the circuit, the workload and the electrical state (e.g., voltage supply and/or droop). One cannot guarantee a bit-flip to be easily reproducible, nonetheless, so long as the validation sequences are long (as it is the case for post-silicon validation) it is expected that the underlying electrical problem will manifest itself as a bit-flip in the logic domain. It is needless to say that our confirmation experiments, even by using emulation platforms, are limited in duration (when compared to post-silicon validation experiments), nonetheless if an injected bit-flip is detected during the confirmation experiments by the subset of selected assertions, we *estimate* that a post-silicon validation experiment would also detect it because of the huge volume of additional clock cycles that are applied in post-silicon validation (approximately three and six orders of magnitude faster than FPGA-based emulation and simulation respectively).

Finally, both these metrics can reveal important characteristics and features about the effectiveness of the embedded DfD logic and can also provide important feedback to different steps of the proposed methodology. For example, the original pool of assertions might need to be expanded if the confirmation experiments indicate that, for some flip-flops, no bit-flips were detected; it is also possible that an insufficient number of preparation experiments have been carried out, in which case the confirmation experiments will indicate that the quality of the violation matrix needs to be improved. Moreover, it can also serve as proof of due diligence when the subset of the selected assertions that were used on the silicon prototype exhibit no failure after extensively long post-silicon validation experiments; in which case, the confidence to commit to high-volume of manufacturing is substantiated by the values captured by the coverage estimates.

10.3 Ranking Algorithm

As emphasized in the previous section, mapping all the discovered assertions from the assertion generation step to hardware for the purpose of low-latency bit-flip detection is impractical due to area and wire usage overheads. Hence, a subset of the discovered assertions must be selected based on the goal of maximizing the flip-flop coverage estimate. In this section, we propose a novel algorithm that works on selecting a set of assertions to maximize the *flip-flop coverage estimate* by taking into account the wire usage constraint. The objective is to maximize the potential coverage for each flip-flop, while at the same time the area should be contained. The pseudo-code for this algorithm is given in Algorithm 1. Before we elaborate on the main steps of the algorithm, we should define a metric that is used in order to weight assertions for the purpose of one-to-one comparisons. The *Detection Potential (DP)* is defined below

$$Assr(i)_{DP} = \frac{(\alpha \times FCov + TotalViolation)}{(\beta \times WireCnt) + Area}$$

The terms in the detection potential and also the different steps in the algorithm are described as follows:

- <u>Initialization</u>: At the beginning, the backbone data structure that is the basis of violation matrix is constructed by reading the associated data from experimental results and wire and area report files (step 1).
- <u>TotalViolation</u>: After creation of the violation matrix, the sum of violation counts of each assertion for each flip-flop is calculated (column sum). The violation count of each assertion for a specific flip-flop is the total number of times that a specific assertion has been violated when a bit-flip has been injected in that flip-flop (step 2).
- <u>FlopCov</u>: This stands for the flip-flop coverage of an assertion, that is, the total number of flip-flops for which the associated violation count entry in the violation matrix is greater than zero (step 3).
- In order to bring `Area` and `TotalViolation` to the same scale, the respective values for these two attributes are scaled linearly in such way that in each itera-

```
input : Violation Matrix, Wire and Area reports
output: Candidate Assertions
initialization;                                                      // step 1
while Used Wires ≤ wire Budget do
  | foreach assertion in assertion list do
  |   | find TotalViolation;                                         // step 2
  |   | find FCov;                                                   // step 3
  |   | scale Area and TotalViolation;                               // step 4
  |   | find α, β;                                                   // step 5
  |   | find DP;                                                     // step 6
  | end
  | Possible Candidates = Assertions with DP within 1% of Max DP;    // step 7
  | foreach assertion_i in Possible Candidates do
  |   | find σ_i;                                                    // step 8
  |   | DP_i = DP_i / σ_i
  | end
  | select candidateAssr;                                            // step 9
  | usedWire += candidateAssr_wire;                                  // step 10
  | usedArea += candidateAssr_area;                                  // step 11
  | foreach FF_j in FlipFlops do
  |   | if candidateAssr_vc of FF_j > 0 then
  |   |   | cover FF_j;                                              // step 12
  |   | end
  | end
  | foreach assertion_i in assertion list do
  |   | update assertion_i wires;                                    // step 13
  | end
  | if AllFlopsCovered then Break                                    // step 14
end
```

Algorithm 1: Ranking Algorithm for maximizing flip-flop coverage estimate.

tion, the minimum and maximum values of these attributes are measured and the associated area and total violation count of each assertion is scaled accordingly in a linear manner (step 4).

- α, β: These coefficients are used to bring different terms in the nominator and denominator of the detection potential formula to the same scale, so that one term is not accidentally given more importance than necessary. However, if the user wants to deliberately adjust the importance of one specific term (i.e., FlopCov) in the detection potential equation, it can be done by multiplying that term further with another coefficient. For the sake of clarity, we have omitted that extra coefficient. The α and β scale factors are defined as:

$$\alpha = \frac{T_{avg}}{F_{avg}} \quad \beta = \frac{A_{avg}}{W_{avg}}$$

where T_{avg} is the average of the total violation counts for all assertions, F_{avg} is the average of the total flip-flop coverage for all assertions and A_{avg} and W_{avg} stand for the area and wire count average of all assertions, respectively. As it will be clarified later, all these coefficients are calculated based on the remaining assertions and their updated wire counts in each iteration of the algorithm (step 5).

- Once all the prior steps are done, the detection potential for each assertion can be calculated according the Eq. 10.1 (step 6). Note that the wire count of each assertion has been determined in step 1.
- At this point, all assertions for which their associated DP is within 1% to the maximum DP are marked as Possible Candidates and are taken for future evaluations (step 7). The selected assertion should not have a significant discrepancy in violation counts for different flip-flops, since it will likely have a higher potential to detect bit-flips for all the flip-flops that it is related to. Therefore, for all assertions in the Possible Candidate list, the violation count standard deviation (σ) of those assertions must be calculated and their respective DP is divided by this standard deviation (step 8). The lower the standard deviation is, the larger the DP will be.
- As soon as the detection potential (DP) values of all assertions are available, the assertion with the highest DP is selected as the candidate assertion (step 9), and the wires that are part of this assertion statement are added to the used wire list (step 10). Equally, the associated area of the assertion is also added to the total area usage (step 11).
- Following to selection of the candidate assertion, all the flip-flops that are potentially covered by it are marked and taken out of further consideration. A flip-flop is potentially covered by an assertion if the corresponding entry of the violation matrix for that flip-flop has a nonzero violation count (step 12).
- The algorithm continues by carrying out an important task that is to update the wire count for each assertion. This is crucial because, in the following iterations when the DP of each assertion is to be determined, the wires that have already been used should not be taken into consideration. For example, during evaluation of assertions DP, an assertion that initially had five wires but three of its wires are already in the used wire list will be treated as an assertion with only two wires (step 13).
- Finally, before moving to the next iteration, though unlikely, it is sensible to check if all the flip-flops have been potentially covered by the already selected assertions so that there is no need to select more assertions (step 14).

In the following section, all steps of the algorithm are reviewed by the aid of an example.

Example

The example in Fig. 10.3 will illustrate the process of assertion selection in each iteration in the proposed algorithm. Initially, the detection potential of each assertion is to be calculated according to formula in Eq. 10.1. Hence,

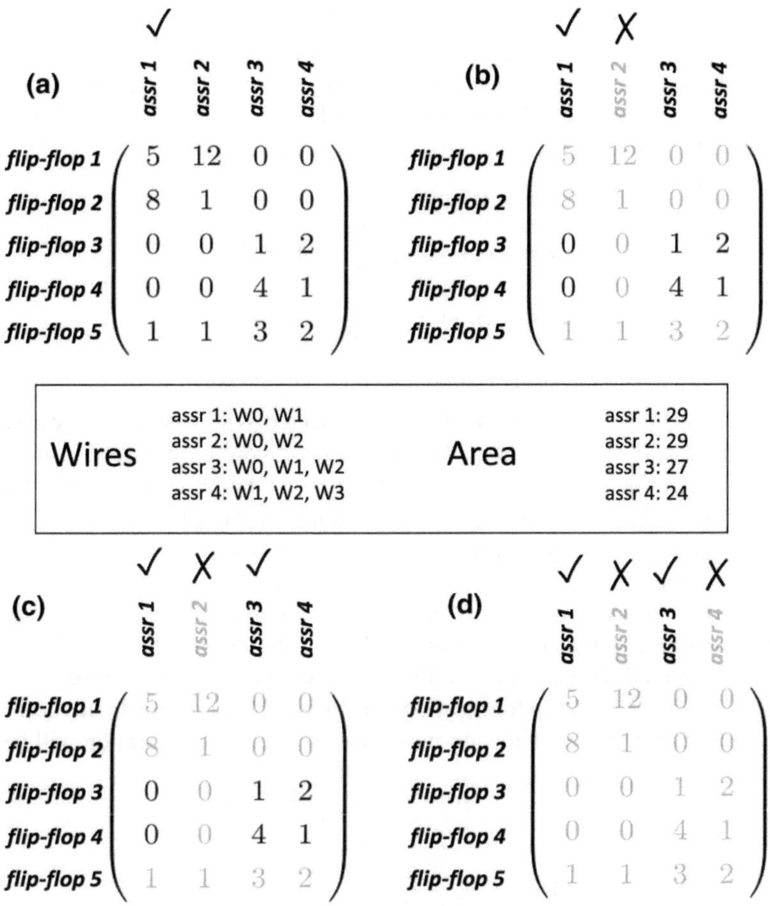

Fig. 10.3 An example showing the different steps on choosing a suboptimal set of assertions to maximize flip-flop coverage is estimated

$$assr(1)_{DP} = \frac{(19.5 \times 3) + 100}{(26.04 \times 2) + 100} = 1.04$$
$$assr(2)_{DP} = \frac{(19.5 \times 3) + 100}{(26.04 \times 2) + 100} = 1.04$$
$$assr(3)_{DP} = \frac{(19.5 \times 3) + 33}{(26.04 \times 3) + 59.4} = 0.66$$
$$assr(4)_{DP} = \frac{(19.5 \times 3) + 1}{(26.04 \times 3) + 1} = 0.75$$

Note that α and β have been calculated based on the scaled values of Total Violation and Area. As it can be seen, the maximum detection potential is 1.04 and is associated with both assertion 1 and assertion 2. This is because both assertions have similar

`TotalViolation`, `FCov`, `WireCnt` and `Area`. Referring to the ranking algorithm (Algorithm 1), the standard deviations for both assertions have to be calculated and their detection potential must be divided by the respective standard deviation. After the division, the assertion with highest DP is selected as the candidate assertion and the other one is dismissed from further evaluation. For our example, the DP values becomes

$$assr(1)_{DP'} = \frac{1.04}{2.867} = 0.362$$
$$assr(2)_{DP'} = \frac{1.04}{5.185} = 0.200$$

Hence, assertion 1 which has a higher DP is selected as the first candidate assertion. Taking standard deviation into consideration ensures avoiding an assertion that covers a dominant flip-flop (that contributes most to its violation count) and potentially misses the other flip-flops. Subsequent to selecting assertion 1, its wires then have to be added to the used wire list. Moreover, all flip-flops that are potentially covered by this assertion are also marked and taken out from further consideration (Fig. 10.3b). Assuming that the wire budget is 6, and also the fact that not all the flip-flops have been potentially covered, the algorithm goes to the second iteration. During the second iteration, all the values in the DP formula must be updated in such way that attributes (Wires, FCov, etc.) for flip-flops that have been covered already should not be counted. Therefore, the detection potentials for the second iteration will be

$$assr(3)_{DP} = \frac{(25.25 \times 2) + 100}{(33.67 \times 1) + 1} = 4.34$$
$$assr(4)_{DP} = \frac{(25.25 \times 2) + 1}{(33.67 \times 2) + 100} = 0.307$$

Since the detection potential of assertion 4 is not within 1% of the maximum detection potential (assertion 3), there is no need to calculate the standard deviation. Hence, assertion 3 is selected as the second candidate assertion, its wires are added to the used wire list and also the flip-flops that are covered by it are marked as potentially covered. As it can be seen in Fig. 10.3d, by selecting assertion 3, all the remaining flip-flops in the design are potentially covered, hence the algorithms stops.

10.4 Assertion Synthesis

As stated earlier, RTL assertions are written in either PSL or SVA formats which cannot be directly synthesized to hardware. Although these high-level assertions monitor design's behavior and provide useful feedback in pre-silicon verification, in order to perform on-chip property checking during post-silicon validation and debug, their equivalent hardware must be generated and integrated to the circuit under validation (CUV).

The problem of automating the generation of equivalent hardware units of pre-silicon assertions has recently been well studied (e.g., [5, 11, 12]). Most of these tools support sophisticated features such as sequences, repetitions, first occurrence matching, assertion covers, etc. However, since the assertions that are employed for bit-flip detection have a simplified structure (an antecedent condition implies a consequence), and in this section, we propose our own custom SVA assertion synthesis algorithm. It is important to note that the resulting circuit incorporates only those operators and constructs that are required by the assertions used for bit-flip detection and therefore additional features that serve purposes other than what is needed in this work are not taken into consideration. One objective of removing the unnecessary features is to reduce the area overhead. In addition to this, the other motivating factor for using an in-house assertion synthesis algorithm is that, we can estimate each assertion's area overhead because we know the exact number of flip-flops, inverters and the size of the AND gate that constructs the assertion. Since the ranking algorithm does relative area comparisons in between the assertions, it is sufficient to work out the estimated area overhead as long as the method is consistent for all the assertions. This will eliminate the need to pass the equivalent hardware unit of the assertions through commercial synthesis tools to achieve their area estimates.

The focus of this section is on assertion synthesis and therefore, for the sake of completeness, we quickly review the structure of SVA assertions. This provides the necessary foundation for understanding the different steps in our assertion synthesis flow. Figure 10.4 represents a type of an SVA assertion that we typically encounter for bit-flip detection. In SVA, the ## construct is called the *cycle-delay* construct [13] and the number followed by ## represents the cycle in which the right-hand side Boolean event must occur with respect to the left-hand Boolean event. In addition, as shown in Fig. 10.4, the signals are grouped based on whether they are at the left side of the implication operator (antecedent signals) or at the right side of it (consequent signals). SVA provides two implication operators |−> and |=>. The former is called the *overlapped implication* operator which means that if the left-hand side prerequisite sequence holds, then the right-hand side sequence must hold. The latter is called the *non-overlapped implication* operator which is similar to the overlapped operator except for the fact that the right-hand side sequence is evaluated in the next clock cycle. Hence *a* |=> *b* and (*a* |−> ##1 *b*) are equivalent. Now, let us assume for the sake of simplicity that we want to generate the hardware circuit for an assertion with all its signals in the same clock cycle (temporal depth of zero), *assert property* ((*a* == 1) && (*b* == 0) |−> (*c* == 1)) . This assertion is read as if signal *a* is 1 and signal *b* is 0, signal *c* is implied to be 1. Now since we know that so long

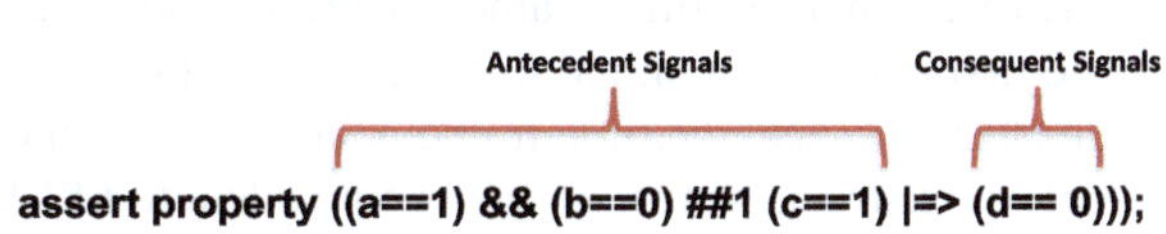

Fig. 10.4 Example of an System Verilog Assertion (SVA assertion)

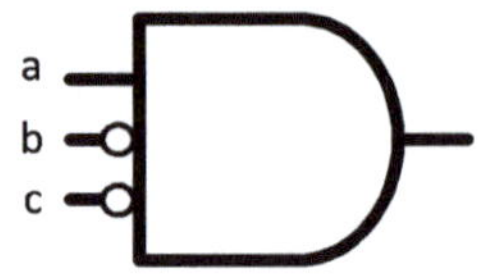

Illegal combination ←

a	b	c	assr
0	0	0	0
0	0	1	0
0	1	0	0
0	1	1	0
1	0	0	1
1	0	1	0
1	1	0	0
1	1	1	0

Fig. 10.5 Example showing the equivalent hardware circuit for the SVA assertion "*assert property* (($a ==$ 1) && ($b ==$ 0) |−> ($c ==$ 1))"

as a is 1 and b is 0, signal c is 1, a hardware circuit that evaluates to logic 1 when the assertion is violated can be constructed using a three-input AND gate as shown in Fig. 10.5. The output of the assertion should go to 1 when at the same clock cycle, signals a and b are 1 and 0 respectively but signal c is 0 instead of 1. Hence, in order to trigger that the assertion is violated, the complement of signal c must be connected to the AND gate. The output of the hardware circuit associated with the assertion will always be 0 as long as a, b and c contain legal values. As shown in Fig. 10.5, only one entry in the associated truth table of the circuit evaluates the output to 1 and that entry is known as the illegal signal combination.

For all the assertions that are to be synthesized, one has to find whether a signal is at the antecedent side or the consequent side, its polarity which determines whether it must be inverted or not when it is connected to the AND gate, and lastly its time frame. For instance, for the assertion that was just discussed, the required synthesis information can be encoded in an statement as $(a_0, -b_0, -c_0)$. The sign before a signal determines if the signal must be inverted or not when connected to the final AND gate. Likewise, the subscript represents the signal's time frame. For assertions with nonzero temporal depth (assertions that span across multiple clock cycles), signals with the highest time frame are directly connected to the AND gate, whereas other signals are buffered accordingly. Figure 10.6 provides three examples for generating hardware circuits for assertions with and without temporal depth of 0. As it can be seen, for each signal, the time frame and the polarity is found and is subsequently connected to the AND gate. The algorithm for finding the required synthesis information for an assertion has several steps which are elaborated below.

1. For each signal s_i in the assertion statement (both antecedent and consequent sides), find its time frame and prepare a time-frame list $T_S = \{t_{s_1}, t_{s_2}, \ldots, t_{s_n}\}$ in such way that the time frame for the leftmost signal is 0. While moving from the first antecedent signal to the last consequent signal, each time a cycle-delay construct is seen, the time frame is increased by the value that follows the cycle-delay construct. Note that, |=> is equivalent to ##1 |−>. For example, in the third

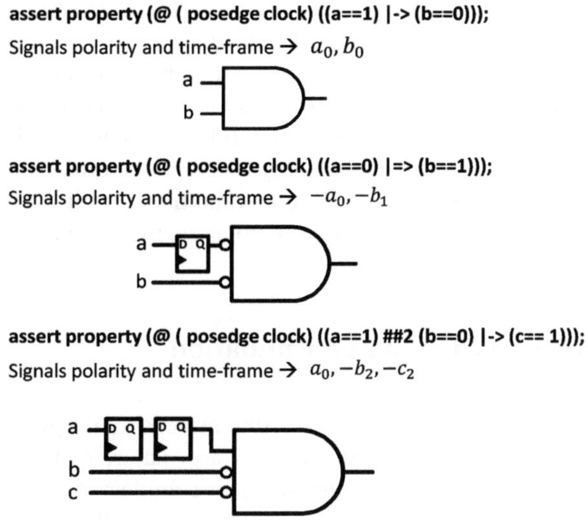

Fig. 10.6 Example showing the equivalent hardware circuit for assertions used in this work

example of Fig. 10.6, the time frame of a is 0 while the time frame of b is 2. The time frame of the consequent signal c is also 2.

2. Prepare the list of signals in the consequent side $C = \{c_1, c_2, \ldots, c_n\}$ and antecedent side $A = \{a_1, a_2, \ldots, c_n\}$.
3. For each signal s_i whose time frame is smaller than the maximum time frame, create a shift register with the size equals to $(max(T_S) - t_{s_i})$. Signals with the maximum time frame are connected directly to the AND gate with the correct polarity which is found in the next two steps.
4. Find polarity of signals in the consequent side $P_C = \{p_{c_i} \mid \forall c_i \in C \; if \; (c_i == 1), \; p_{c_i} \leftarrow \texttt{false}, \; else \, p_{c_i} \leftarrow \texttt{true}\}$. If polarity is `false`, the signal must be complemented and if it is `true`, it is connected as is. For instance, for the third assertion in Fig. 10.6, the consequent signal c must be a 1, therefore its polarity is negative thus, its complement is connected to the AND gate.
5. Find polarity of signals in the antecedent side $P_A = \{p_{a_i} \mid \forall a_i \in A \; if \; (a_i == 1), \; p_{a_i} \leftarrow \texttt{true}, \; else \, p_{a_i} \leftarrow \texttt{false}\}$. For instance, for the third assertion in Fig. 10.6, the antecedent signal b must be a 0, therefore its complement is connected to the AND gate. Note, whenever the output of the AND gate goes to 1, it means that the assertion has been violated.

The area estimate for each assertion is estimated based on

$$Assr(i)_{area} = (\alpha \times F) + (\beta \times In) + (\gamma \times Inv) \tag{10.1}$$

where F represents the total number of flip-flops, Inv is the total number of inverters and In represents the number of inputs to the AND gate. α, β and γ are technology independent coefficients that reflect the relative differences in the area of flip-flops, inputs to an AND gate, and inverters. These three coefficients can be customized by the user based on the specific standard cell library that is employed.

Finally, it is important to note that the very purpose of finding the area estimates is to provide necessary information for the cost function of the heuristic ranking algorithm detailed in Sect. 10.3. It by no means provides the exact area overhead of the assertions, nonetheless rapid computation of estimates still captures the size of assertions relative to each other and therefore it avoids the need to use third-party commercial synthesis tools to compute the exact area.

10.5 FPGA-based Emulation

In Sect. 10.2, we elaborated different steps for the generation and selection of embedded hardware assertions. Due to slow nature of functional simulators with respect to the actual silicon, attempting to complete the preparation and the confirmation experiments using functional simulators will lead to a short snapshot of design's behavior being captured. This results in inaccurate assessment of the potential of the assertions for detecting bit-flips and limits the number of error injections that can be performed, which indirectly affects the accuracy of selecting the embedded assertions checkers that will be committed to silicon for bit-flip detection. Since field-programmable gate arrays (FPGAs) emulators are capable of verifying logic designs at clock speeds of at least three orders of magnitude faster than a software simulator, they have been widely used to narrow the large gap between simulation and silicon speed [14]. Moreover, FPGA-based emulation platforms have recently been employed for various post-silicon validation purposes. For instance, the idea presented in [15] is to use emulation to evaluate the critical-path timing coverage of a validation plan. In this section, we present an automated methodology to design emulation-ready hardware architectures that can be merged into the methodology discussed in Sect. 10.2 to improve the accuracy of selecting assertions that are going to be instrumented as on-chip monitors. An important benefit of emulation-based experiments is to improve the density of the violation matrix, as illustrated in Fig. 10.7. We have observed that the small number of error injections, which is limited in simulation-based experiments by their slow speed, affects the accuracy of the violation matrix. By increasing the number of error injections, the nonzero elements in the violation matrix will increase which clearly provide a more meaningful input to the assertion ranker.

10.5.1 Hardware Architecture and Tool Flow

Figure 10.8 overviews the architecture and the details for each block are elaborated below. As shown later, the internals of the subblocks in Fig. 10.8 are different for

	assr 1	assr 2	assr 3	assr 4	assr 5
flip-flop 1	0	2	0	0	0
flip-flop 2	0	0	1	0	0
flip-flop 3	0	0	0	0	0
flip-flop 4	3	1	2	0	4

(a) Simulation-based Experiments

	assr 1	assr 2	assr 3	assr 4	assr 5
flip-flop 1	0	3	1	0	1
flip-flop 2	1	0	2	0	0
flip-flop 3	0	0	0	1	0
flip-flop 4	3	1	2	0	4

(b) Emulation-based Experiments

Fig. 10.7 Violation Matrix for a circuit with four flip-flops and five assertions. The red elements are in **b** represents the improvement in accuracy that is due to a higher number of error injections in Emulation-based experiments

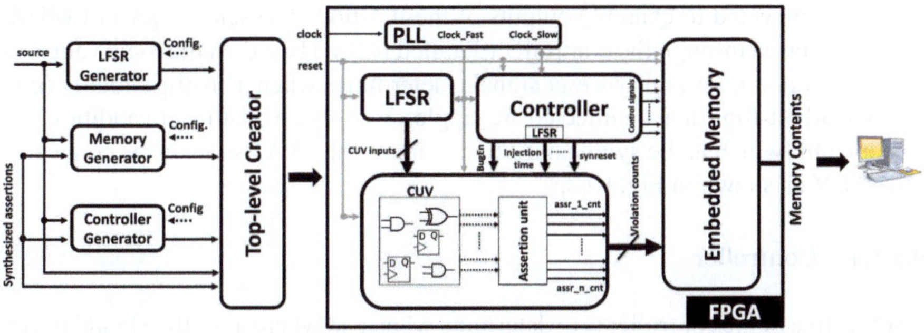

Fig. 10.8 Tool flow for automatic generation of emulation-ready hardware architecture to accelerate error injection experiments in post-silicon validation

preparation experiments where the focus is on grading all the assertions and confirmation experiments where the focus is on measuring the coverage metrics.

10.5.1.1 Phase-Locked Loop

For reasons that will be clarified in Sect. 10.5.1.4, one of the proposed architectures operates using two-clock domains. We have decided to use phase-locked loops (PLLs) mainly because of the following reasons: first, using divide-by-n clock division is prone to timing closure problems and secondly, the logic resources that are required to implement divide-by-n clock division can be saved by utilizing embedded PLLs that are prefabricated on most FPGAs. Our current implementation will instantiate a PLL that generates a fast and a slow clock signal. The ratio between the fast and slow clocks are configurable by the user (dependent on the specific FPGA device that is used).

10.5.1.2 Input Stimuli Generation

Since the input stimuli used in both preparation experiments and confirmation experiments are random, a *PI*-bit linear feedback shift register (LFSR) is created by our tool where *PI* is the total number of primary inputs in the design. The characteristic polynomials for LFSRs are primitive and irreducible in order to support the maximal-length sequences.

10.5.1.3 Circuit-Under-Validation and Assertion Unit

Prior to connecting the CUV to the assertion unit, each flip-flop in the design is instrumented with a 2-to-1 MUX to facilitate error injections. The inputs of the MUX are connected to Q and $\bar{Q}$ outputs of the flip-flop. The select signal of MUX, which is asserted through the controller (detailed below) based on the logic values of *BugEn*, *injection time*, and *synreset* signals, determines when a bit-flip should occur. Note that all bit-flips that are injected are single-clock-cycle bit-flips. In addition, the assertion unit which is the synthesized hardware of the SVA assertions is connected to the CUV as shown in Fig. 10.8.

10.5.1.4 Controller

The objective of the controller is to determine when and where a bit-flip should occur, monitor the status of assertions after bit-flip injections and initiate a burst of writes to the memory. The injection time is determined by a 20-bit LFSR, inside the controller whose initial state is configured randomly by our proposed tool prior to compilation. Once the entire platform starts running, the controller will inject *E* number of errors in the first flip-flop at different times which have been determined by the LFSR within the control unit. *E* is selected by the user and it is passed as an option to the proposed toolflow. After each error injection (a single cycle bit-flip), the circuit continues to run for a predefined number of clock cycles (given by the user) during which the controller activates the assertion unit to monitor if there are any violations. Due to the evident differences in the objective of preparation experiments and confirmation experiments, the sequence of writes to the memory are done through two different architectures.

1. **Preparation Experiments architecture**: For each assertion, there is a counter which determines the number of times the respective assertion has been violated. Once all the errors are injected in a flip-flop, the controller stops the circuit's operation. At this time, the controller will check the violation count register of each assertion and organizes memory writes in case a nonzero value is observed. The controller's interface to the memory can be configured in two different ways. The first configuration utilizes two-clock domains in such way that the sequence of checks for each assertion is done through a second finite-state machine (FSM) which works based on a slower clock as shown in Fig. 10.10. The main reason

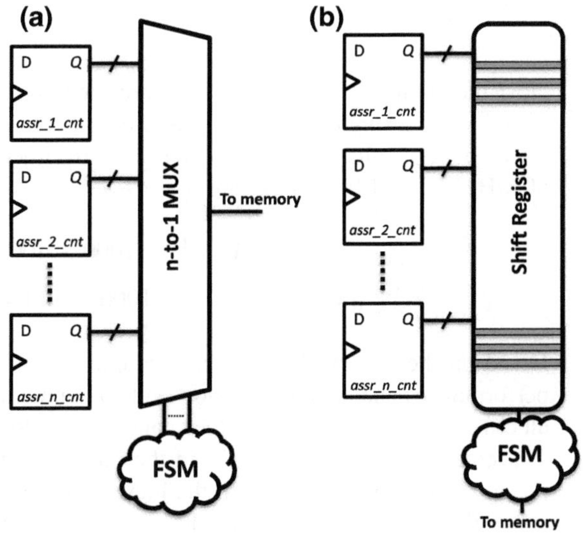

Fig. 10.9 Memory interface for two-clock domain (**a**) and single-clock domain (**b**) architectures in preparation experiments. Note that flop ID and assertion ID are concatenated to the output of the MUX which then will be written to the memory

for using a slower clock for this part of the design can be explained as follows. For the sake of argument, assume that we inject E number of bit-flips in one of the flip-flops in a design and that the design has M number of assertions. After all the injections have been done in a flip-flop, the controller will have to check the violation counter of every single assertion. Since a fairly large number of assertions are instrumented to the CUV, which need to be checked one by one, and also the fact that there is a single embedded memory that can accommodate one write at a time, the multiplexer that determines the words that are to be written to the memory can become reasonably large as shown in Fig. 10.9a. This can result in a timing closure problem due to long propagation delay in this multiplexer. One has to note that the number of clock cycles spent on checking the status of each assertion counter is bounded by the number of assertions (in this case M), which is significantly less than the number of clock cycles needed for error injections ($E\times$ number of clock cycles between error injections). Therefore, it is sensible to keep the clock frequency of error injection block (CUV and assertion checker unit) as high as possible and adjust the clock frequency of the assertion checking FSM as shown in Fig. 10.10.

The following example shows that using a slow clock to write the assertion counts to memory will have a negligible impact on the overall emulation time. Assume that we want to run emulation experiments for a circuit with 1000 flip-flops (F) and 400 assertions (A) and that 1024 error injections (E) are scheduled to be injected in each flip-flop. The insertion time is generated by a 20-bit LFSR. Let us assume that on average, 2^{10} clock cycles (CC) are elapsed for each error injection. Hence the total time spent for error injections based on a 50 MHz clock (F_{50}) is:

$$T = F \times E \times CC \times \frac{1}{F_{50}} = 10^3 \times 2^{10} \times 2^{10} \times 20\,\text{ns} = 20.9\,\text{s}$$

Now, the number of clock cycles that are spent for checking assertions after error injections in a single flip-flop is based on the total number of assertions. Hence, the amount of time spent in this step for error injection in all flip-flops (F) for a 50 MHz and a 1 MHz clocks are

$$\begin{aligned} T^{Assr}_{clock50} &= F \times A \times \tfrac{1}{f_{50}} = 1000 \times 400 \times \tfrac{1}{50\times10^6} = 8\,\text{ms} \\ T^{Assr}_{clock10} &= F \times A \times \tfrac{1}{f_1} = 1000 \times 400 \times \tfrac{1}{01\times10^6} = 400\,\text{ms} \end{aligned}$$

Based on the observations in the example above, it can be concluded that the performance penalty of using a slower clock for memory writes (in this case 1 MHz clock) is only 1.8%. On the other hand, if the clock frequency is slowed down to accommodate the slow paths in the memory writing FSM, the performance penalty will scale linearly with the increase in the clock period. For a circuit of a size like the one in this example, we have observed that the clock period can become three times longer.

It is worth mentioning that so long as the ratio between the two clocks is a natural number, synchronizers will not be needed between the two-clock domains. This is because the handshaking signals (dotted arrows in Fig. 10.10) that facilitates the communication between the faster clock FSM and the slower clock FSM are not acknowledged until they are captured by the slower clock FSM and the other way round. One might argue that the slower clock can be avoided by pipelining this large multiplexer in such way that one write is performed in each clock cycle. Although this will lead to faster memory updates, using this approach will result in significant on-chip area usage associated with pipelining registers and logic (recall that the number of inputs to the multiplexer is in the range of hundreds). Therefore, we choose to operate this FSM at a lower frequency (thus avoiding the registers needed for pipelining).

For applications where only one clock domain is available, the architecture can be configured in such way that, once all the errors are injected in a flip-flop, the contents of each assertion's violation counter is stored in a shift-register structure as shown in Fig. 10.9b. Afterwards, contents of this shift-register are offloaded to the memory one-by-one at every clock cycle. The small boost to the run-time is significantly outweighed by the resulting resource usage overhead imposed by the size of shift-register that has to account for the worst-case scenario (many violations per error injections). Therefore, the authors prefer the fast/slow clock method, especially because the performance penalty is negligible.

2. **Confirmation Experiments architecture**: As explained in Sect. 10.2.5, the focus of the confirmation experiments is on the assessment of the assertions that have been selected by the ranking algorithm. *Flip-flop coverage estimate* is used as a metric that determines whether the selected assertions fulfill the required expectations or not. For computing the flip-flop coverage estimate the number of times

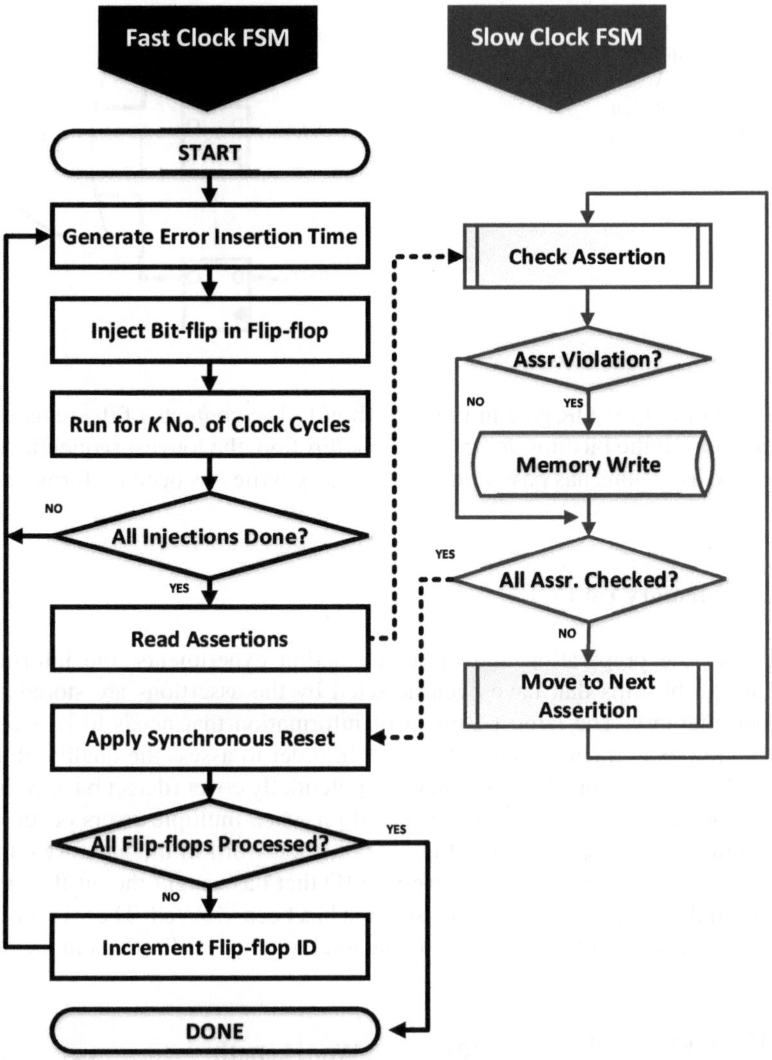

Fig. 10.10 Controller FSM showing two different state machines for fast and slow clocks. Note that *k* is user defined and represents the number of clock cycles for which the circuit runs after each bit-flip injection

these assertions have been violated is of no importance, unlike preparation experiments where for the sake of grading, the number of times an assertion has fired will increase its bit-flip detection potential and hence is important. As a result, the violation counters of the assertions are removed and the assertion outputs are connected to a register that becomes 1 (and stays 1) if at least one out of all the bit-flips in the target flip-flop violates that assertion. A memory write is organized

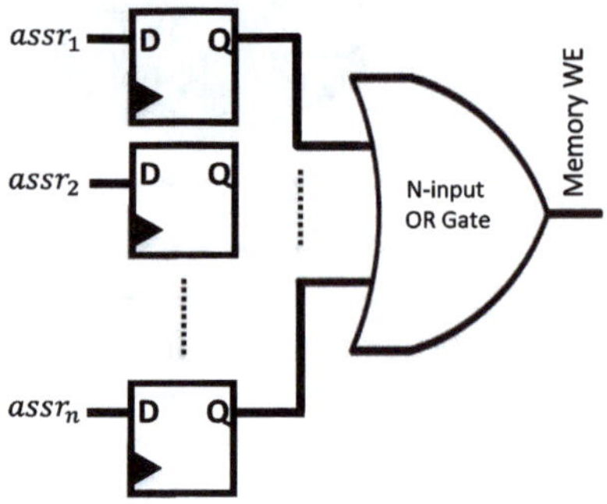

Fig. 10.11 Memory interface for flip-flop coverage estimate. When write enable is granted, a single-bit 1 is written to the memory

if the output of the OR gate in Fig. 10.11b is 1. The contents of the registers are reset when all the bit-flips are injected in a flip-flop, the longest sequential depth of all the assertions has passed and the memory write has been performed.

10.5.1.5 Memory Unit

During both the preparation and the confirmation experiments, the information related to the bit-flips that have been detected by the assertions are stored in the embedded memory. The largest amount of information that needs to be stored is in preparation experiments. This is because, in order to assess the quality of assertions, one has to know the flip-flops they can potentially cover (detect bit-flips in that flip-flop) and the number of times they can do it when multiple errors occur in the same flip-flop. Hence, as shown in Fig. 10.12a, each word in the memory contains information about the flip-flop, the assertion ID that has caught the bit-flips in that flip-flop and the number of times that assertion has been violated. The width of Flop ID and Assr. ID are configured based on the total number of flip-flops in the design

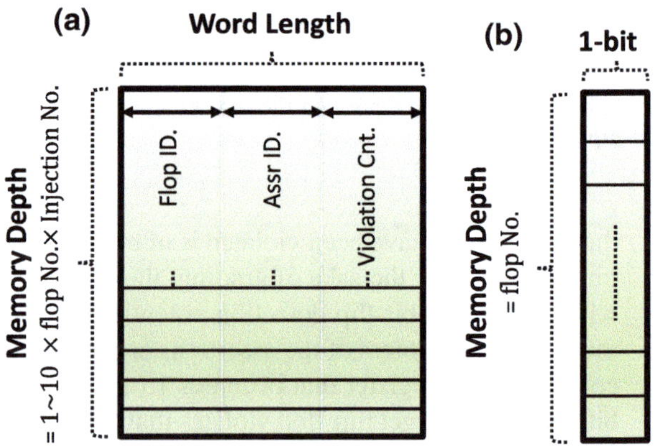

Fig. 10.12 Memory layout for **a** preparation experiments and **b** flip-flop coverage estimate in confirmation experiments

and the total number of assertions that are embedded in hardware. The width of Violation Cnt. segment is determined by our tool based on the maximum number of error injections. In order to determine the memory depth, we used our findings from simulation-based experiments. Based on those observations, the expected number of assertions that would fire for each bit-flip injection is set to 10. Therefore, we used this coefficient for accommodating enough space in the available physical memory. Though, for debug sessions that have a large number of assertions and using 10 as the coefficient would result in the memory not fitting on the target FPGA device, smaller coefficients in the range of 1–10 is used. The tool flow has an automatic mechanism such that if memory overflow occurs before all the flip-flops are evaluated, the session is divided into multiple sessions, with less assertions in each session. Clearly, this coefficient and the number of assertions that can be mapped to the device in a single session depends on the capacity of the target device.

During confirmation experiments, only the information needed to determine bit-flip coverage estimate and the flip-flop coverage estimate are stored. For bit-flip coverage estimate, every time a bit-flip is injected in a flip-flop and the largest sequential depth of all assertions is passed, the output of the OR gate in Fig. 10.11a is evaluated such that if it is 1, it implies that the bit-flip has been detected and a single-bit 1 is written to the memory in a sequential manner. In an ideal case where all bit-flips are detected, the maximum required memory depth equals the total number of flip-flops multiplied by the number of errors that are injected in each flip-flop. For instance, for a design with 20 flip-flops, if 5 bit-flips are set to be injected, the required memory size would be 100 bits.

On the other hand, for flip-flop coverage estimate, after all the bit-flips are injected in a flip-flop, if the output of the OR gate in Fig. 10.11b is 1, then that flip-flop is marked as potentially covered. Since we want to determine how many flip-flops are potentially covered, it would be sufficient to store a single-bit 1 every time a flip-flop is marked as covered. Therefore, the maximum required memory depth is the total number of flip-flops. Clearly, the memory address at which a 1 is written corresponds to the ID of the covered flip-flop. Memory layouts for confirmation experiments are shown in Fig. 10.12b and c.

10.6 Results and Discussions

The proposed tool flow has been implemented on an Intel core i7 machine with 32GB of RAM using GCC 4.8.4 for compiling C++ source codes and Tcl 8.5 and Python 2.7.6 for scripting. For assertion discovery, we have used GoldMine [7], an automatic assertion generation tool which has multiple built-in data mining engines for finding the likely design invariants. These likely invariants are passed through a commercial formal verification tool that filters out the incorrect invariants leaving out true invariants as design's final assertions. Since GoldMine operates based on simulation traces, results from both the random input (Goldmine's default) stimuli and the deterministic vectors produced by Validation Vector Generator tool from

Virginia Tech [16] have been provided to GoldMine's mining engine. Assertions from these two methods have been merged as are the assertions from different mining engines of GoldMine. Details about the specifics of GoldMine are out of the scope of this work and the interested reader is referred to [7].

In order to generate the equivalent hardware circuits of the high-level SystemVerilog Assertions (SVA assertions), we have implemented the algorithm detailed in Sect. 10.4 in such way that, initially, all the assertions that have been found using GoldMine are added to the source code and passed to our tool to produce their equivalent hardware description language (HDL) description and their area estimate (required by the ranking algorithm). Once a subset of these assertions are selected through the ranking algorithm, namely *selected assertions* in Fig. 10.1, we will instrument them to the original design and pass them through Synopsys Design Compiler to find the accurate area overhead, which takes into account the logic sharing in between the assertions. This is a more realistic measure of the area overhead (due to logic sharing between assertions) than the trivial method of summing up the area overheads of the individual assertions.

To support the random occurrence of bit-flips in post-silicon validation, we have used random input stimuli throughout both the preparation and confirmation experiments. As explained in Sect. 10.5.1.2, the LFSR unit is designed to produce random vectors whose bit-width is equal to the number of CUV inputs.

We have deployed our architecture on an Altera DE2 Cyclone IV device with a 50 MHz reference clock. All memory dumps have been done using the in-system memory content editor feature of Quartus that operates through the JTAG port. Finally, the faster clock in our controller in preparation experiments is a 50 MHz clock and the slower one is a 1 MHz clock (both coming from a PLL unit), and the confirmation experiments operate on a single 50 MHz clock.

As explained, generating a more accurate violation matrix after preparation experiments is the main motivation for using emulation-based experiments instead of simulation-based ones which in turn, results in an improvement in the selection of assertions that are to be mapped to hardware, which eventually leads to more precise coverage estimates. Therefore, we have integrated our proposed tool flow and architecture that detailed in Sect. 10.5 to the methodology shown in Fig. 10.1 and ran preparation experiments with 256 error injections (discussed later) per flip-flop. Afterwards, we have passed the resulting violation matrix to the assertion ranker, which selects a different number of assertions by varying the wire count constraint. Following to this, we ran confirmation experiments on each set of these selected assertions and varied the number of error injections from 5 to 5000, something that is infeasible to do in simulation-based experiments. In addition, we have run a limited number of simulation-based experiments to justify the benefits of emulation-based experiments. In the following subsections, we will provide and discuss our experimental findings.

10.6.1 Preparation Experiments

Due to the limited capacity of FPGAs, it is not feasible to instrument all the assertions discovered by GoldMine and map them to the FPGA for the preparation experiments. Hence, assuming that only one FPGA board is available, depending on the number of assertions and the FPGA capacity, the preparation experiments are divided into multiple emulation sessions as seen in Fig. 10.13. It is worth noting that, although one can have a single emulation session by using a high-capacity FPGA that can accommodate all the assertions that are mined, it is more practical to employ multiple FPGA boards (with devices of lower capacity) and run experiments in parallel because there is no dependency in between the different emulation sessions. Though in this work, we have used a single FPGA board and all our comparisons are based on measuring the total time spent in all the emulation sessions.

Table 10.1 shows the comparison between the total amount of time spent in simulation and emulation for running preparation experiments for the ISCAS89 s38584 benchmark circuit. The reason behind choosing this circuit for comparison is that it has the highest number of assertions compared to the other two ISCAS circuits that have more than 1000 flip-flops. Therefore, it will have the worst run-times both in emulation-based (larger number of emulation sessions) and simulation-based (larger number of assertions) experiments. Note that, a similar trend also holds for s38417 and s35932 circuits from the ISCAS89 benchmark set [10].

As stated earlier, for the same number of error injections, the simulator run-times are longer for preparation experiments in comparison with the confirmation experiments. This is because of the fact that in preparation experiments, a large number of SVA assertions are added to the the design and the associated overhead caused by the concurrent property checking makes the simulators run much slower when a large number of assertions are added to the design. This is however not an issue in emulation because the assertions are synthesized to their hardware equivalent circuit and all the units (CUV and assertion units) are working concurrently at the same clock speed. Nevertheless, as shown in Fig. 10.13, due to the limited FPGA device capacity and the large number of assertions that need to be evaluated in preparation experiments, the experiments are done in multiple sessions. For ISCAS s38584, the entire experiment was divided into 45 sessions. The reported run-time is the sum of time spent in each session. Due to long run-times, authors did not

Fig. 10.13 Running preparation experiments in multiple emulation sessions

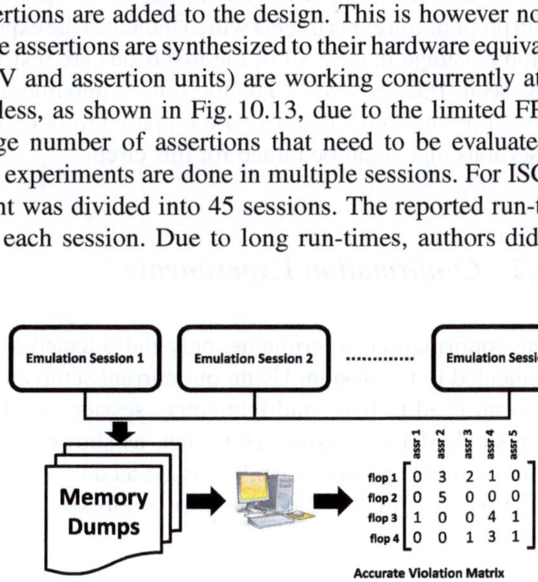

Table 10.1 Comparison of run-time and accuracy improvements for ISCAS s38584 circuit. The emulation (total) represents the total time spent on creating the top-level design, preparing the bitstream, compilation and on-board execution together with memory dumps

Preparation experiments			
No. of error injections	Run-time		Flip-flop coverage estimate (%)
	Simulation (hours)	Emulation (hours)	
2	26.3	$45 \times 0.001 = 0.045$	79.6
8	105	$45 \times 0.12 = 5.4$	86.15
64	–	$45 \times 0.89 = 40.05$	88.56
256	–	$45 \times 1.64 = 73.8$	88.94

Table 10.2 Maximum observed flip-flop coverage in preparation experiments for three largest ISCAS benchmark circuits

Circuit	Flip-flop coverage (%)
s35932	98.6
s38417	54.7
s38584	88.9

carry out simulation-based experiments beyond 20 error injections per flip-flop for preparation experiments.

In addition to the significant run-time improvements, it can be seen that the flip-flop coverage estimate also increases by 9%. This means that the ranker will benefit from a more accurate relation between flip-flops, assertions and their error space (Violation Matrix) which results in the selection of more accurate assertions. Table 10.2 represents the maximum observed flip-flop coverage with 256 number of error injections in preparation experiments which indicates the experimental maximum possible flip-flop coverage in case all of the assertions are instrumented to the design. As it can be seen, ISCAS s38417 has the lowest maximum flip-flop coverage which is explained by the lower than usual (compared to other benchmark circuits) number of assertions that could be mined for this circuit.

10.6.2 Confirmation Experiments

During confirmation experiments, only the selected assertions from the ranker are instrumented to the design. Using our current setup and the selected FPGA device, there is no need to have multiple debug sessions as the entire architecture can be fit in a single debug session. Note that, as shown in Fig. 10.12, the memory unit of confirmation experiments and its associated FSM are much simpler than those of preparation experiments. During confirmation experiments, the quality of the selected

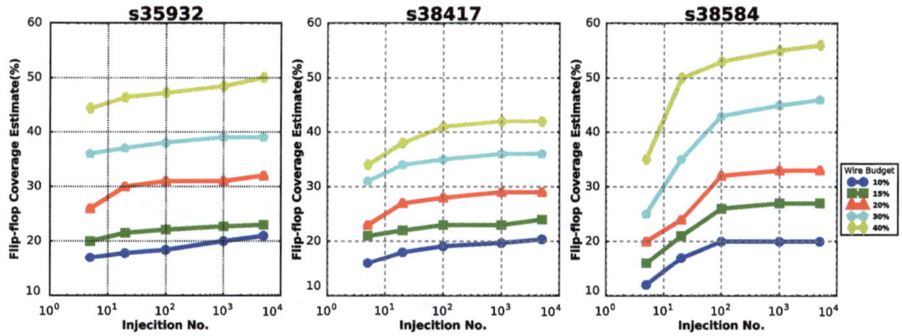

Fig. 10.14 Evaluation of flip-flop coverage in confirmation experiments as the number of error injection increases for different wire counts

assertions are assessed using the flip-flop coverage estimate. Figure 10.14 shows the flip-flop coverage estimate for different wire budgets when the number of injections is increased from 5 error injections to 5000 error injections. As it can be seen, with small number of injections, there is a possibility that some assertions are never sensitized which leads to the inability to cover the flip-flops that are monitored by those assertions. There is a steep increase when changing the number of injections from 5 to 20, however the slope of the curves decreases as the number of error injections increases further. This is because for the majority of the flip-flops a bit-flip is detected in one of its first 20 occurrences in the respective flip-flops. Nonetheless, because bit-flip detection is dependent on the state of the circuit when the bit-flip occurs, for some flip-flops it might take a larger longer time until one of those corner states is reached that will cause the assertions to detect the bit-flip. Note, while the generation of long validation sequences for post-silicon validation is an important problem, it is beyond the scope of this study. We rely on extensive randomized validation sequences for our experiments and it is a normal expectation that focused sequences (that push the circuit in its corner states faster) will detect bit-flips even sooner.

Finally, the area overhead for assertions that maximize flip-flop coverage estimate is shown in Fig. 10.15. As explained in Sect. 10.4, the main motivating factors for designing our custom synthesis tool are (1) to abandon the need to pass all the discovered assertions through a commercial synthesis tool to achieve their area estimate and (2) to eliminate unnecessary features that are not needed in this work from the synthesized model (i.e., violation counter for assertions in confirmation experiments). Authors would like to emphasize that the focus of this work is not on assertion synthesis and generic assertion synthesis tools such as MBAC [5], support most features in SVA and PSL that we do not use and therefore our tool does not support.

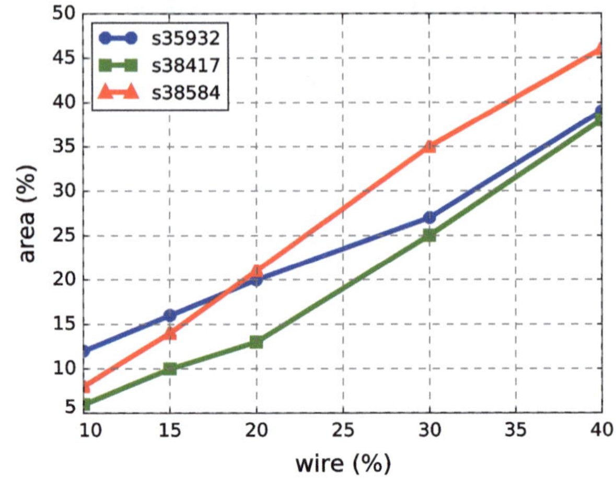

Fig. 10.15 Evaluation of area overhead with respect to different wire budgets for the largest ISCAS circuits when the ranker is set to maximize flip-flop coverage

Table 10.3 Evaluation of the running time of different steps of the tool flow in Fig. 10.1 for ISCAS s38584

Task	Configuration	Time (hour)
Assertion generation	Miner (Single core), Formal verifier (Multi core)	228
Preparation experiments	FPGA (45 sessions), 256 injections	74
Assertion mapper	Single core	0.17
Assertion ranker	Single core	0.04
Confirmation experiments	FPGA (1 session), 5000 injections	16.2

10.6.3 Run-Times for Different Steps of Our Methodology

Finally, we will provide the run-times of each box of the methodology in Fig. 10.1. Results provided in Table 10.3 are for the case where ranker has been constraint with 40% wire usage for ISCAS s38584 where the total time spent for finding assertions as well as the total number of assertions is the largest (worst-case scenario). As for assertion generation, GoldMine has been configured to generate assertions both by coverage miner engine and decision forest miner engine. A combination of random stimuli as well as deterministic stimuli produced by [16] has been used and the formal verifier has been set to provides 6–10 counter examples which are fed back to the miner for refining the original trace. For preparation experiments and confirmation experiments, a total of 28365 and 633 assertions were assessed, respectively. As for assertion mapper, the reported run-times are the sum of the run-time for assertion synthesis using our proposed algorithm in Sect. 10.4 as well as the run-time for Synopsys Design Compiler for measuring the area overhead for the final selected assertions.

10.7 Summary

In this chapter, we discussed the idea of using hardware assertions for reducing error detection latency and improving internal observability in post-silicon validation. We proposed algorithms and hardware architectures that facilitate the selection of the most suitable assertions to be embedded on-chip for real-time monitoring of the bit-flips during post-silicon validation.

Acknowledgements The authors wish to thank University of Illinois at Urbana-Champaign for GoldMine [7] (automatic assertion generation), Virginia Tech for the Validation Vector Generator [16] (used for deterministic stimuli for Goldmine) and McGill University for providing us with MBAC [5] for assertion synthesis.

References

1. J.H. Barton, E.W. Czeck, Z.Z. Segall, D.P. Siewiorek, Fault injection experiments using FIAT. IEEE Trans. Comput. (TCOMP) **39**(4), 575–582 (1990)
2. P. Taatizadeh, N. Nicolici, Automated selection of assertions for bit-flip detection during post-silicon validation. IEEE Trans. Comput. Aided Design Integr. Circuits Syst. (TCAD) **35**(12), 2118–2130 (2016)
3. H.F. Ko, N. Nicolici, Automated trace signals identification and state restoration for improving observability in post-silicon validation, in *ACM/IEEE Design, Automation and Test in Europe (DATE)* (2008), pp. 1298–1303
4. Y.-S. Yang, N. Nicolici, A. Veneris, Automated data analysis solutions to silicon debug, in *ACM/IEEE Design, Automation Test in Europe Conference Exhibition (DATE)* (2009), pp. 982–987
5. M. Boulé, Z. Zilic, *Generating Hardware Assertion Checkers: For Hardware Verification, Emulation, Post-Fabrication Debugging and On-Line Monitoring* (Springer, Berlin, 2008)
6. S. Hangal, S. Narayanan, N. Chandra, S. Chakravorty, IODINE: a tool to automatically infer dynamic invariants for hardware designs, in *42nd ACM/IEEE Design Automation Conference, 2005. Proceedings* (2005), pp. 775–778
7. S. Hertz, D. Sheridan, S. Vasudevan, Mining hardware assertions with guidance from static analysis. IEEE Trans. Comput. Aided Design Integr. Circuits Syst. **32**(6), 952–965 (2013)
8. B. Vermeulen, S.K. Goel, Design for debug: catching design errors in digital chips. IEEE Design Test Comput. **19**(3), 35–43 (2002)
9. M. Gao, K.-T. Cheng, A case study of time-multiplexed assertion checking for post-silicon debugging, in *IEEE International High Level Design Validation and Test Workshop (HLDVT)* (2010), pp. 90–96
10. F. Brglez, D. Bryan, K. Kozminski, Combinational profiles of sequential benchmark circuits, in *IEEE International Symposium on Circuits and Systems (ISCAS)*, vol. 3 (1989), pp. 1929–1934
11. C. Fibich, M. Wenzl, P. Rssler, On automated generation of checker units from hardware assertion languages, in *Microelectronic Systems Symposium (MESS), 2014* (2014), pp. 1–6
12. M. Wenzl, C. Fibich, P. Rssler, H. Taucher, M. Matschnig, Logic synthesis of assertions for safety-critical applications, in *IEEE International Conference on Industrial Technology (ICIT)* (2015), pp. 1581–1586
13. H.D. Foster, A.C. Krolnik, D.J. Lacey, *Assertion-Based Design*, Information Technology: Transmission, Processing and Storage (Springer, Berlin, 2004)
14. C.-Y. Huang, Y.-F. Yin, C.-J. Hsu, T.B. Huang, T.-M. Chang, SoC HW/SW verification and validation, in *ACM/IEEE Asia and South Pacific Design Automation Conference (ASP-DAC)* (2011), pp. 297–300

15. K. Balston, A.J. Hu, S.J.E. Wilton, A. Nahir, Emulation in post-silicon validation: it's not just for functionality anymore, in *IEEE High Level Design Validation and Test Workshop (HLDVT)* (2012), pp. 110–117
16. A Parikh, W. Wu, M.S. Hsiao, Mining-guided state justification with partitioned navigation tracks, in *IEEE International Test Conference (ITC)* (2007), pp. 1–10

Part IV
Post-Silicon Debug

Chapter 11
Debug Data Reduction Techniques

Sandeep Chandran and Preeti Ranjan Panda

11.1 Drinking from the Fire-Hose

The on-chip Design-for-Debug (DFD) hardware has the single objective of maximizing the visibility of the internal functioning of the chip. However, the severe area and resource constraints that these structures operate under, make achieving this target extremely challenging. In order to appreciate the magnitude of the challenge, let us consider an example where an erroneous behavior of the chip was exposed by a simple directed test that takes 1 second to execute on the sample chip at a frequency of 3 GHz. The near-native operating frequency results in a span of 3 billion cycles. In order to gain some visibility into the chip, let us assume that a simple on-chip DFD hardware generates a trace of 128-bits (or 16 bytes) per clock cycle. The simple test would therefore generate a trace of size $16 \times 3 \times 10^9$B, or approximately 45 GB, which is larger than the DRAM present in most computers. Transferring this data off-chip would require a transmission link with a whopping bandwidth of 384 Gbps. The other alternative of halting the chip frequently, so as to restrict the rate at which the traces are generated is not desirable for various reasons such as the perturbations it causes to the test environment, the difficulty in pausing the execution without introducing inconsistencies, and the loss of near-native execution speeds offered by the sample chip. Clearly, managing such large volumes of execution trace under severe area constraints requires some ingenious engineering.

We summarize the goals that DFD hardware has to meet so as to set the context under which design decisions are made:

- Maximize visibility into the internal functioning of the chip.
- Minimize the area overhead due to the DFD hardware.

S. Chandran (✉) · P. R. Panda
Indian Institute of Technology Delhi, Hauz Khas, New Delhi, India
e-mail: sandeep@cse.iitd.ac.in

P. R. Panda
e-mail: panda@cse.iitd.ac.in

P. Mishra and F. Farahmandi (eds.), *Post-Silicon Validation and Debug*,
https://doi.org/10.1007/978-3-319-98116-1_11

- Minimize interference by DFD hardware on the functionality of the underlying hardware. It is best if the DFD hardware is transparent to the underlying architecture.

Clearly, these objectives are orthogonal to each other, and balancing them is difficult. An example DFD hardware that halts the chip and exercises the debug link to examine the chip every cycle gives very good visibility and occupies very less area, but is intrusive and hampers debug efficiency. On the other hand, using a large on-chip buffer to record all the internal activity of the chip gives good visibility and is nonintrusive, but incurs a prohibitively large area overhead.

These factors have resulted in the DFD hardware in modern chips being characterized by very limited on-chip storage and low bandwidth off-chip transmission links. The techniques we discuss here improve the efficiency of debug on such architectures by reducing the overall data volume by processing them as they are generated.

11.2 Popular Debugging Methodologies

11.2.1 Decoding a "Bug"

An erroneous behavior of the chip that adversely impacts a target application is loosely called a "bug". The error in the behavior arises either from an erroneous input to a module within the chip, or through a corruption of its internal state, or a combination of the two. The state of the system is a snapshot of all the memory elements on the chip, and is the result of a sequence of the inputs the circuit has received in the past.

The corruption in the state is possible for several reasons such as: (i) the sequence of inputs given in the past was illegal, and the circuit did not gracefully raise an exception when it encountered it; (ii) the designer did not envisage all possible corner cases that the circuit would be subject to, and hence designed the circuit incorrectly; and (iii) errors due to manufacturing defects.

Any data captured from within the chip that gives insights into the functioning of the chip, (internal state and how it changes over time when subjected to different stimuli) is called *Debug data*. Different techniques are used to capture different debug data; for example, recording a few signals from within different modules, recording only transactions over the interconnect, capture signatures representative of a region of execution, transfer the state off-chip, and so on. This debug data is then carefully analyzed by the validation engineers to localize bugs and find the root cause of their occurrence.

11.2.2 Overview of Debugging Methodologies

Traditionally, scan chains were used to debug chips where a subset of critical flip-flops and registers were connected to form a long shift-register. Then, these chains were loaded with test vectors, a string of 0 s, and 1 s, and finally unloaded after execution for further analysis. Since the chip would have to be tested against numerous vectors, loading, and unloading these vectors became the bottleneck. This is analogous to the challenges faced by the DFD hardware in post-silicon validation where the off-chip link is the bottleneck. Therefore, several techniques were proposed to compress these input vectors before transferring them over the debug link. These techniques used compression techniques such as (i) code-based compression schemes that use various methods to encode the vectors, (ii) linear-decompression based compression schemes where decompression requires only linear operations such as XOR, and (iii) broadcast-scan based schemes that merge values present across multiple vectors and send them out only once [1]. The increasing complexity of the chips have made scan chains inefficient to debug functional errors detected during post-silicon validation. However, concepts such as compression that were used to overcome the off-chip communication bottleneck have inspired some more recent post-silicon debug strategies.

Modern debugging paradigms such as run-stop debug and at-speed debug have been proposed to meet the debugging requirements of complex chips. Under the run-stop debug paradigm, the chip is allowed to execute normally for a certain duration before it is paused to collect its internal state. Under the at-speed debug paradigm, instead of capturing the entire state of the chip, only a continuous trace of a few signals are recorded in an internal buffer of limited size. The contents of this buffer have to be transferred off-chip so as to preserve it for further analysis as and when it overflows. With increasing complexity of the chips, the volume of debug data that is required to be processed under each of these paradigms increases significantly. Therefore, it is important to restrict the size of the debug data to manageable limits such that it does not hamper the efficiency of post-silicon debug. This chapter discusses techniques to reduce the debug data volume, while attempting to not compromise the quality of the captured data.

11.3 Reduction Schemes for Run-Stop Debug

Under the run-stop method of debug, the chip is executed for a certain duration and is paused to examine the state, before proceeding again. In order to examine the state, an atomic snapshot of all the on-chip memory elements including the register-file, buffers, and caches, is transferred off-chip at regular intervals. The large size of the last-level cache (LLC) dominates the time it takes to transfer the state off-chip. Let us take an example of an 8 MB LLC. In addition to this, there is a 1 MB L2 cache per-core. A 4-core machine would have over 12 MB of debug data to transfer over the debug

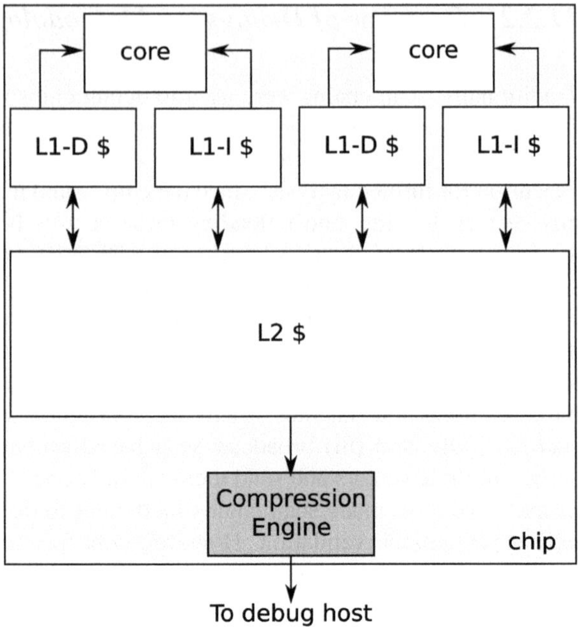

Fig. 11.1 High-level architecture of a multi-core processor where the DFD hardware is a compression engine that compresses the contents of the last-level cache (L2 here)

link. Considering a 100 Mbps JTAG link, capturing each snapshot and transmitting it off-chip will take around 120 ms, which is equivalent of 120 million clock cycles when the core is running at a modest frequency of 1 GHz. This becomes extremely inefficient as the dumping frequency increases. Therefore, several techniques have been proposed to reduce the size of the debug data when capturing the state of the LLC.

11.3.1 Compression Using LZW

Figure 11.1 shows the outline of an architectural proposal [2] to reduce the size of the debug data by compressing the last-level cache before transferring it off-chip. The debug data that is transferred off-chip in this case includes the tag array, control bits, and the Error Correction Codes (ECCs) in addition to the data stored in the cache. This technique exploits the cache organization of the debug data to achieve superior compression ratio as compared to the brute-force method of compressing the entire debug data in a structure-agnostic manner. In order to exploit the cache organization of the debug data for compression, the contents of each of the cache fields such as tag, control-bits, data, and ECC, are compressed separately.

When compressing the contents of the tag array, spatial locality of references causes the higher order bits of the adjacent cache lines to be similar, and hence allows for better compression. Similarly, the control-bits of the LLC change less

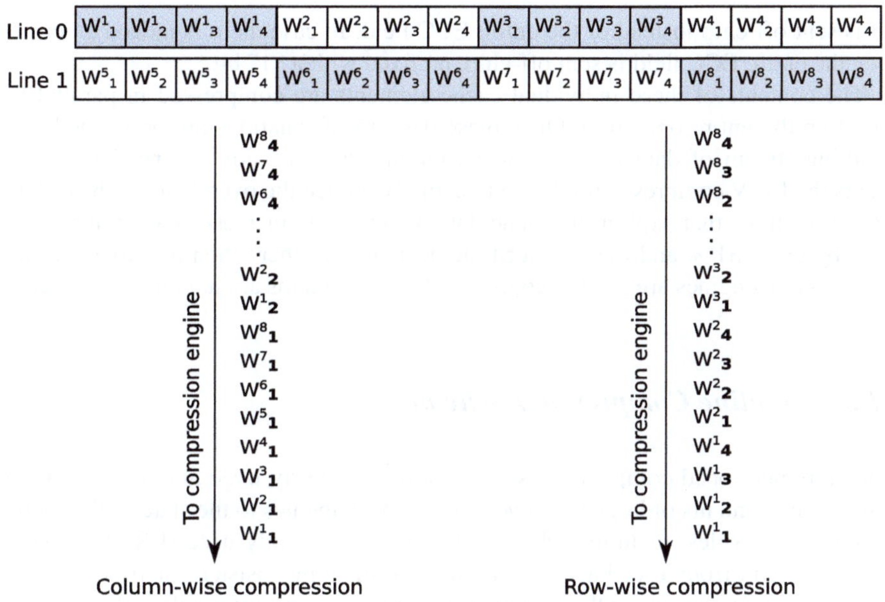

Fig. 11.2 Column-wise and Row-wise compression of the data array

frequently based on whether it holds code or data. For example, the dirty bit for lines containing code is unlikely to be set. The data array is the largest element within the cache and also is the most diverse.

In order to compress the data array, a key observation is made on the data types of the values it stores: in applications that work largely on integer data, the upper bytes of the integer data usually change less frequently. Hence, the upper bytes of each word in the data cache are compressed before the least-significant bytes (LSBs) of a word. This is *Column-wise compression*. For applications dominated by floating-point and image data, words are grouped sequentially one line after another and compressed. This is called *Row-wise* compression.

Figure 11.2 explains these two methods with an example. Here, each byte in the cache is denoted by w_y^x where x is the word this byte belongs to, and y is the position of the byte within each word. Therefore, w_1^6 indicates this byte is the most-significant byte of the sixth word in the cache. Under column-wise compression, the upper bytes of each word are compressed before the lower words. Hence, the order of compression is w_1^1, w_1^2, w_1^3,...,w_2^1, w_2^2, w_2^3, and so on, until finally the LSBs (w_4^1, w_4^2, w_4^3,...) are sent. Under row-wise compression, each word is sent one after another, as w_1^1, w_2^1, w_3^1, w_4^1, w_1^2, w_2^2, w_3^2, w_4^2, and so on.

For compressing the ECC bits, the fact that there would be very few errors at any given point is exploited. The ECC bits are sent off-chip only when there is indeed an error in a particular cache line. The ECC of each cache line is recomputed during the dumping phase, to see if there is an error in the line. If there is no error, a single "0" bit is sent off-chip instead of the entire ECC field. In case there is indeed an error, then the entire ECC field is sent off-chip, prefixed with a "1" bit.

The contents of these individual cache elements are compressed in parallel to speed up the entire operation. Once these data transformations are performed, the resulting stream of data is passed to an on-chip compression engine that implements the LZW compression scheme to actually reduce the overall size. These data transformations that exploit the cache data organization increases the compression ratio by up to 31%, and the transfer time decreases further when multiple parallel compression engines are used to compress the data streams arising out of the cache.

11.3.2 Online Compression Scheme

The aforementioned compression scheme still takes fairly large amount of time to transfer the contents of large last-level caches. Any updates to the state of the cache during these cycles, including changes to the least-recently used (LRU) counters that may result from a read access, would result in an inconsistent snapshot. Hence the processor is paused for this duration which is a significant number of processor cycles.

Figure 11.3 shows a technique to overcome this where the cache is compressed in an online manner [3]. An additional hardware unit present on-chip keeps track of the *dumped*, and *non-dumped* regions of the cache. When a request to update

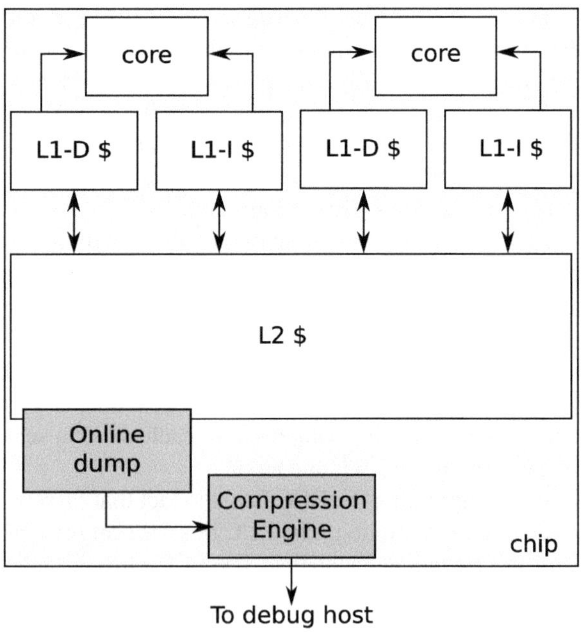

Fig. 11.3 Online cache compression

the cache state arrives, it is first determined whether the cache line that needs to be updated belongs to the dumped or the non-dumped region. If the update is to the dumped region, it is allowed to complete as usual. However, if the update is to the non-dumped region, the cache line is first dumped before allowing the request to complete. When such a cache line is transferred out of sequence, along with the cache line, the way number is also transferred off-chip so as to identify the way and the line during off-chip reconstruction of the state.

It is possible that the state of the same cache line is updated multiple times before it is dumped, and consequently the same cache line would be transferred multiple times off-chip. In order to avoid this, a *Dump History Table* is maintained. The Dump History Table maintains one bit per cache line to indicate if the line has already been transferred off-chip. All updates to the non-dumped region refers to this bit. The setting of this bit is an indication that the cache line has already been transferred off-chip by a previous update, and hence the current update could proceed as usual. If this bit is not set, then the cache line is transferred off-chip first before the update to the cache line proceeds, and the bit is set on completing the update. This technique manages to reduce the effective L2 cache dump time to between 0.01 and 3.5% of the time taken in the previous case.

11.3.3 Reduction Using Update History Table

The size of the debug data to transfer off-chip can be further reduced through incremental dumping of updated cache lines [4]. Instead of transferring all the cache lines, it is sufficient to transfer only those cache lines that are updated after the previous dump. The information on the cache lines that are updated since the previous dump is maintained in an *Update History Table (UHT)* where a bit is stored per cache line. Along similar lines, one bit can be used to track the update status of multiple cache lines instead of just one cache line. This reduces the area required to track the update information. In this case, a bit is set if any one of the cache lines tracked by the particular bit is updated, but all cache lines tracked by the particular bit are dumped if the bit is set.

Figure 11.4 shows an example where cache lines 6, 10, and 11 have been updated since the previous dump. In the case of a UHT where each bit tracks the updates to a single cache line, only these 3 lines are transferred off-chip. However when a bit in UHT tracks the updates to two cache lines, the cache line 7 is also dumped although it was not updated.

In order to dump the cache in an online manner under this scheme, a pointer to the UHT entry of the cache line that is currently being dumped is maintained. The entries before this pointer correspond to the region of the cache that is already dumped, and the later entries correspond to the non-dumped region. Two UHTs are maintained: *Update History Table Current (UHT-C)* and *Update History Table Previous (UHT-P)* to support online dumping of cache lines. Similar to the previous technique, if a line is updated in the non-dumped region of the cache, the cache line is

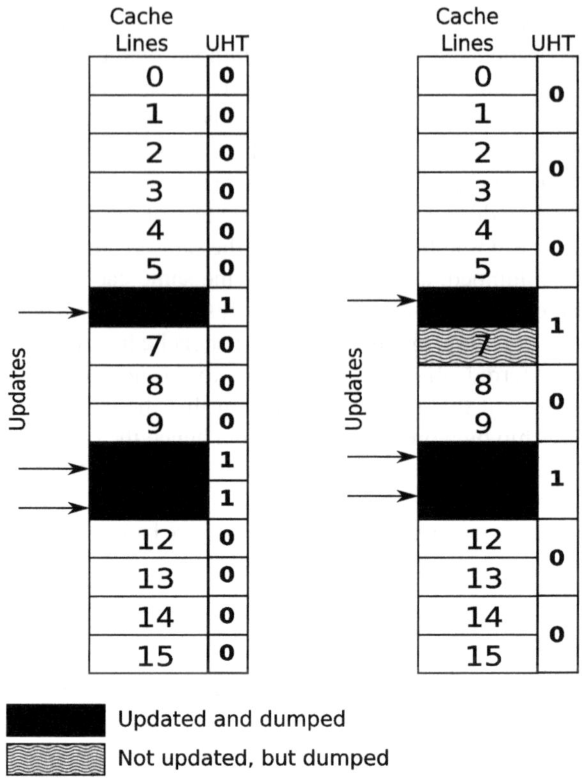

Fig. 11.4 Update history table where a bit tracks 1 and 2 cache lines, respectively

first transferred offline, and the corresponding bit in the UHT-C is cleared. If a cache line that is already dumped is updated, the update request is allowed to complete, and the information is recorded in UHT-P. At the end of the dumping phase, the UHT-C and UHT-P are interchanged. Such cycling of these two tables continues with every dump. Such incremental dumping reduces the number of cache lines dumped by 64% and the dump time is reduced to just 16% of the original dump time. In case the incremental dumping is done in an online manner, the dump time is reduced to mere 0.0002% of the original dump time.

11.3.4 Reduction Using Interval Table

For large LLCs found in the modern day processors, the UHT may incur prohibitively high area overheads. For example, a nominal cache of size 2 MB with 128 bits per line and eight-way associativity requires a UHT of 16 k bits. With shrinking technology sizes, the size of on-chip caches are bound to further increase.

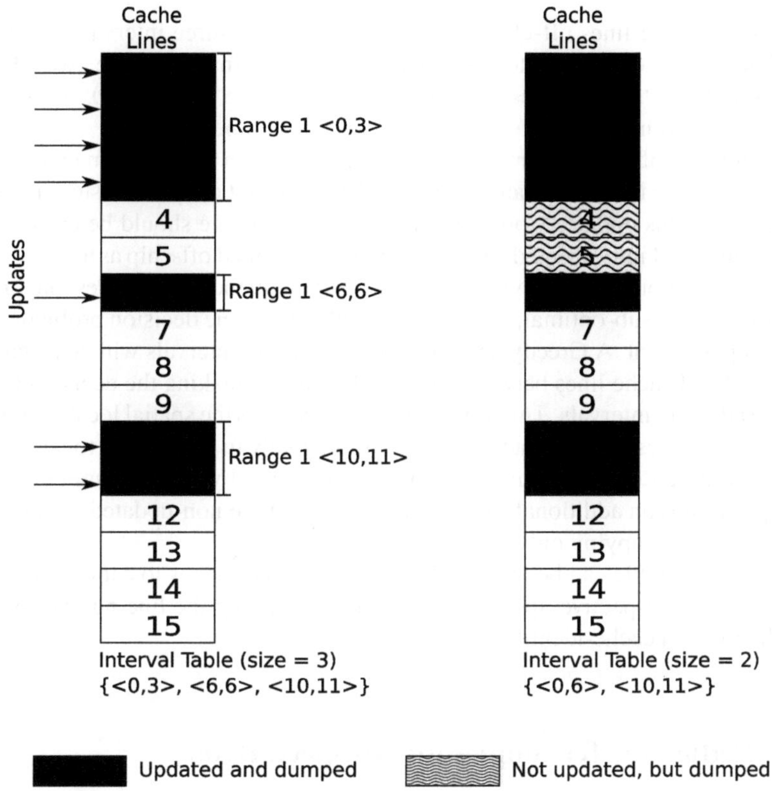

Fig. 11.5 Interval table

An alternative technique for incrementally dumping the cache is to maintain only the starting and ending addresses of a continuous range of cache lines whose state is updated since the previous dump instead of maintaining the updated information of each cache line separately [5]. Each range of adjacent updated cache lines is called an *Interval*, and a fixed set of such intervals is called an *Interval Table*. There is an upper limit on the number of intervals that can be maintained in the Interval Table due to area constraints.

The actual number of ranges of updated cache lines at any given time depends on the cache accesses thus far. It is possible that the number of ranges of updated cache lines is higher than the number of intervals that can be stored in the Interval Table. Since information on cache lines that are updated cannot be missed, adjacent intervals of the Interval Table have to be merged into a single interval so as to accommodate the excess ranges into the Interval Table. Once two intervals of the Interval Table are merged, the non-updated cache lines between the two intervals become part of the merged interval and are considered as updated cache lines and dumped off-chip. Therefore, there is an additional overhead incurred due to transferring the non-updated cache lines off-chip. Figure 11.5 shows the three intervals of updated cache lines. When only two intervals can be stored in the Interval Table, two adjacent intervals (<0, 3> and <6, 6>) are merged into a single interval <0, 6> so as to accommodate all information on all the updated cache lines.

The merging of adjacent intervals has to take place in an online manner as and when a cache line is updated, and the Interval Table is out of space to store information regarding this update. Moreover, the intervals to merge should be chosen such that the number of non-updated cache lines that are dumped off-chip as a result of it is minimal. This choice is nontrivial because a future sequence of accesses may render a previous choice sub-optimal, as is the case with any online decision problems such as page replacement. A Greedy heuristic where adjacent intervals with least number of non-updated cache lines between them at the time of making the decision is used to merge adjacent intervals. This simple heuristic exploits the spatial locality of reference seen in caches where adjacent lines are quite likely to be updated together. This online algorithm deviates from the theoretical optimum by a maximum of 2x. This technique dumps an additional 11.5% cache lines that are non-updated as compared to UHT, while occupying only less than 10% of its area.

The Interval Table can be also be shared by multiple caches to capture the information on their respective updated cache lines by tagging the line number of each cache line with a cache-identifier bit.

11.4 Reduction Techniques for At-Speed Debug

An alternative paradigm for post-silicon validation is *at-speed debugging* where instead of pausing the operation of the chip to collect information on the state, a small buffer called the *Trace Buffer* is built into the chip, into which changes in internal signals over multiple cycles are recorded. Due to severe area constraints on the DFD hardware, this trace buffer is usually very small (only a few KB). Moreover, this trace buffer is generally organized as a circular buffer, and is configured to overwrite older entries when it runs out of space. If the execution trace captured within the trace buffer is from a region of execution that is of interest currently, the execution is paused briefly and the trace buffer contents are transferred off-chip before overwriting it. The duration for which the execution is halted is much shorter as compared to the aforementioned run-stop debugging techniques.

The window of visibility that the trace buffer offers through the execution trace stored in it is determined by two factors: width and depth, where width is the number of signals that are recorded in the trace buffer, and depth is number of clocks for which the behavior of internal signals are recorded. In order to maximize the visibility obtained from a trace buffer of limited size, the input trace is compressed before storing it into the buffer, thereby increasing the effective capacity of the buffer.

Trace compression can be achieved either through *Width compression* under which the width of each entry in the trace buffer is reduced, or through *Depth compression* under which the number of entries stored in the trace buffer is increased. The trace signal selection techniques discussed in previous chapters 3–6 are examples of width compression techniques. Two popular depth compression techniques are: (i) differential encoding where an entry is written into the trace buffer only when the input changes and (ii) event triggering, where execution traces are recorded only in a particular region of interest demarcated by occurrence of pre-specified events. These two compression schemes are orthogonal to each other and hence can be combined together to achieve superior compression. For example, trace signal selection techniques can be used in conjunction with differential encoding.

Compression schemes can also be classified as either *Lossless* or *Lossy*. Lossless compression schemes can reproduce the input stream from the compressed contents. The differential encoding under depth compression technique mentioned above is an example of a lossless compression scheme. Similarly techniques that cannot reproduce the input stream from the compressed contents are lossy compression techniques. Recording input traces in response to occurrence of events is an example of a lossy technique because the traces discarded before the occurrence of the event cannot be reproduced.

Domain-Specific Considerations

Since the nature of the input stream is known apriori, compression techniques can exploit them to achieve better compression ratios. Some domain-specific considerations that significantly help improve the compression ratio are:

- *Deterministic replay of the test case.* If the erroneous behavior observed can be deterministically repeated over several executions, the compression ratio that can be achieved is significantly higher. In such a case, the visibility into the internal functioning of the chip can be gathered over several runs, one small window at a time.
- *Availability of "golden" output as a reference.* A compression engine that knows the reference behavior of the chip can exploit this information to achieve a superior compression ratio through lossy methods where execution traces that match with reference behavior can be simply discarded.

The schemes that exploit these considerations stand the risk of performing poorly when the erroneous behavior do not match their expectations. A validation engineer would have to resort to other debugging methods in order to meet the expectations of these techniques before it can be of value to him.

11.4.1 Lossless Compression

Lossless compression techniques *encode* a symbol or a sequence of symbols in the input stream using a possibly shorter symbol [6]. A fundamental building block of such techniques is a dictionary of symbols that is constructed as it processes the input stream. The differences arise from how this dictionary is constructed and maintained by each technique. The compression engine implements the dictionary

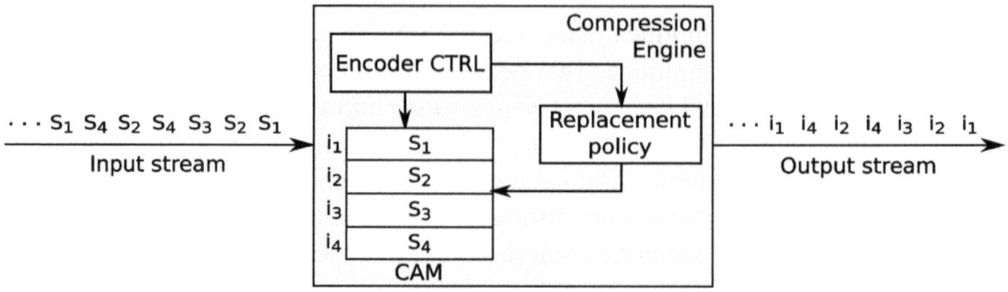

Fig. 11.6 Lossless compression technique

using a Content-Addressable Memory (CAM) so as to support compression of the input stream in an online manner. If the input symbol already exists in the CAM as determined by a lookup, the index of the entry in the CAM that matches with the input symbol is sent on the output stream as the encoded symbol. When a new symbol is discovered in the input stream, it is sent as is on the output stream before adding it to the dictionary. Figure 11.6 shows a high-level architecture of a lossless compression engine. Since the width of the indexes i_x is smaller than that of the input symbols S_x, the total size of the output stream will be less than that of the input stream.

Since the number of entries in the CAM is limited, some symbols would have to be replaced by newly discovered symbols as and when the CAM runs out of space. The replacement scheme determines the quality of the dictionary and hence significantly affects the overall compression ratio. We briefly discuss a few replacement schemes here.

- FIFO replacement: Under this replacement policy, the oldest symbol is replaced as and when a symbol in the CAM has to be replaced.
- Random replacement: Here, a symbol is picked at random from the CAM to be replaced.
- Modified Least-Recently Used (LRU) scheme: An implementation of a true LRU replacement policy incurs a significant area overhead because of all the counters it has to maintain and update. Therefore, an approximation of the same is implemented to reduce the area it occupies and the time it takes to update its state. One such approximation is to use a combination of FIFO and true LRU. Under this scheme, the CAM is divided into multiple address segments and the pointers to the start of each segment is maintained separately. The first pointer in the array of the segment pointer is the most recent segment, and the last pointer is the least-recently used segment. When an entry in the CAM is accessed, the pointer to this segment is moved to the front, and other segment pointers are shifted down by one entry in the next cycle. Within a segment, the entries follow a FIFO replacement strategy. This LRU scheme is beneficial, because the number of registers to update the LRU state on each access is restricted to the segment pointers only instead of the total number of CAM entries.

- Modified Least Frequently Used (LFU) scheme: The drawback of the LRU scheme is that it only considers the recency of access and not the frequency of matches of an entry in the CAM. Therefore, it is possible that a frequently accessed entry that was not accessed recently (within the depth of the CAM), was removed from the dictionary altogether. The LFU policy avoids this by using an additional counter with each entry to keep track of the number of matches. However, just as the LRU scheme, an implementation of true LFU scheme is costly and hence similar modifications as the aforementioned Modified LRU are used to achieve an area-efficient Modified LFU policy.

Some compression schemes can encode consecutive symbols simultaneously instead of just one symbol. There are several implications of doing so on the size and structure of the CAM that has to be considered. If the number of consecutive symbols that are considered for encoding is fixed, then the size of each entry in the CAM would have to be increased, which would increase the latency of each lookup. Moreover, as the length of consecutive symbols under consideration together increases, its frequency of repetition decreases, as a result of which the compression ratio may decrease. Therefore, finding the right balance between the number of consecutive symbols to consider for encoding, and limiting the latency of the CAM lookup so as to improve the compression ratio is nontrivial.

A popular compression scheme that uses fixed-width, multi-symbol dictionary entries is BSTW, that is named after its inventors. A modified version of this is proposed to suit the area constraints of post-silicon validation without compromising the compression ratio. Under this proposal, each CAM entry continues to remain one symbol wide, and consecutive symbols are stored in successive CAM entries. Each output symbol is now tagged to indicate the number of successive entries of the input stream that match in the dictionary. For example, let us consider a compression scheme that uses a 2-symbol dictionary and 2-bit tags in the output stream. The 2-bit tags can be encoded to indicate the following: (i) 00—no match and that the output symbol is same as the input symbol, (ii) 01—a single symbol match only and the output symbol is the index of the matching entry in the dictionary, (iii) 10—two successive symbols of the input stream are exactly the same as the entry in the matching index, and (iv) 11—two successive symbols of the input stream match consecutive entries of the CAM and that the index of the first entry the output symbol.

Figure 11.7 shows a technique to improve upon the compression ratio achieved through the aforementioned technique by using a "golden" trace as a reference to construct the dictionary statically offline [7]. Since, the chip is well-tested and only very few functional errors remain, the footprint of the bug could be very small. Hence, an optimized dictionary can constructed for a representative trace statically in an offline manner, and loaded into the CAM before running the test case. Such a statically constructed dictionary does not need a replacement policy because all possible input symbols and their expected frequencies have already been taken into consideration when constructing the dictionary. Only the most frequent symbols are loaded into the on-chip CAM for online compression of the execution trace. The maximum possible deterioration in the compression ratio is bounded by the observed error rate.

11.4.2 Lossy Compression Schemes

Lossy compression schemes require a reference behavior to be provided to them in order to compress the input stream. Since the reference behavior is established to be correct, only regions of the execution trace that deviate from the reference provided need to be captured for further analysis. The lossy compression schemes vary based on the kind of reference behavior they take as input, and how they use it to achieve better compression ratios.

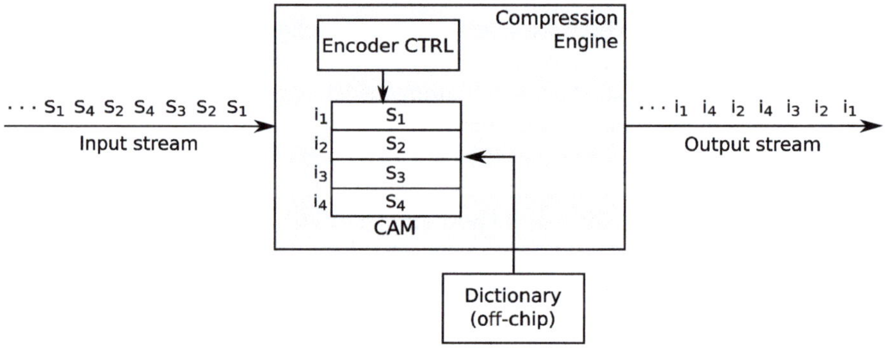

Fig. 11.7 A static lossless compression scheme

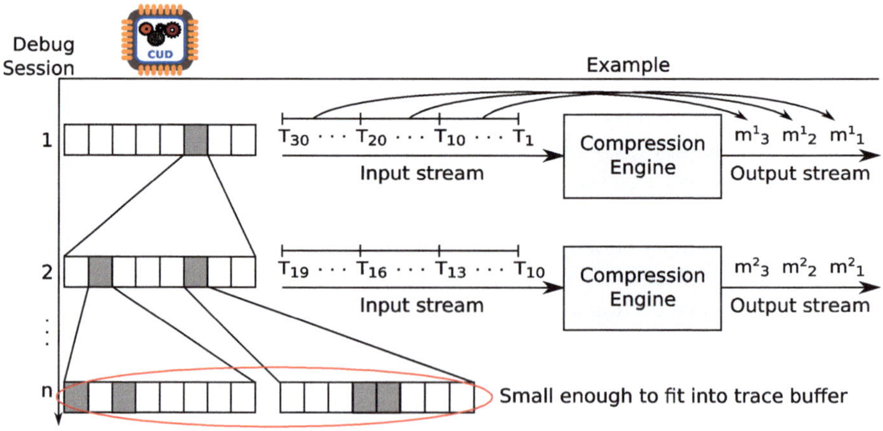

Fig. 11.8 Lossy compression scheme

11.4.2.1 Lossy Compression Using MISRs

Figure 11.8 highlights a lossy compression technique that was proposed for a deterministically repeatable functional error [8]. This technique uses the traditional event triggers to demarcate pre-specified execution intervals. A Multi-Input Signature Register (MISR) generates a signature of this execution interval and stores it into the on-chip trace buffer. Thus captured MISRs are then transferred off-chip and are compared against the corresponding MISRs generated from the golden trace to identify intervals of execution where the actual functioning of the chip has deviated from its expected functioning. The golden trace is obtained through simulations or through emulations on FPGAs, and the intervals of execution are demarcated through event triggers.

In the subsequent debug session, the intervals corresponding to the mismatching MISRs are further divided into smaller intervals, and MISR signatures are generated for those smaller intervals. This is again compared with its corresponding golden MISR signatures. Therefore, it iteratively zooms in to the erroneous regions, until an interval is obtained where the execution traces from it completely fit into the trace buffer, thereby significantly reducing the amount of traces to be transferred off-chip.

The overall efficiency of this can be improved further through various methods. One such method is to use variable sized intervals instead of fixed sized segments to better capture bursty behavior of functional errors, where segments with a large number of errors can use the space of segments with fewer errors to store the captured MISRs. Another improvement to this technique is to eliminate the need to transfer the actual MISRs off-chip by storing the golden signatures into an empty area of the trace buffer. Therefore, intervals can be checked for errors as and when the corresponding MISRs are generated by comparing them with that stored in the trace buffer.

11.4.2.2 Lossy Compression Using 2D Compaction

A drawback of the above technique is that it may take several iterations to sufficiently zoom into an erroneous region of execution. To overcome this, a technique was proposed that takes exactly three debug sessions to capture traces only from the erroneous regions of execution [9]. Figure 11.9 gives an overview of this technique. During the first debug session, the parity of each input trace (T_1–T_9) is computed and stored into the trace buffer. This is then checked with the "golden" parity generated from simulation data to determine the error rate offline. The error rate thus computed is used to determine the size of the intervals over which signatures should be generated in the second debug session.

The second session uses a 2D compaction technique to determine the exact cycles that were erroneous. Under this scheme, two signatures are generated—MISR signature and Cycling Register (CR) signature. An MISR generates a signature for k consecutive trace messages, where k is determined through the first debug session and the number of MISRs present on-chip. Figure 11.9 shows an example where $k = 3$. The input traces $T_1, T_2, T_3, \ldots, T_9$ are generated in 9 consecutive cycles. The first

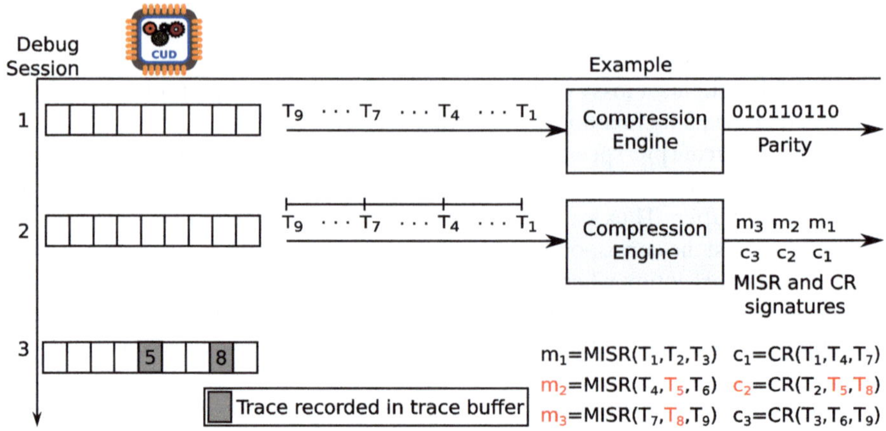

Fig. 11.9 2D compaction scheme

MISR generates a signature for T_1, T_2, T_3; the second MISR generates a signature for T_4, T_5, T_6; and the third MISR generates a signature for T_7, T_8, T_9. The cycling registers however, cycle through the input traces before generating their signatures. Therefore, the first CR generates the signature for T_1, T_4, T_7, the second generates for T_2, T_5, T_8, and the third for T_3, T_6, T_9. Through this method, the input trace stream is spread over two dimensions before generating the signatures. Therefore, when the trace deviates from the golden trace available off-chip in a particular cycle, there are exactly two signatures (an MISR signature and a CR signature) that mismatch with the corresponding golden signatures. The intersection of the mismatching MISR and CR signatures determines exact cycles to trace in the third debug session.

In the third debug session, the cycles to capture the execution trace are loaded into the trace buffer as *cycle tags*. These cycle tags are used by the trace buffer controller to record the input trace into the trace buffer, thereby achieving significantly higher compression ratio.

11.4.2.3 Lossy Compression for Multi-core Systems

The brute-force method to compress traces from multiple cores is to sequentially capture traces from each core. A technique to capture only erroneous traces from multiple cores in exactly three debug sessions [10] also uses MISRs generated from a golden trace obtained from simulations of a behavioral model or emulations on an FPGA in the first debug session to determine the erroneous intervals in each core. The golden MISRs are preloaded into the system and used to detect erroneous intervals as and when the actual MISRs are generated. In the second debug session, the golden data is streamed into the trace controller in order to detect the clock cycles where there is a mismatch between the golden data and the actual trace. Since streaming in large volumes of debug data is resource-intensive, this technique is applicable on modern chips with on-chip DRAM that support very high bandwidths. Since the golden data needs to be streamed into all the cores, it is possible that this method becomes very intrusive. In order to avoid this, traces from a core that has been certified as error-free in the first session are used as a reference to check the traces generated by the erroneous cores. Golden traces are streamed into the core for the second session only for those intervals in which signatures from all cores mismatched in the first session. In the third debug session, traces from only those cycles where the actual trace message mismatched with that of the reference trace are recorded into the trace buffer.

11.4.2.4 Lossy Compression Through Invariants

All the aforementioned techniques for lossy compression require the reference behavior to be given as a golden trace, and generating the golden trace through simulations or emulations could be time-consuming. Similarly, the reliance on a golden trace requires the functional error to be deterministically repeatable. However, in practice these requirements could be limiting. For example, it is difficult to deterministically reproduce a multi-threaded test case because its behavior is dependent on the order of acquiring and releasing locks and other synchronization constructs. A technique that addresses this concern requires the reference behavior to be specified as an invariant that has to be followed by the execution at all times [11]. An execution trace is recorded into the trace buffer only if it violates the specified invariant.

This technique proposes a specification language called *Storage Specification Language (SSL)* that borrows the semantics of a subset of *Property Specification Language (PSL)* operators so as to implement a set of logical and temporal operators that are tailored for recording execution traces. Such properties, defined through logical and temporal relationships between various signals, are extensively used by formal verification tools and assertion-based verification methodologies. This technique extends it to the domain of post-silicon validation.

The invariants specified using SSL operate on the granularity of fields within a trace, where a field is a group of bits that are captured atomically to give meaningful information about the execution. For example, the program counter (`PC`) of an instruction trace is an example of a field within the trace. Width compression is achieved when an SSL is specified to store only a select few fields from within the trace. The subset of fields to record in the trace buffer in the current debug session is specified by the validation engineer based on the error observed, and the information that is of value to him. Further, an event is triggered when the field takes a pre-specified value. Traces can be recorded into the buffer in response to logical or temporal relationships between various events.

Figure 11.10 shows a working example of an SSL statement that records a trace only when the field `LDST` becomes `1` (indicating a store operation) exactly two cycles after `PC` takes the value of `0xaa`. In this example, the construct *after* is a temporal expression. This SSL statement performs both width and depth compression on the input trace.

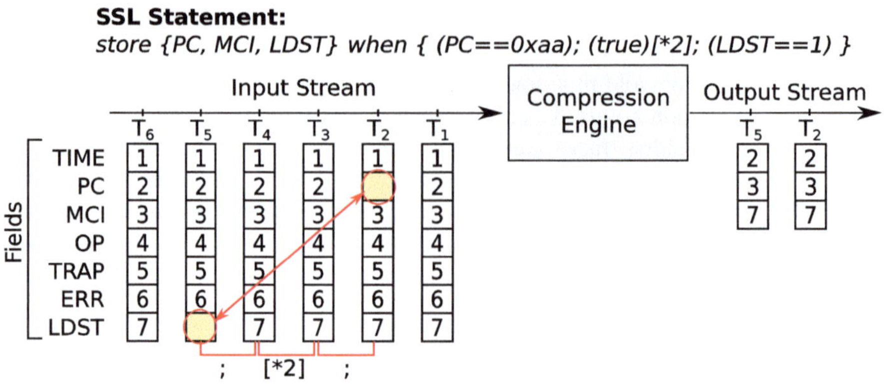

Fig. 11.10 An overview of compression through different invariants

Experimental results show a reduction in transfer times of up to 70% when only width compression is used, and negligible transfer time is incurred when width and depth compression are achieved through appropriately specified properties.

11.5 Summary

The on-chip DFD hardware handles large volumes of debug data in order to maximize the visibility into internal functioning of the chip. However, the resources under the disposal of DFD hardware are often constrained, in order to limit the area overhead due to their presence. Therefore, capturing and transferring information on the internal state becomes a bottleneck to efficient debugging of functional errors observed during post-silicon validation. This is alleviated by compressing the debug data before storing or transferring them off-chip. Several techniques have been proposed to do so based on the nature of debug data being captured by the post-silicon validation methodology.

Under the run-stop debug methodology, where the off-chip transfer time is dominated by the data present in large last-level caches, a modified version of the LZW compression algorithm that compresses the tag, data, control, and ECC bits separately in parallel is known to achieve superior compression ratio in an area-sensitive manner. Also, techniques have been proposed to transfer only the changes to the cache state in an incremental manner in order to reduce the volume of debug data.

Under the at-speed debug methodology, compression schemes have been proposed to reduce the width of each execution trace, and increase the number of traces that can be stored into the on-chip trace buffer. Several lossless compression schemes have been proposed that replace each input symbol in the execution trace with an index into a dictionary that is smaller in size. The techniques vary based on how this dictionary of symbols is created and maintained during the execution of the test case.

Such schemes are suitable when the execution trace is from a region of interest, and therefore, all the information captured in the trace is valuable.

Several lossy compression schemes have also been proposed for scenarios where the footprint of the observed error has not been localized to a particular region of execution. These schemes rely on providing a golden behavior as a reference input against which the actual behavior is compared. Only traces from regions of execution where the actual behavior deviates from the reference behavior is captured, and the rest is discarded, thereby significantly reducing the volume of debug data. The proposed techniques have used varied reference inputs such as traces generated from behavioral simulations and FPGA emulations, specifications of invariants that must be followed, and traces from other cores that are established as non-faulty previously.

With evolving architecture of modern systems, some of the resource constraints on the DFD hardware may be relaxed, thereby increasing the scope of further improving the efficiency of post-silicon validation. One such direction is that of reusing architectural elements such as cache and on-chip DRAM as a backing store of the trace buffer.

References

1. N.A. Touba, IEEE Des. Test Comput. **23**(4), 294 (2006). https://doi.org/10.1109/MDT.2006.105
2. A. Vishnoi, P.R. Panda, M. Balakrishnan, in *2009 Design Automation Test in Europe Conference Exhibition* (2009), pp. 208–213. https://doi.org/10.1109/DATE.2009.5090659
3. A. Vishnoi, P.R. Panda, M. Balakrishnan, in *2009 46th ACM/IEEE Design Automation Conference* (2009), pp. 358–363. https://doi.org/10.1145/1629911.1630007
4. P.R. Panda, A. Vishnoi, M. Balakrishnan, in *2010 18th IEEE/IFIP International Conference on VLSI and System-on-Chip* (2010), pp. 55–60. https://doi.org/10.1109/VLSISOC.2010.5642623
5. S. Chandran, S.R. Sarangi, P.R. Panda, IEEE Trans. Very Large Scale Integr. (VLSI) Syst. **24**(5), 1794 (2016). https://doi.org/10.1109/TVLSI.2015.2480378
6. E.A. Daoud, N. Nicolici, IEEE Trans. Comput. Aided Des. Integr. Circuits Syst. **28**(9), 1387 (2009). https://doi.org/10.1109/TCAD.2009.2023198
7. K. Basu, P. Mishra, in *29th VLSI Test Symposium* (2011), pp. 14–19. https://doi.org/10.1109/VTS.2011.5783748
8. E.A. Daoud, N. Nicolici, IEEE Trans. Comput. **60**(7), 937 (2011). https://doi.org/10.1109/TC.2010.122
9. J.S. Yang, N.A. Touba, IEEE Trans. Very Large Scale Integr. (VLSI) Syst. **21**(2), 320 (2013). https://doi.org/10.1109/TVLSI.2012.2183399
10. H. Oh, I. Choi, S. Kang, IEEE Trans. Comput. **66**(9), 1504 (2017). https://doi.org/10.1109/TC.2017.2678504
11. S. Chandran, P.R. Panda, S.R. Sarangi, A. Bhattacharyya, D. Chauhan, S. Kumar, IEEE Trans. Very Large Scale Integr. (VLSI) Syst. **25**(6), 1881 (2017). https://doi.org/10.1109/TVLSI.2017.2657604

Chapter 12
High-Level Debugging of Post-Silicon Failures

Masahiro Fujita, Qinhao Wang and Yusuke Kimura

12.1 Motivation and the Proposed Post-silicon Debugging Flow

With the increase of the VLSI design complexity and the demand for shortening the time to market, the design processes should be started from higher abstracted levels such as the design descriptions in C language. High-level synthesis has been developed and can automatically transform a given behavioral description in C into its corresponding RTL structural implementation. By starting the design process from high-level, advantages such as faster simulation, smaller amount of codes in design descriptions, better readability of the design descriptions, and others can significantly enhance the design productivity.

It is said that for complicated and large chip designs, the majority of efforts are spent to make sure the correctness of designs rather than synthesizing and optimizing them. In general, verification processes become more efficient in high-level, as the amount of descriptions to be analyzed can be much smaller than those of implementations. Therefore, the debugging of bugs must be performed as much as possible in high-level so that fewer bugs may remain the later design stages. Unfortunately, however, some logical bugs may still escape from all of the verification efforts including formal methods used in high-level and RTL designs. Some bugs can only be recognized after the chip has been fabricated and run in the field. As the actual chip runs much faster than logical simulation in high-level, bugs which need very long sequences of error traces may be found with the actual chip runs. As a result, it is

M. Fujita (✉) · Q. Wang · Y. Kimura
University of Tokyo, Bunkyo-ku, Tokyo, Japan
e-mail: fujita@ee.t.u-tokyo.ac.jp

Q. Wang
e-mail: wang@cad.t.u-tokyo.ac.jp

Y. Kimura
e-mail: kimura@cad.t.u-tokyo.ac.jp

P. Mishra and F. Farahmandi (eds.), *Post-Silicon Validation and Debug*,
https://doi.org/10.1007/978-3-319-98116-1_12

extremely important to realize an efficient verification and debugging environment for the chip in post-silicon assuming that some sorts of high-level design descriptions are available.

Another problem in C-based design is the fact that the synthesis and optimization processes could introduce logical non-equivalency between the designs before and after such processes are applied due to the bugs and/or inappropriate uses of the tools. This is practically an unavoidable situation, and the design flow should be ready for such situations. As the design becomes more complicated and larger, it is very hard to completely check the equivalence between the final implementation and the high-level design. It is highly preferred for the design flow to have mechanisms by which the implementation can be modified in post-silicon. That is, the introduction of some amount of programmability to the design can be essential. In this chapter, we discuss such mechanisms assuming the design starts in high-level.

The rest of the chapter is organized as follows. In the next section, the proposed design flow for C-based design is introduced and discussed, In the following section, issues to introduce small amount of programmability into a given design is discussed. Then, the methods to analyze and debug the implementation in post-silicon using such programmability are discussed in the following two sections. The methods to reproduce the C design descriptions which are equivalent to the actual chip implementation are introduced next. The final section gives concluding remarks.

12.2 Proposed C-Based Design Flow Supporting Post-silicon Analysis and Debugging

The proposed design flow is shown in Fig. 12.1. We start with a high-level design in C which should be intensively verified as much as possible before going to the next design process. The high-level design after verification is considered to be the golden model in our proposed design flow, and all the other designs in later design stages should be behaviorally equivalent to it, although some designs in later design processes may not be so simply due to the inclusion of some logical bugs through the synthesis and optimization processes.

In our proposed design flow, when synthesizing lower and more concrete designs from the high-level ones, some amount of programmability realized by LUT (Lookup Table) is introduced into the designs as shown in the figure. This is discussed in the next section. The implementation designs in our proposed design flow have some amount of programmability by which functionality of the implementation can be modified by some amount.

The introduced programmability is utilized when analyzing the implementation, or the actual chip, which is discussed in Sect. 12.4. With the programmability in the implementations, we can analyze the root cause of errors in more efficient and effective ways. For example, the flipflops used in LUTs are connected to the chip scan chains in order to reprogram LUTs in post-silicon, and so the functionality and

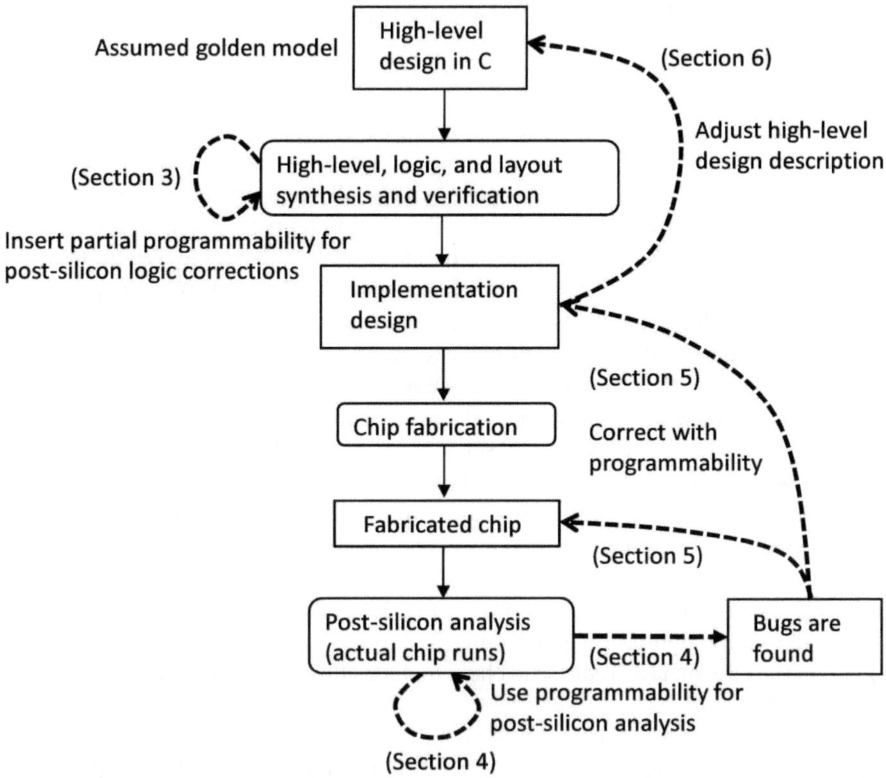

Fig. 12.1 Proposed C-based design flow supporting post-silicon analysis and debugging

the values of LUTs can be dynamically monitored and modified for efficient analysis, which is demonstrated in Sect. 12.4.

After the root cause of the bugs or errors are identified, actual design corrections are tried to be performed, which is explained in Sect. 12.5. If the correction does not need much changes in the designs, there may be no need of re-fabrication, as the reprogramming LUTs can be sufficient. However, if the bug corrections need large changes in the implementation, changing the functionality of the LUTs may not be sufficient, and the design must be revised and the chip must be re-fabricated.

Finally, after the implementations are corrected, their corresponding high-level designs in C are reproduced based on the template-based designs and the CEGIS [1] methods, which is explained in Sect. 12.6. Those templates can be efficiently generated with the help of designers who have actually corrected the bugs in the implementations, as they know which kind of changes are required in the high-level designs. In general, there can be two different situations, first one is the case where high-level and implementation are equivalent and the second case where they are not equivalent (bugs are introduced during synthesis and optimization). The reproduction of the high-level designs in C is very useful in both situations.

12.3 Introduction of Small Amount of Programmability into Designs

In our proposed methods, programmability in the implementations is introduced in two ways, one is for control part and the other is for datapath as shown in Fig. 12.5. Control parts of RTL designs are typically represented in FSMs (Finite State Machines) in HDLs (Hardware Description Languages) which can be compiled into synchronous sequential circuits automatically by logic synthesis tools. Here we insert a set of LUTs to the synthesized sequential circuits in the processes of high-level and logic synthesis. For the datapaths, not only LUTs but also additional or spare functional unites with multiplexers are inserted (Fig. 12.2).

For example, for the sequential circuit shown in the upper half of Fig. 12.3, a gate or a set of gates are replaced with LUTs as shown in the lower part of the figure. Here two LUTs replace a gate and a set of three gates. The selection of gates to be replaced with LUTs are based on the suggestions by designers. The inserted LUTs are connected also to the scan chain of the chip so that the LUTs can be programmed and the values of LUTs can be monitored and manipulated from the outside of the chip through the scan chain of the chip. For example, the LUTs in the circuit shown in Fig. 12.3 can be connected as shown in Fig. 12.4. The scan chain is shown as the dashed line in the figure. It is connecting flipflops as well as LUTs as shown in the figure. In the case of a LUT, the scan chain includes flipflops for the truth table values for the LUT.

In general, the data transfers in the datapath in a RTL design are controlled by FSM (Finite State Machine). Figure 12.5 is an example of a data flow graph and its RTL implementation with the control part and datapath. Data flow graphs, such as the ones in Figs. 12.5 and 12.6 can be obtained from high-level design in C. By setting the values of control signals, p, q, r, s, t, and u as shown in the figure, the target computation shown in the data flow graph can be computed with the datapath of the

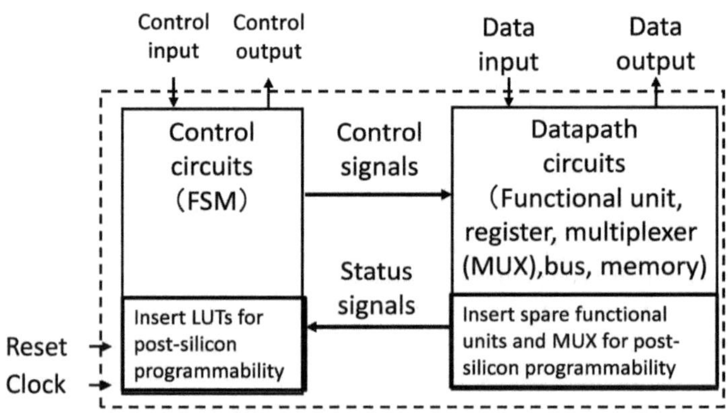

Fig. 12.2 Partial programmability for control part and datapath in RTL designs

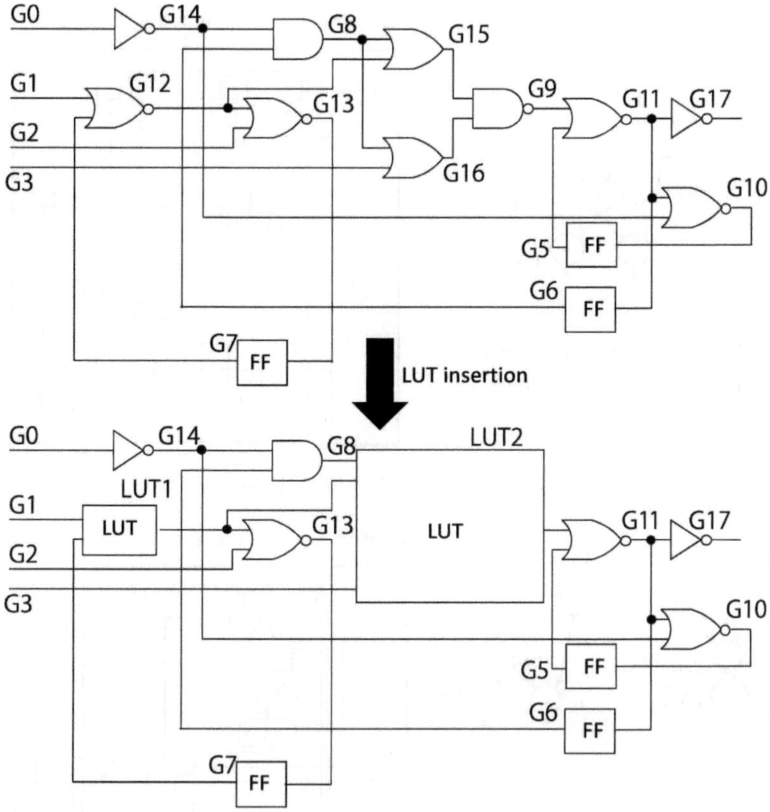

Fig. 12.3 Gates are replaced with LUTs for partial programmability in post-silicon

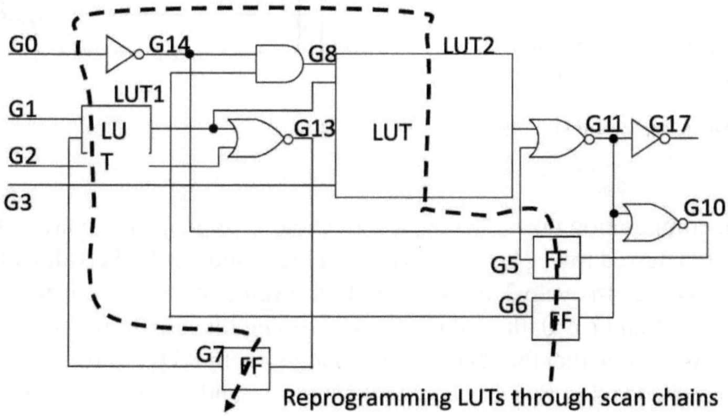

Fig. 12.4 Inserted LUTs are connected to the scan chain of the chip for controllability and observability

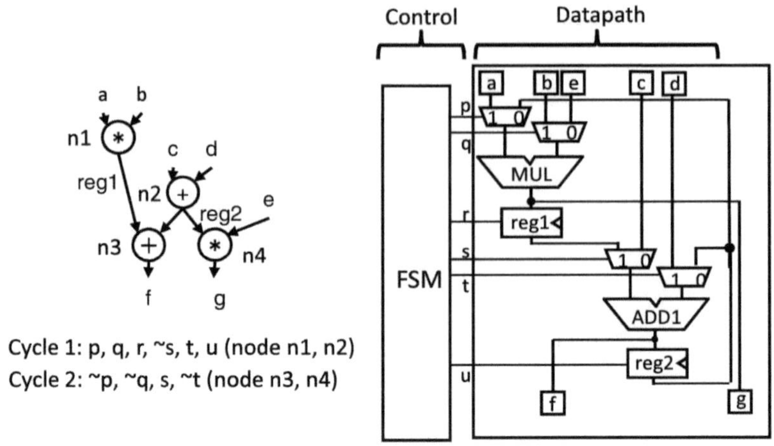

Fig. 12.5 An example of data flow graph and its RTL implementation

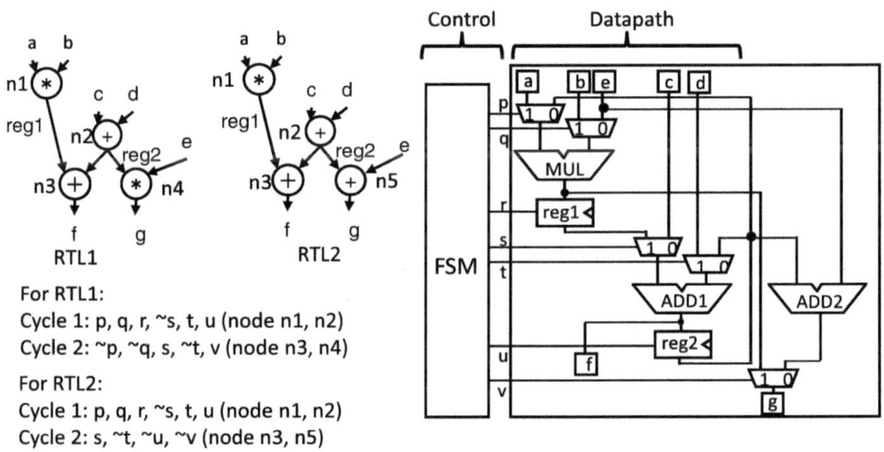

Fig. 12.6 An example of RTL changes

figure. The computation takes two cycles. In cycle 1, as p, q, and r are 1, the value of $a * b$ is transferred to $reg1$. Also, as t and u are 1 and s is 0, the value of $c + d$ is transferred to $reg2$. In cycle 2, as p, q, are 0, the value of $reg2 + e$ is transferred to g. Also, as s is 1 and t is 0, the value of $reg1 and reg2$ is transferred to f.

Now, let us assume that the RTL design changes from RTL1 to RTL2 as shown in Fig. 12.6. The node n4 in RTL1 becomes the node n5 and computes addition instead of multiplication. Then the datapath shown in Fig. 12.5 cannot perform the modified computation in two cycles. However, the datapath shown in Fig. 12.6 can compute both RTL1 and RTL2 in two cycles as shown in Fig. 12.6. This datapath is somehow redundant if the computation is only the one shown in Fig. 12.5, but that redundancy can be utilized for the computation of RTL2 as shown in Fig. 12.6.

Our design methodology is to have some redundancy in datapaths and a set of LUTs in control part so that some modified RTL designs can also be incorporated keeping small numbers of execution cycles.

12.4 Post-silicon Analysis Using Programmability in Implementation Designs

Here, we present methods for efficient and effective post-silicon analysis, verification and debugging which utilize the programmability inserted into the control parts of circuits. That is, the problem is to identify the buggy locations and understand the reason of the bugs. We call the programmability inserted into control parts of circuits "patch logic" [2]. With patch logic, various methods for post-silicon verification and debugging can be implemented in the fields. For example, we can trace all state transition information for control parts and periodically trace registers so that chip error traces can be monitored and reproduced with computer simulations. Once we successfully reproduce the error trace by some sorts of simulation, analysis and debugging processes can become much more easier. Basically we can repeat tracing and analysis/debugging processes until we finally locate the buggy portions by repeatedly changing the patch logic. Although repetition of tracing and analysis is essentially the same as other post-silicon supporting methods, for each tracing, the patch logic may change the circuit functionality and is utilized to monitor different registers and memory elements for more precise and effective analysis.

With the patch logic, we can also compare the results of duplicated computations dynamically in order to see if there is any electrical error. This is a kind of assertion checking and a very effective and efficient way to detect electrical errors if the portions for duplicated computations are dynamically adjusted in the field, which can be easily implemented with the patch logic.

Using the patch logic, we can add additional tracing mechanisms in the fields which transfer internal values of datapaths as well as control states to the outside of the chip and/or to some internal memories. When to trigger tracing can also be defined as part of patch logic. Therefore, assertions can be defined and modified in the fields in such a way that appropriate sets of registers are traced in appropriate timing. This patching mechanism can be used repeatedly in the post-silicon verification and debugging processes.

For example, in the first round, registers R1 and R2 are programmed for tracing when an assertion A is satisfied. After getting the corresponding trace and performing its analysis, in the second round, registers R2 and R3 with another assertion A1 becomes the target of tracing. This process is thoughtfully continued until some sort of conclusions in the bugs is obtained.

Here, we assume that the target design to be analyzed/debugged is implemented in such a way that its controller part and datapath part can be clearly separated as shown in Fig. 12.5. In other words, we consider a target design can be represented by

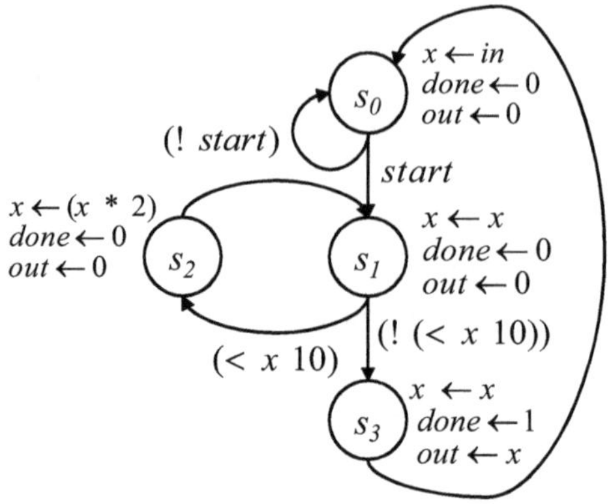

Fig. 12.7 An Example of FSMD

an FSMD (Finite State Machine with Datapath), which is a representation of RTL designs. In this paper, we use the definition of FSMD described in [3]. Figure 12.7 shows an example design represented in FSMD. In an FSMD, a state transition from a state to one of the next state is executed at every clock cycle.

In order to reproduce the chip error trace by simulation, control state information, i.e., state-transition sequences in the control FSM of accelerators is essential and so, it is better to be always traced. Basically, 1-bit data is required to store each state transition as shown in the following. Each state in control FSMs which are synthesized from high-level descriptions basically has either only one successor state or at most two. A state may have a second successor state if that has a branch as well as its immediate successor in the state transitions. So, if the initial state is known, which is mostly the case, in order to trace all state transitions, we have only to record 1 bit for each cycle which shows either state transition to the next, or state transition to the branch.

Therefore, even recording 1 million state transitions may not be so difficult in terms of capacity of trace buffers. Moreover, the patch logic can be programmed in such a way that contents of internal trace buffers can be dumped into off-chip memories, when the buffer becomes full. In the presence of electrical errors, it may be better to periodically record the current state identifier (name or number) as well as state transitions so that we can follow execution sequences more correctly, even if some electrical error happens. If a state name and the computed one from the state transition do not match, that is an indication of possible electrical errors. As a summary of discussions so far, the overall flow of our proposed post-silicon verification and debugging becomes the one shown in Fig. 12.8.

Figure 12.9 shows the number of states used in the high-level synthesized hardware for several high-level synthesis benchmark circuits. As can be seen from the figure, the numbers of states in the control parts are in the order of hundreds and most of the state transitions are just to the next states. This indicates that the state-transition information can be further compressed assuming that most of transitions are only

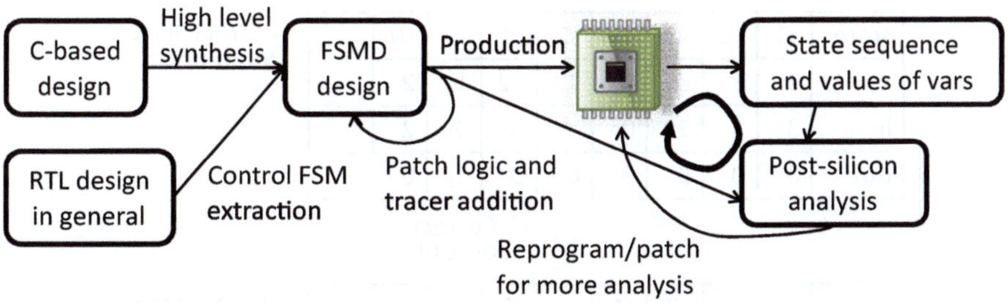

Fig. 12.8 The proposed post-silicon verification and debugging flow

Example	All states	States whose next states are unique	States whose transitions have branches
idct	72	70	2
mpeg_pred	160	149	11
bdist2	102	100	2
adpcm_decoder	80	75	5

Fig. 12.9 Numbers of states in control parts of synthesized accelerators

to their next state. By utilizing lightweight data compressions methods, the required amount of memory for tracing state-transition sequences can be dramatically reduced.

Although state-transition sequences are very useful, it does not provide information on the values of signals in the circuits. Figure 12.10 shows execution sequences based on the FSMD in Fig. 12.7. The first table shows a correct execution where the value on "in" is received by the variable "x" at the first cycle and then it is doubled twice to generate the output at the seventh cycle.

If an electrical error happens in the fourth cycle on the value of "x" and its fourth bit is flipped, the execution becomes the second table which is different from the correct one (the first table), and the output is generated at the fifth cycle with the wrong value. In this case, by tracing only sate-transition sequences, we can have some idea of what is going on in the circuit with an electrical error.

In other cases, however, state-transition sequences may not change at all even with electrical errors. The third table in the figure shows such a case. Here, an electrical error happens at the sixth cycle on the second bit of "x". Although the output value generated is wrong, there is no change in the state-transition sequence. In these cases, it is essential to also trace values of internal signals which can be easily implemented with patch logic.

One way to detect possible electrical errors is to compute some portions more than once, and compare those results. By computing the same register transfer statements (possibly with arithmetic computations) twice and comparing the results, we can detect if there is any electrical error (or some soft error) have happened or not. In order not to degrade the performance of hardware much, it is preferred to apply duplicated

state	s0	s1	s2	s1	s2	s1	s3	s0	...
x	3	3	6	6	12	12	12	2	...
in	3	2	1	0	-1	-2	-3	2	...
out	0	0	0	0	0	0	12	0	...

Correct state sequence

Correct output

state	s0	s1	s2	s1	s3	s0	s1	s2	...
x	3	3	6	14	14	-2	-2	-4	...
in	3	2	1	0	-1	-2	-3	2	...
out	0	0	0	0	14	0	0	0	...

Wrong state sequence

Strategy:

$x \leftarrow x+8$ (*bitflip*) Wrong output at wrong timing

state	s0	s1	s2	s1	s2	s1	s3	s0	...
x	3	3	6	6	12	14	14	2	...
in	3	2	1	0	-1	-2	-3	2	...
out	0	0	0	0	0	0	14	0	...

Correct state sequence

$x \leftarrow x+2$ (*bitflip*) Wrong output at correct timing

Fig. 12.10 Execution examples of the FSMD in Fig. 12.7

computations to very small portions of the whole register transfer operations. With patch logic, the portions to be duplicated can be dynamically decided based on the previous analysis in the post-silicon verification and debugging cycles shown in Fig. 12.8, which is very essential for efficiency. We have implemented the duplicated computation based error detection with patch logic for a benchmark circuit, 8×8 IDCT. Here, we show two example cases of duplicated executions.

The first one is just to duplicate one register transfer statement as shown below:

Original statement: x8 = 565 * (x4 + x5);
After duplication: x8 = 565 * (x4 + x5); x8_ = 565 * (x4 + x5); check(x8 == x8_);

For this, we only need small modifications in the control part of the target hardware, which can be arranged by changing the functionalities of the LUTs inserted into the control part.

In the second case, a computing sequence for a memory access is computed twice as shown below:

$$Memorystoring(blk[8*i])sequence:$$
$$x1 = blk[8*i+4] << 11; x2 = blk[8*i+6]; x3 = blk[8*i+2];$$
$$x4 = blk[8*i+1]; x5 = blk[8*i+7];$$
$$x6 = blk[8*i+5]; x7 = blk[8*i+3];$$
$$x0 = (blk[8*i+0] << 11) + 128;$$
$$x8 = W7*(x4+x5); x4 = x8 + (W1 - W7)*x4;$$
$$x8 = W3*(x6+x7); x6 = x8 - (W3 - W5)*x6;$$
$$x8 = x0 + x1; x1 = W6*(x3+x2);$$
$$x3 = x1 + (W2 - W6)*x3; x1 = x4 + x6; x7 = x8 + x3;$$
$$blk[8*i] = (x7 + x1) >> 8;$$

All of the above are duplicated. In order to realize the duplication, we need to use 20 more states in the control part of the hardware. Basically unused states, which are unreachable states in the original design, are used by changing the functionalities of LUTs in the control part.

As can be seen from the above, the amount of changes required in patch logic is not so large for partial duplication. By repeatedly applying partial duplicated computations in the proposed flow shown in Fig. 12.8, efficient localization of electrical errors can be expected.

12.5 Correction of Implementation Designs with Programmability

This section presents automatic correction methods for logical bugs, and the correction is based on changing the functionalities of the LUTs in the circuit. As discussed above, some amount of programmability is inserted into the implemented and fabricated chips. Here, we are using such programmability for post-silicon debugging. The debugging method is based on templates which are generated from the normal designs by making some portions vacant.

The debugging and correction problem can be formulated as Quantified Boolean Formula (QBF) problem as follows. The combinational part of the control circuit with partial programmability can be shown as Fig. 12.11. Here, In is the set of primary inputs and Out is the output of the combinational part of the control circuit. Although there are multiple outputs, here for simplicity, we assume the single output circuit. By interpreting the function as the vector of functions, it is straightforward to be extended for multiple outputs.

Most portions of the circuits are completely fixed as they are already in silicon. There are, however, a small number of LUTs which are inserted in the design process and actually implemented in silicon. In Fig. 12.11, there are m of such LUTs. The behavior of k-input LUT can be completely described in truth table

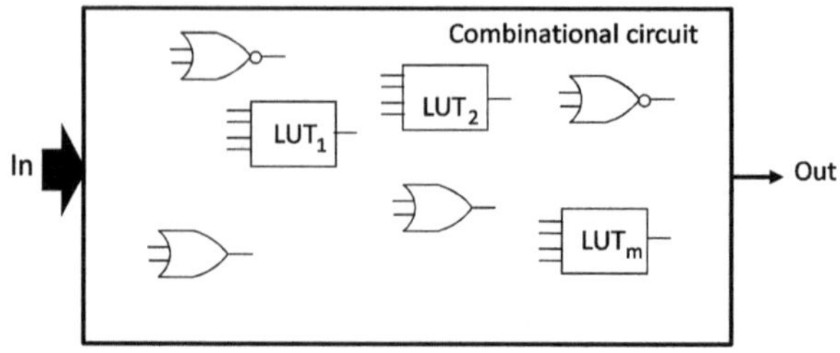

Fig. 12.11 Target circuit to be debugged

using 2^k rows. In our formulation, $x_1, x_2, \ldots, x_{2^k}$ represent the values of the truth tables. That is, by assigning values to them, the k-input logic function is completely defined. As there are in total m LUTs, there are m of such set of variables $x_1^1, x_2^1, \ldots, x_{2k}^1, x_1^2, x_2^2, \ldots, x_{2k}^2, \ldots, x_1^m, x_2^m, \ldots, x_{2k}^m$. Now, the debugging or correction problem is to make the entire circuit in Fig. 12.11 to be equivalent to its specification by assigning values to $x_1^1, x_2^1, \ldots, x_{2k}^1, x_1^2, x_2^2, \ldots, x_{2k}^2, \ldots, x_1^m, x_2^m, \ldots, x_{2k}^m$.

Let $Spec(In)$ be the given specification for the circuit. Also, let $Circuit(In, X)$ be the function realized by the combinational circuit where X is a vector of variables, $x_1^1, x_2^1, \ldots, x_{2k}^1, x_1^2, x_2^2, \ldots, x_{2k}^2, \ldots, x_1^m, x_2^m, \ldots, x_{2k}^m$ Then, the problem to be solved is $\exists X.\forall In.Spec(In) = Circuit(In, X)$.

This is a QBF problem but can be solved by repeatedly applying SAT solvers [4]. In order to solve the problem with SAT solvers, the variables under universal quantifiers are given constant values. Once constant values or a set of constant values are given to the universally quantified variables, the condition becomes just a necessary condition not a sufficient condition. By solving such a necessary condition with SAT solvers, a solution candidate can be generated. If the SAT problem becomes UNSAT, that means there is no way to correct the circuit by changing the functionalities of LUTs, and so we have to give up the correction and must redesign the circuit.

Once a solution candidate is generated, it can be formally checked to equivalent to the specification with SAT solvers. Please note that equivalence checking of two combinational circuits can be formulated as a SAT problem by connecting the outputs with an exclusive-OR gate. If they are equivalent, we have found a solution and finished. If they are not equivalent, a counterexample under which the two circuits behave differently is generated. Then the generated counter example becomes the additional constraints. The universally quantified variables are assigned according to the counter example, and the circuit must be correct under that inputs. This process is continued until we find a correct solution, or we prove the nonexistence of solutions.

The basic ideas discussed above on the solving process is illustrated in Fig. 12.12. It is an iterative process and in each iteration, a new constraint which says the output must be correct for one particular input is added. In the figure, for an iteration, there are three input–output value constrains, i.e., the cases where the inputs are (0, 1, 0), (0, 0, 0), and (1, 1, 0). Based on these constraints, SAT solvers check if there is any satisfying assignments of X variables. If there is no solution, i.e., UNSAT, there is no way to make the circuit equivalent to the specification, and the process finishes without success. If there is a solution, then the circuit is programmed

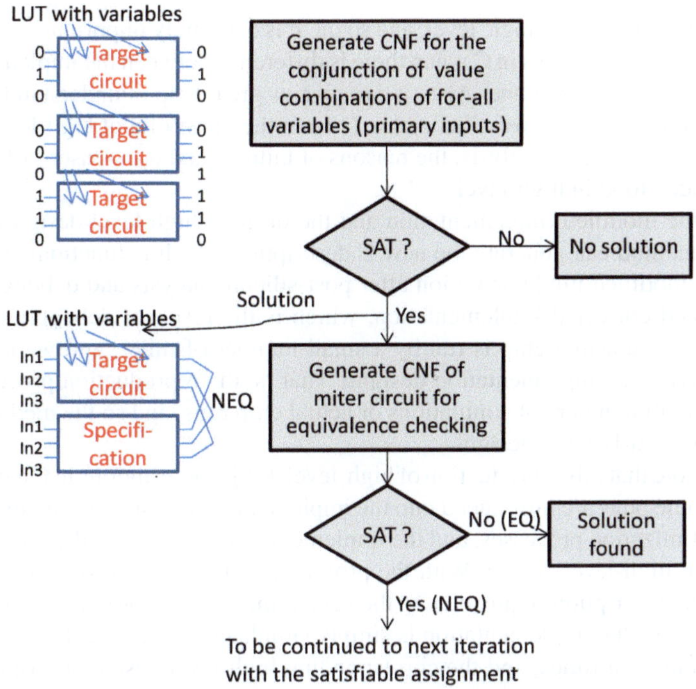

Fig. 12.12 Solving QBF problem with repeated applications of SAT solvers

according to the solution, and it is checked to be equivalent to the specification. The equivalence checking is formulated as the search problem of counterexamples for the equivalence. If there is no counterexample, the current circuit is the solution. If there is a counterexample, it is added as a new constraint and the process continues.

The correction method discussed above is scalable to circuits having hundreds of thousands of gates as long as the numbers of LUTs are not large compared to the total numbers of gates in the circuit [4]. For all the experiments we have conducted so far, the numbers of iterations have always remained small, e.g., just hundreds even if there are more than 500 primary inputs in the circuits.

12.6 Automatically Adjusting High-Level Design When Post-silicon Bugs Happen

The previous sections discuss the methods for post-silicon verification, analysis, and debugging, focusing on programmability of the implemented circuits. In this section, a method for reproducing the C description from the implementation is discussed and presented. In order to keep the consistency of the designs in various abstraction levels such as high-level, logic level, and so on, it is extremely important to adjust the original high-level designs in C when there is difference between the implementation design and the high-level one. Also, it can give us great help in understanding what is actually happening in post-silicon clearly from the viewpoint of high-level design descriptions. With such analysis, the reasons of failures and root reasons of the bugs may be understood in high-level.

Given the modified implementation and the original high-level design in C, the proposed method can generate the new C description which is functionally equivalent to the modified implementation after post-silicon analysis and debugging. The modified and corrected implementation, which is the result of debugging, is only simulated or the actual chip is run by a small number of times, and we never formally analyze the implementation designs. That is, the reproduction processes are based on small numbers of simulations or actual chip runs, and so the method could be applied to fairly large designs.

Please note that this reproduction of high-level design descriptions in C can be very useful if some bugs are introduced into the implementation design during the synthesis and optimization processes, and the implementation is not logically equivalent to the original high-level design. With the proposed method, the corresponding high-level design descriptions equivalent to the actual implementation can be obtained. In such situations, the implementation is simply simulated or the actual chip is run by a small number of times, and the corresponding high-level design description in C can be obtained.

The proposed methods are based on template-based synthesis, [5] which has been utilized in partial synthesis of hardware and software [1, 4, 6, 7]. In our method, a template-based synthesis method is used to generate the new high-level description in C only by repeatedly simulating implementation designs or actually running the chips repeatedly.

The overall flow of the template-based method for reproducing the C description from the (modified) implementation is shown in Fig.12.13. Given a high-level design in C (1) and its modified implementation (2), the goal is to generate a revised high-level design in C (4) which is logically and behaviorally equivalent to (2). Our method first generates a template-based design (3) by making a set of small portions in (1) vacant. This process is an interactive one, and users/designers specify where to be made vacant and which kind of parametrized statements should be used in those vacant portions. The details of how to generate the template, in other words, where to be made vacant and which parametrized statements are supposed to be there, are discussed in the next subsection. The problem to determine exactly how the vacant portions are filled is formulated as Quantified Boolean Formula (QBF) problem, and Counter Example-Guided Inductive Synthesis (CEGIS) based method is applied to reconstruct the missing portions in the template [4].

In the CEGIS process, the modified implementation (2) is not at all formally analyzed. Instead it is only simulated by a number of times. If a solution is found, we have successfully refined the template so as to be equivalent to (2). If there is no solution found through the CEGIS process, the template used cannot represent the behavior of (2), and other templates are to be generated and analyzed.

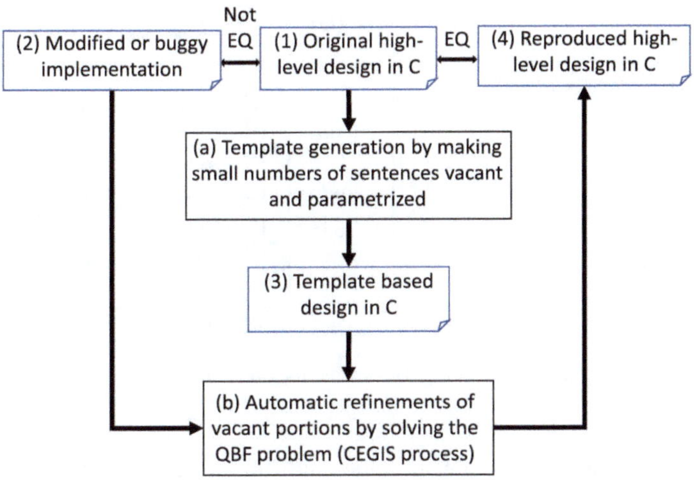

Fig. 12.13 The proposed flow

12.6.1 Template Definition

A template here is a description in C where small numbers of sentences are intensionally made vacant from a complete program. Those vacant portions can be replaced with parametrized statements having symbolic variables or expressions. The symbolic statements may contain constants, program variables and a set of arithmetic or logical operators. Please note that, the designers/users are supposed to be able to understand the implementation for coming up with the appropriate operators and variables in the template. Otherwise, the template may not cover the target behaviors. If there is no way to refine the template to be equivalent to the target behaviors, template generation and refinement processes are repeated. The actual reasoning is based on repeated applications of SAT solvers.

In order to illustrate the proposed method, we give a simple example as shown in Fig. 12.14a. Here, (a) is the original C code computing the smallest power of 2

```
uint32_t largest_power(uint32_t x){
  uint32_t num = 0xffffffff;
  while(num!=0){
    if(num<x) break;
    num = num>>1;
  }
  return num;
}
```

(a) Original C codes

```
uint32_t largest_power(uint32_t x){
  uint32_t num = /*** block1 (const0) ***/;
  while(num!=0){
    if(num<x) break;
    num = /*** block2(>>, |, num,const1) ***/;
  }
  return num;
}
```

(b) Template

Fig. 12.14 An original high-level design and a generated template from **a**

(a)

```
uint32_t largest_power(uint32_t x){
  uint32_t num = /*** block1(const0) ***/;
  while(num!=0){
    if(num<x) break;
    num = /*** block2 (>>,|, num,const1) ***/;
  }
  return num;
}
```

(b)

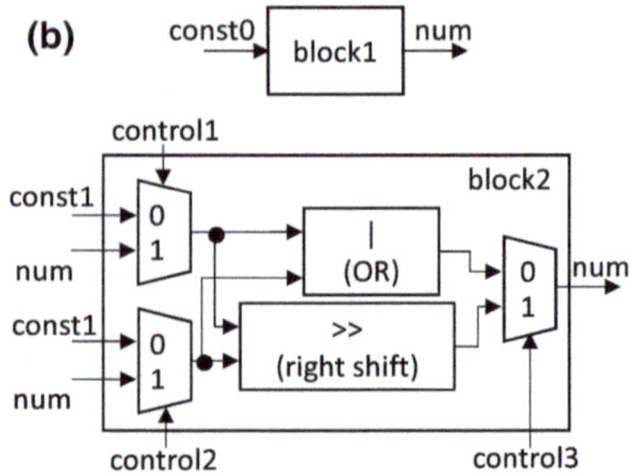

Fig. 12.15 Template generated from Fig. 12.14a and its schematic

which is larger than a given integer value, and (b) is a template-based description which is generated from (a) by making some vacant portions. In this template, the initial value of num is replaced by block1 with const0 which represents a symbolic constant, and the computation of the next value for num is replaced by block2 with symbolic expression including operators >> and |, a variable num and a constant const1. Block2 consists of more than one multiplexers and various operators, whose connections are controlled by control signals of the multiplexers. These are basically parametrized circuits.

For example, block1 in Fig. 12.14b (also shown in Fig. 12.15) is a simple circuit whose input is const0 and output is num. In block2, there are two multiplexers at its input and one multiplexer at its output as shown in Fig. 12.15. The multiplexers in the input part select incoming signals from const1 and num, and the selected ones are the actual inputs to the two functional units which compute OR and sight shift operations as shown in the figure. The multiplexer in the output selects which output of the two functional units are actually connected to the output of the parametrized circuit. The values for the control signals of the multiplexers and constants are the target of the synthesis for the template.

In our method, a number of behaviors can be represented by one template. Then, refinement of the template is formulated as a QBF problem and solved by the CEGIS-based algorithm which is discussed next.

12.6.2 Basic Algorithm

The flow of template-based method for reproducing C description and the method for generation of template has been presented in the last section. In this section, CEGIS-based method is presented.

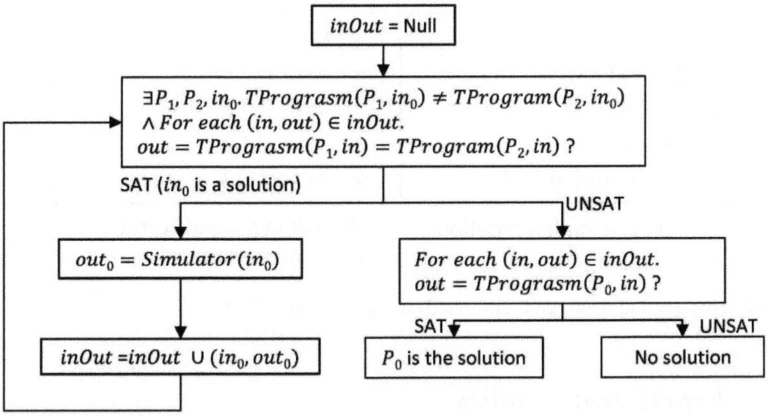

Fig. 12.16 The CEGIS flow used in the proposed method

Given a defined template with parameters and a correct simulator for implementation, how to refine the template by fiiling out the vacant portions, P, in the template can be formulated as a QBF problem as following:

$$\exists P\ \forall in.\ Tprogram(P, in) = Simulator(in) \tag{12.1}$$

Here, P represents the parameters in the template, in represents the input variables, $Tprogram$ is the output of the template, and $Simulator$ presents the results of simulation of the implementation, Please note that there is no logical specification required in this method. This QBF problem can efficiently be solved by incremental SAT solvers, as the problem can be incrementally reasoned as shown in citeProcFujita.

Figure 12.16 shows the algorithm and the flow to find a final solution. A set of input and output values are stored in $inOut$. ii and oo represent a set of input and output values. $inOut$ is initially empty. First, find two sets of parameter values $P1$ and $P2$ which satisfy the relationships in inOut, If we can find two solutions, generate another input $in0$ under which $P1$ and $P2$ generate different output values. If the entire condition is SAT, we add $in0$ and its corresponding output$out0$ to $inOut$. We repeat this process until the condition becomes UNSAT. Then, check if P_0 satisfy $TPrograsm$ for all possible inputs and outputs in $inOut$ as the following formula:

$$\exists P_0, Tprograsm(P_0, ii) = oo. \tag{12.2}$$

If it is SAT, this means there is a parameter value P_0 as the solution. Otherwise, no solution is generated, and the flow terminates with no solution. In such cases, another template may be tried.

(a) A buggy implementation	(b) Debugged and correct implementation
x = (x >> 1) | x;	x = (x >> 1) | x;
x = (x >> 2) | x;	x = (x >> 2) | x;
x = (x >> 4) & x;	x = (x >> 4) | x;
x = (x >> 8) | x;	x = (x >> 8) | x;
x = ((x >> 16) | x) >> 1;	x = ((x >> 16) | x) >> 1;

Fig. 12.17 Buggey and debugged (corrected) design for Fig. 12.14

12.6.3 Experiments Studies

We have implemented the proposed method and conducted a set of experiments. In our experiments, CBMC [8] is used as a bounder model checker for C, which is primarily used to convert the template description into its CNF formula. All the experiments ran on Linux kernel 4.6.4 on Core i7-3770(3.4 GHz) with 16GB memory. CBMC tool version is 5.4. The timeout is set to 24 hours. All variables appearing in the examples are 32-bit integers.

The first experiment is to apply an ECO to the example given in Fig. 12.14a. Here, we assume that a bug exists in the original implementation, such as one operator "|" becomes "&", due to errors of manual design or automatic synthesis as show in Fig. 12.17. Figure 12.14a is a buggy implementation and the bug is the use of "&" in line 3. It should be "|" as shown in (b). Here, we assume that after the bug is detected and eliminated by some debugging methods such as the one shown in the previous Sect. [4], a correct implementation becomes the one shown in (b).

Following the approach in the previous section, a template is generated from the original high-level design which described as a C code as shown in Fig. 12.15a. The method shown in the previous section is applied to this template, and `0x7fffffff` and `num>>1` for the two blanks are found in 23.7s with 4 iterations as shown in Fig. 12.18. Please note that the first blank gets a slightly different expression `0x7fffffff` from the original C description `0xffffffff` in Fig. 12.14a, that is also a correct implementation under the template. This can be made sure by checking the equivalence between the C descriptions of Fig. 12.14a and the refined template. In this case, CBMC can check the equivalence.

In the second experiment, four designs in C are used, and their basic functionality is to count the numbers of 1's in a given integer value. Figure 12.19a shows the original design in C and the variable x is an 32-bit integer in this program. Instead of scanning the integer value bit by bit by 32 times, one possible schematic implementation which is easier to be implemented as hardware is shown in Fig. 12.19b.

According to the high-level design and above implementations, four experiments are performed. In the first one, we assume that the bugs were in the implementation, but they have been debugged. The debugged implementation is shown in Fig. 12.19b,

```
uint32_t largest_power(uint32_t x){
  uint32_t num = 0x7fffffff;
  while(num != 0){
    if(num < x) break;
    num = num >> 1;
  }
  return num;
}
```

Fig. 12.18 Refined template for Fig. 12.14b

(a)

```
uint32_t bit_count(uint32_t x){
   uint32_t num=0;
   uint32_t mask=1, i;
   for( i=0; i<32; i++){
        if( x&mask ) num++;
        mask = mask<<1;
  }
        return num;
}
```

(b)

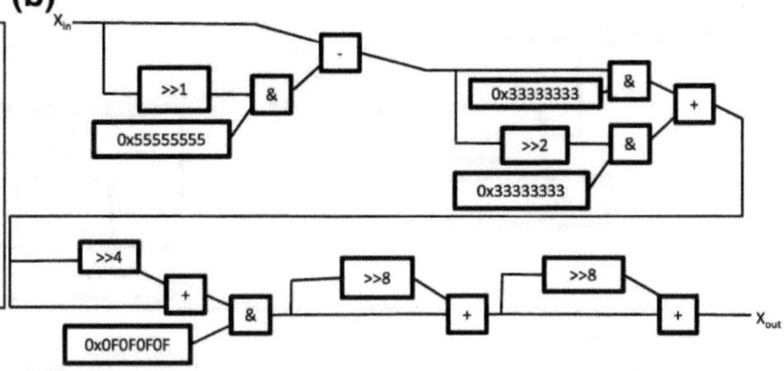

Fig. 12.19 High-level design and its possible implementation for counting numbers of 1's

which is the same situation in the previous experiment. In the second one, the correct implementation should compute the number of 0's in an integer value instead of 1's. In the third one, function that is computing the number of 1's only in the first 16 bits should be implemented. In last experiment, the implementation should compute the 1's in the first 16 bits, and the number of 0s in the second 16 bits. The corresponding modified implementations for these experiments are shown in Fig. 12.20a–c, respectively. As can be seen that an inverter is inserted in Fig. 12.19b and in Fig. 12.20c.

According the proposed method, a template is generated from the high-level design in Fig. 12.19a, which is shown in Fig. 12.21a. In order to generate the template, three potions are selected and replaced with parametrized and programmable expressions. In this template, A is only a symbolic constant. B in condition statement is a symbolic expression consisting of x, mask, and constant with three optional operators &, | and >>. C is a symbolic expression including num and a constant with four optional operators $+$, $-$, &, and |.

The experimental results are shown in Table 12.1. The columns represent the example number as explained above, the symbolic expressions, numbers of iterations, and runtime in seconds, respectively.

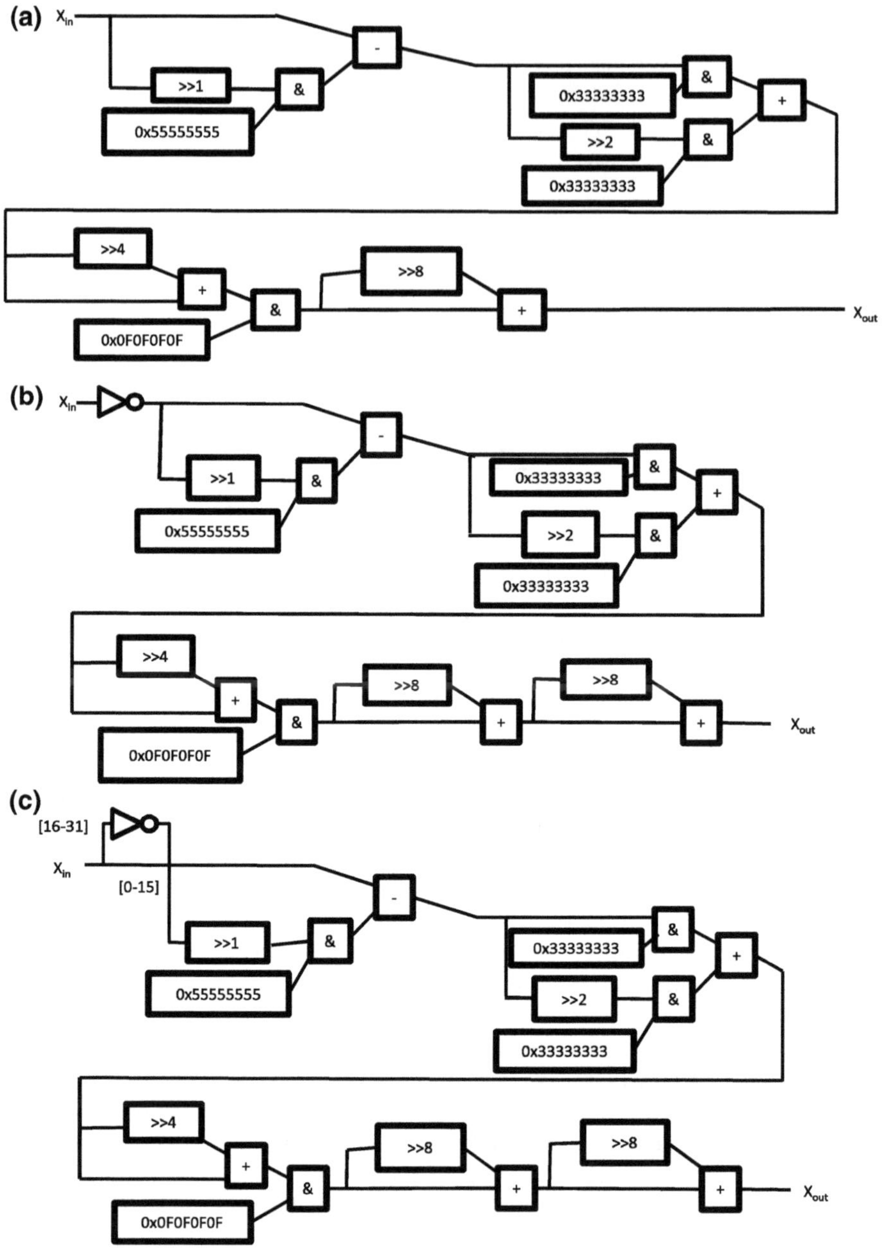

Fig. 12.20 Four implementations for different functions

(a)

```
uint32_t bit_count(uint32_t x){
   uint32_t num=0;
   uint32_t mask=1, i;
   for( i=0; i< A ; i++){
         if( B ) num++;
         mask = mask<<1;
   }
         num = C ;
         return num;
}
```

(b)

```
uint32_t bit_count(uint32_t x){
   uint32_t num=0;
   uint32_t mask=1, i;
   for(i=0; i< A1 ; i++){
         if( B1 ) num++;
         mask=mask<<1;
   }
   for(i=0; i< A2 ; i++){
         if( B2 ) num++;
         mask=mask<<1;
   }
         num= C ;
         return num;
}
```

Fig. 12.21 Two templates generated from the C description in Fig. 12.19a

Table 12.1 Experimental results in the second experiment

	A	B	C	Iterations	Time (s)
Example 1	32	x&mask	num	4	0.6
Example 2	32	x&mask	(32-num)&63	6	2.2
Example 3	16	x&mask	num	4	0.6
Example 4	N/A	N/A	N/A	0	0.2

In Example 1, the solution is A = 32, B = x&mask, C = num, and the refined template is exactly the same as the description in Fig. 12.19a. In Example 2, A = 32, B = x&mask, and C = (32-num)&63 are found in 2.2s with six iterations. Furthermore, another solution A = 32, B = !mask, C = num is generated with the same template which is also a correct solution. In Example 3, B=x&mask, C = num and A =16 are found within 1s only by four iterations. In Example 4, no solution is found by the template, which implies that the template has no way to be refined to be a consistent description with the modified implementation design. That is, we need another template for the reproduction of a correct C description.

Multiple templates can be defined and used. In this case, we redefine the template by adding more symbolic variables or expression as shown in Fig. 12.21b. Here, A1 and A2 are symbolic constant, and B1, B2 and C are symbolic expressions, which are the same definitions as A and C in template1. In this case, A1 = 16, A2 = 16, B1 = x&mask, 2 = !x&mask, and C = num are successfully found in 1.3 s with 4 iterations. It is noteworthy to say that there may be multiple solutions some of which are somehow redundant, such as num−1+1. Here, such solutions can be regarded as the same solution in our results.

In the third experiment, a typical high-level benchmark, the AES encryption algorithm is used for studying the scalability of our proposed method. The original C code has about 200 lines and over 10,000 statements must be executed when running an encryption process. After synthesizing from the high-level design, the generated implementation has around one hundred thousand of gates. We assume that bugs exist in the original implementation and can be debugged by some methods. Here, we try to reproduce the C description by our template-based method.

The template is generated by replacing the constants in some basic functions from the original C description with symbolic variables. In this case, refining the template takes around 960 min for three blanks in the function. The reason for long run time would be that CBMC must analyze and process a larger number of statements for every input and output values, thus, around 10 million CNF clauses are received by SAT solvers.

12.7 Concluding Remarks

We have discussed issues on dealing with post-silicon failures from the viewpoints of high-level designs such as C-based designs. Use of programmable circuits such as LUT (Lookup Table) in control parts and datapaths of hardware designs have been explored. The partial programmability inserted into silicon is utilized for analysis and debugging in post-silicon phases. With those partial programmability, bugs can be efficiently analyzed and corrected. We have also discussed reproduction methods of equivalent high-level design descriptions after the post-silicon debugging processes are applied. After debugging in post-silicon, the functionality of the debugged implementation is different from the original high-level design description, which must be adjusted. We have proposed an automatic method by which high-level descriptions are generated as modifications of the original high-level descriptions by simulating the debugged implementations.

Future research directions include more experimental analysis of the proposed methods with large and industrial environments, establishing template libraries for the analysis and debugging in post-silicon, and others.

References

1. A. Solar-Lezama, Program Synthesis by Sketching, Ph.D thesis, University of California, Berkeley, 2008
2. M. Fujita, H. Yoshida, Post-silicon patching for verification/debugging with high-level models and programmable logic, in *ASP-DAC* (2012)
3. D. Gajski, N. Dutt, A. Wu, S. Lin, *High-Level Synthesis: Introduction to Chip and System Design* (Kluwer Academic Publisher, Dordrecht, 1992)
4. M. Fujita, Toward unification of synthesis and verification in topologically constrained logic design. Proc. IEEE **103**(11), 2052–2060 (2015)
5. M. Fujita, Y. Kimura, Q. Wang: Template based synthesis for high performance computing, in *IFIP WG10.5 VLSI-SoC* (2017)

6. P. Subramanyan, Y. Vizel, S. Ray, S. Malik, Template-based synthesis of instruction-level abstractions for SoC verification, in *Formal Methods in Computer-Aided Design (FMCAD)* (2015)
7. J. Matai, D. Lee, A. Althoff, R. Kastner, Composable, parameterizable templates for high-level synthesis, in *Design, Automation & Test in Europe Conference & Exhibition (DATE)* (2015)
8. http://www.cprover.org/cbmc/

Chapter 13
Post-Silicon Fault Localization with Satisfiability Solvers

Georg Weissenbacher and Sharad Malik

13.1 Introduction

Owing to the high cost of recalling and replacing faulty circuits, verification, and validation have a particularly high significance in hardware design [1]. Chip manufacturers were the first to adopt automated verification techniques such as model checking in their development process (cf. our survey [2]). These verification algorithms can be readily applied to hardware designs (provided in hardware description languages such as VHDL or Verilog) or logic net lists, ensuring the correctness of the design at early "pre-silicon" development stages.

Functional correctness of the high-level hardware design, however, does not guarantee the absence of bugs in the chip prototype or the final integrated circuit. Electrical faults introduced during the manufacturing process are not reflected by the high-level model and need to be caught in the "post-silicon" stage of hardware development (Fig. 13.1). The "Rowhammer" security vulnerability is a prominent example of a bug not reflected in a high-level model. The vulnerability is a result of the increasing density of integrated circuits; the physical proximity of individual DRAM cells result in an undesired correlation between signals (so-called bridging faults). Consequently, signals in one row of the DRAM circuit influence cells in adjacent rows, allowing the attacker to draw conclusions about regions of the memory that should remain inaccessible.

The cost of post-silicon validation is high: 35% of the development cycle of a new chip being is spent on debugging hardware prototypes [3]. The fact that test scenarios can be executed at full speed (unlike in model checking or simulation of

G. Weissenbacher (✉)
TU Wien, Vienna, Austria
e-mail: georg.weissenbacher@tuwien.ac.at

S. Malik
Princeton University, Princeton, NJ, USA
e-mail: sharad@eprinceton.edu

P. Mishra and F. Farahmandi (eds.), *Post-Silicon Validation and Debug*,
https://doi.org/10.1007/978-3-319-98116-1_13

the high-level model) allows for extensive testing of the prototype and can result in extremely lengthy erroneous execution traces. Locating faults in such a trace is particularly challenging due to limited observability of signals in hardware. Only a small percentage of the state space traversed by an erroneous execution can be recorded using trace buffers and scan chains [4].

In this chapter, we discuss fault localization techniques based on satisfiability solvers [5]. Satisfiability solvers (or SAT solvers for short) are efficient tools to determine whether a given propositional logic formula is satisfiable. Modern SAT solvers can deal with formulas containing millions of variables, allowing us to analyze large hardware designs.

Section 13.2 provides an abstract view of the fault localization process, describing how a "golden model" of the circuit combined with observations recorded during the execution of a prototype of the circuit can be used to derive fault candidates. Section 13.3 describes the implementation of this localization technique using symbolic encodings and satisfiability solvers. Moreover, Sect. 13.3 discusses a number of variations and optimizations of the basic SAT-based fault localization approach.

13.2 An Abstract View of Post-Silicon Fault Localization

To locate faults in a prototype of an integrated circuit, we juxtapose a formal model of the circuit with the information logged during the execution of the actual physical chip. The formal model is assumed to be "golden", i.e., a reference model that is functionally correct and serves as a specification of the correct behavior of the chip. The circuit prototype itself may contain faults introduced during manufacturing; the goal is to locate the temporal and spatial location of the malfunction in the (execution of the) circuit. We describe this setting below in more detail.

Figure 13.1 illustrates the development process of a digital circuit. The developer creates a design using a hardware description language (HDL) and subsequently derives a register-transfer level (RTL) model of the circuit. The RTL model of the circuit contains declarations of registers and a description of the combinational logic describing the flow of signals between these registers. The combinational logic is described using familiar programming language constructs such as conditional statements and arithmetic instructions (see for instance, Fig. 13.6a). Synthesis tools then turn the RTL representation into lower-level representations and ultimately into a design implementation in terms of logic gates and wires.

In our setting, we assume that the HDL and RTL representations have been extensively tested and verified using techniques such as simulation and model checking [1]; this part of the verification process is called *pre-silicon* verification. Pre-silicon verification is hampered by the fact that simulation and model checking are computationally expensive techniques. Once a prototype of the digital circuit has been manufactured, the verification process enters its *post-silicon* phase, where test cases can be executed at-speed, allowing for longer and more elaborate test scenarios. The advantage of increased speed comes at the cost of observability— while simulation

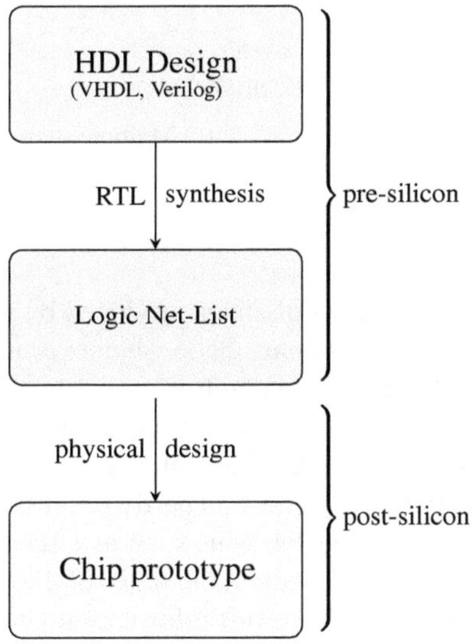

Fig. 13.1 Hardware development

allows the developer to stop and resume tests at will and thus inspect the full state of the circuit model at any time, post-silicon debug relies on logging mechanisms (such as trace buffers and scan chains [4]) that provide only limited access to the execution history.

Bugs caught at the post-silicon level are either *logic bugs* (caused by logic design errors that were missed in the pre-silicon verification phase) or *electric bugs* introduced during the manufacturing process. Electric bugs are caused by unintended behavior of elements of the physical circuit that is not reflected at the HDL and RTL abstraction levels such as crosstalk, power supply noise, thermal effects, or manufacturing process variations. Consequently, electrical bugs cannot be caught by analyzing the HDL and RTL descriptions. The techniques presented in this chapter focus on locating the cause of electrical bugs in physical circuits, i.e., the temporal and spatial locations where the behavior of the manufactured circuit deviates from its RTL specification (the "golden model").

In the following, we assume that the golden RTL model of the circuit is given in terms of a transition relation T relating states s_i to their successor states s_{i+1} (Fig. 13.2a). States represent the data stored in the latches or registers of the circuit, (i.e., a mapping of register names to values $\langle x \mapsto 1, y \mapsto 0\rangle$). The transition relation corresponds to a combinational circuit relating inputs, outputs, and states and their successor states. Each iteration of T corresponds to the execution of a single cycle of the circuit (for simplicity, we are considering only synchronous circuits with a single clock). We use s_0 to refer to the starting state of the circuit. A number of n steps of the transition relation T produces a sequence of states $s_0, \ldots, s_n$ (where $T(s_i, s_{i+1})$ is true for $0 \leq i < n$), as indicated in Fig. 13.2b. If at any point there is

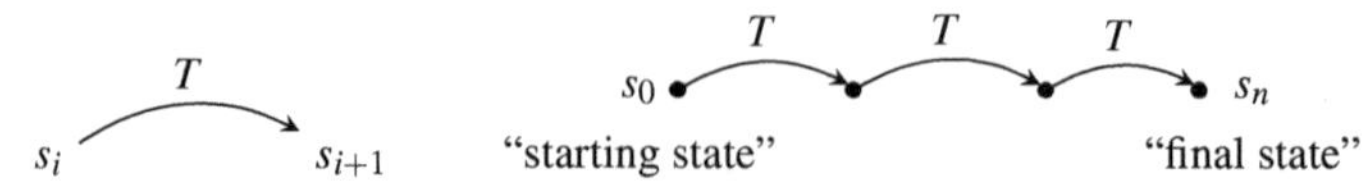

(a) Transition relation T (b) Multiple steps of T

Fig. 13.2 Steps of the transition relation T

a discrepancy between the sequence of states produced by the transition relation T of the golden model and the observations made when executing the physical circuit, then the implementation of the circuit is faulty.

Locating the exact point where such a discrepancy arises can be challenging in practice, partly because of long error detection latencies and the resulting length of the execution trace that needs to be analyzed, and partly because of limited observability of the circuit at runtime. Debug mechanisms such as trace buffers and scan chains allow us to record *partial* information about the states of the circuit during execution. Typically, we can only assume to have full information about the terminal state of the execution: scan chains and scan-based debug provides greater observability than trace buffers (which with their size typically measured in kilobytes, can only record a fraction of the content of the registers of the circuit), but require halting the system to perform a scan dump. Consequently, we are unable to draw conclusions about the exact state of the circuit at a given execution cycle; at best, we can narrow it down to the *set of states* consistent with our partial observations at this point of the execution. Note that, not all states contained in that set are necessarily visited or even reachable, since no assumptions about the values of the registers not recorded in the trace buffer can be made.

While a concrete execution trace of the transition relation T corresponds to a sequence of states $s_0, \ldots, s_n$ as described above, it is useful to generalize executions to sets of states. Rather than considering states and their successors individually, one can consider sets of states and their post-image under the transition relation T (see also Fig. 13.3b):

$$S_{i+1} \quad = \quad T(S_i) \stackrel{\text{def}}{=} \{s_{i+1} \mid T(s_i, s_{i+1}) \wedge s_i \in S_i\} \tag{13.1}$$

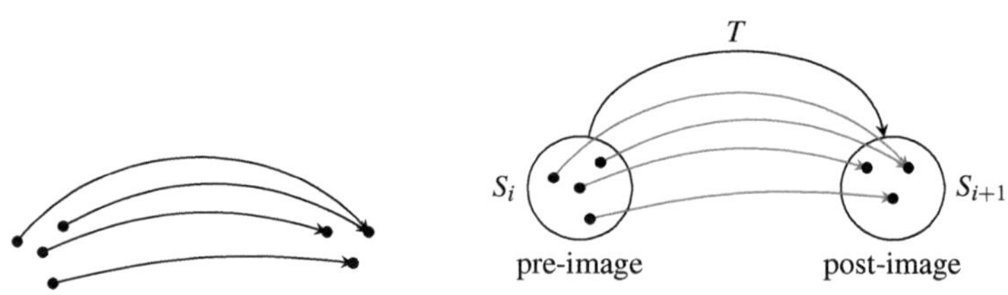

(a) Single transitions and explicit states (b) Binary relation T over sets of states

Fig. 13.3 Explicit states versus image computation

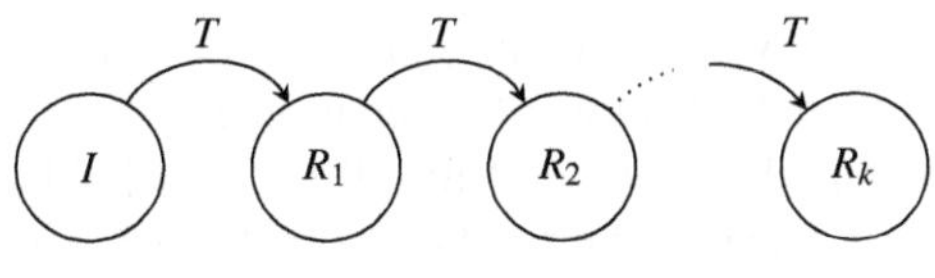

Fig. 13.4 A k-step exploration of reachable sets of states

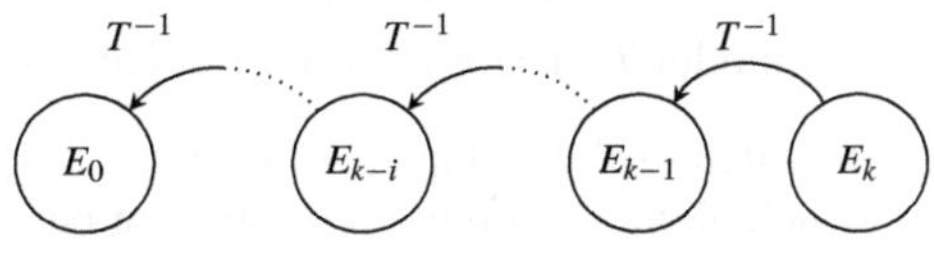

Fig. 13.5 A k-step backwards exploration starting from states E_k

The post-image of S_i under T is illustrated in Fig. 13.3.[1] Similarly, we can define the pre-image of S_{i+1} with respect to T as

$$S_i \quad = \quad T^{-1}(S_{i+1}) \stackrel{\text{def}}{=} \{s_i \mid T(s_i, s_{i+1}) \wedge s_{i+1} \in S_{i+1}\}\,. \tag{13.2}$$

In this setting, a sequence of computations of post-images of T starting from an initial set of states I comprises a set of concrete executions. The set representation allows us to consider a set of starting states I rather than a single state s_0, which is useful if the values of some registers are undetermined on power up in the RTL model. Figure 13.4 illustrates the exploration of the state space up to a depth of k execution steps: each set R_i represents the states reachable after exactly i steps or the transition relation T.

Conversely, we can also reason in a *backwards* manner about the execution of the transition relation T by computing the pre-image (given in Eq. (13.2)) starting from a terminal state. When considering backwards executions, sets of states arise naturally since a given state might s_{i+1} have more than one potential predecessor state s_i. A given value of x in target state s_{i+1}, for instance, does not uniquely determine the values of x and y in the corresponding starting state s_i of an execution of the assignment statement `x <= x && y`.

While it is not physically possible to execute the physical circuit in a backwards direction, a backwards analysis of the transition relation T starting from a terminal error state is useful in practice, since the cause of the error is often located closer to the end of the execution trace. A forwards analysis starting from a set of initial states I (as in Fig. 13.4) yields the sets R_i of states reachable in i steps. Conversely, in Fig. 13.5, E_k represents a set of terminal states, and E_{k-i} the states from which E_k is reachable within i steps of T.

Now, consider a k-cycle execution of the physical circuit under test resulting in an error state s_k observed a the end of the execution. Since s_k is an erroneous state, it cannot be contained in the set R_k of states reachable after k steps of the supposedly error-free golden model T. Similarly, a k-step backwards analysis of T starting from $E_k \stackrel{\text{def}}{=} \{s_k\}$ will yield a set E_0 which does not overlap with the set of initial states I, since otherwise s_k would be a valid reachable state in the golden

[1]Note that S_i and S_{i+1} are not necessarily disjoint, as the domain and co-domain of T coincide.

model. More generally, given a forwards execution $R_0, R_1, \ldots, R_k$ (with $I \stackrel{\text{def}}{=} R_0$) of T and a backwards execution $E_0, \ldots, E_k$ of T where E_k is a set of error states, we have $R_i \cap E_i = \emptyset$ for all $i \in [0, k]$. Clearly, none of the states in E_i (and E_k in particular) can be reached from I via T. Consequently, since s_k was observed when executing the manufactured prototype, the circuit's behavior must deviate from the golden model T. The goal of fault localization is to determine

1. in which of the k steps (or cycles),
2. which element of the circuit did not behave in accordance with T.

We refer to the first aspect of the problem as *temporal* localization and to the second aspect as *spatial* localization of the fault.

Faults can be permanent (i.e., occur in every cycle), or transient or intermittent (occurring only in some cycles). In order to locate the cycle in which a single fault is triggered, we need to determine at which point the circuit behavior deviates from T, or, in other words, at which cycle i the transition from the "correct" states R_{i-1} to the "erroneous" states E_i occurs. The observations stored in the trace buffers facilitate the localization. If we had perfect information about the execution of the physical chip (i.e., if all states were fully observable), temporal localization would amount do simply determining the (first) location i in the sequence of observed states $s_0, \ldots, s_k$ for which $T(s_{i-1}, s_i)$ evaluates to false. In general, however, our observations are incomplete, i.e., after any cycle i we only know that the circuit was in one of a potential set of states S_i. The smaller the information recorded in the trace buffers, the larger this set becomes. Let $S_0, \ldots, S_k$ be the sets of states derived from the observations logged during the execution of the manufactured circuit. If we assume that the erroneous terminal state is the result of the fault being triggered in a single specific cycle (an assumption known as single-fault hypothesis), then all execution steps prior to that cycle i were error-free (consistent with T). By combining our observations $S_0, \ldots S_{i-1}$ and the specification T, we obtain a sequence of sets of states $C_0, \ldots, C_{i-1}$ defined inductively as

$$C_0 \stackrel{\text{def}}{=} I \cap S_0 \quad \text{and} \quad C_j \stackrel{\text{def}}{=} T(C_{j-1}) \cap S_j \tag{13.3}$$

(with $C_j \subseteq R_j$ for $0 \leq j < i$). Similarly, all execution steps after the ith cycle are consistent with T as well, but propagate the error until the terminal error state $s_k \in E_k$ is reached. We obtain a sequence of "bad" states $B_i, \ldots, B_k$ (for $i < j \leq k$)

$$B_k \stackrel{\text{def}}{=} E_k \cap S_k \quad \text{and} \quad B_{j-1} \stackrel{\text{def}}{=} T^{-1}(B_j) \cap S_{j-1} \tag{13.4}$$

(with $B_j \subseteq E_j$ for $i \leq j \leq k$). The fault triggered in the ith step of the transition relation T allows a transition from C_{i-1} to B_i. In practice, i is unknown, but the insights above allow us to narrow down the temporal location of the fault: if C_i is empty (i.e., the reachable states do not intersect with our observations), then the fault must have been triggered in cycle i or before. If B_i is empty at a given cycle i (i.e., our observations rule out the existence of a bad state at cycle i), then the fault has not been triggered yet. The more precise our observations $S_0, \ldots, S_k$ are, the easier it is to narrow down the exact cycle in which the fault was triggered.

To perform *spatial* localization, we need to define what exactly we mean by "element of the circuit" and in which way such an element can malfunction (i.e., deviate from its specified behavior in T). The appropriate granularity for localizing electric faults is at the level of single gates. Manufacturing defects of single gates (or wires) in integrated circuits can then be represented using fault models, i.e., a formal model of how a component may fail. A single stuck-at fault, for instance, is a simple and widely-used gate level fault model representing the situation that a single input or output of a gate is stuck at a constant logical value (0 or 1). Let T_f be the transition relation obtained by injecting a fault f (affecting a single gate) into the golden model T. If a state in B_i can be reached from C_{i-1} via T_f, then the fault f triggered in cycle i is a *fault candidate* consistent with our observations that explain the observed erroneous behavior of the manufactured circuit. Several fault models and numerous faults are viable and needs to be considered. Consequently, the approach yields a (potentially large) range of fault candidates. As before, the accuracy of the spatial localization is contingent on the quantity and quality of the observations made during the execution of the physical circuit.

In practice, the effectiveness of the approach described above is hindered by the large number of potential fault candidates, the length of execution traces in post-silicon validation, as well as limited observability of signals. The following section describes how these problems can be mitigated using symbolic encodings and satisfiability solvers.

13.3 SAT-Based Fault Localization in Practice

The success of set-based reasoning as described above hinges on a compact representation of sets of states as well as on efficient decision procedures enabling us to reason about these representation. In the following, we describe symbolic encodings of transition relations, a consistency-based approach to diagnose faults, as well as the use of satisfiablity solvers to perform this diagnosis.

13.3.1 Symbolic Encodings and Execution

Symbolic model checking [6, 7], an approach that marked a major breakthrough in the scalability of hardware model checking algorithms, uses Binary Decision Diagrams (BDDs) [8] (a graph-based data structure to represent propositional formulas) to encode sets of states. Similarly, first-order predicates provide a compact symbolic representation of sets of program states [9] frequently used in model checking of software and word-level representations of circuits (e.g., [10, 11]). A predicate P over the variables (or registers) V of a design encodes all states in which P evaluates to true. For instance,

$$(x \neq 0) \;\; \text{represents} \;\; \{s \mid s(x) \neq 0\}\,,$$

i.e., the set of states s in which x is not equal to zero and all other registers have arbitrary values. Thus, a single predicate can encode a large set of states.

Transition relations T are encoded as relations over the variables V_i and a set of variables $V_{i+1} \stackrel{\text{def}}{=} \{v_{i+1} \mid v_i \in V_i\}$ representing successor states. For instance, for a simple instruction with a single register $V \stackrel{\text{def}}{=} \{x\}$ and an input d, the statement `x <= d` is encoded as follows:

$$\underbrace{(x_{(i+1)} = d_i)}_{\texttt{x <= d}} \text{ represents } \{\langle s_i, s_{(i+1)}\rangle \mid s_{(i+1)}(x) = s_i(d)\}$$

In case, V is not a singleton set, the constraint $\bigwedge_{y \in (V \setminus x)} (y_{i+1} = y_i)$ overall remaining registers is added to ensure that the values of the variables unaffected by the statement `x <= d` remains unchanged.

Figure 13.6b shows the symbolic encoding of the simple code fragment in Fig. 13.6a, which can be readily obtained from the source code. Each implication in Fig. 13.6b encodes a branch of the conditional statement in Fig. 13.6a; the premise determines which branch is taken. Assignment statements update the state of the register. The initial states can be encoded as a simple predicate over V.

Since we aim at performing localization at the gate level, we require an RTL representation of the circuit. A gate level representation can be obtained via RTL synthesis as in Fig. 13.1, resulting in a Boolean circuit as in Fig. 13.7.

Simple sequential circuits as in Fig. 13.7 (in which the next state Q is determined by the current state Q and the current input signals set and d) are encoded in a similar manner. The propositional variables V represent latches as well as input (and output) signals of the circuit

Fig. 13.6 A code fragment and its symbolic encoding

```
reg x;
always@ (posedge clk)
begin
  if (set)
    x <= d;
end
```

(a) Verilog code

$$(set_i) \Rightarrow \quad (x_{i+1} = d_i)$$
$$(\neg set_i) \Rightarrow \quad (x_{i+1} = x_i))$$

(b) Symbolic encoding

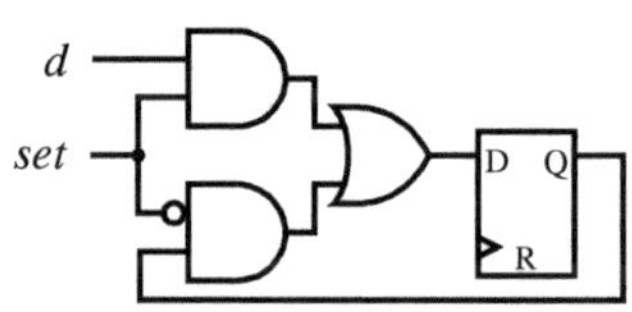

Fig. 13.7 A simple sequential circuit

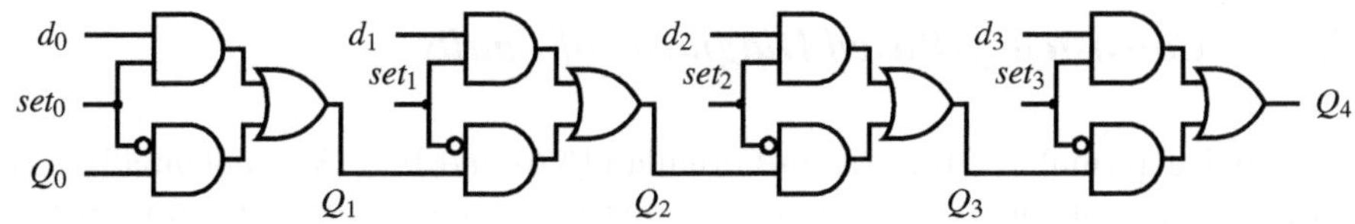

Fig. 13.8 Four-cycle unfolding of circuit in Fig. 13.7

$$Q_{i+1} \Leftrightarrow ((\neg set_i \wedge Q_i) \vee (set_i \wedge d_i)) \tag{13.5}$$

Note that, the counterparts d_{i+1} and set_{i+1} of the input signals are unconstrained, since the inputs of the next cycle are determined externally rather than by the circuit.

As before, a transition $T(V_i, V_{i+1})$ represents the execution of a single cycle of the circuit. An execution of k cycles can be encoded by means of k copies of the transition relation T over $k+1$ versions of the variables V

$$I(V_0) \wedge \left(\bigwedge_{i=1}^{k} T(V_{i-1}, V_i) \right) \tag{13.6}$$

The resulting Formula (13.6) represents all executions of length k (as illustrated in Fig. 13.4). A corresponding "unfolding" of the sequential circuit in Fig. 13.7 is shown in Fig. 13.8. This so-called *iterative logic array* (ILA) [4] is a representation of four execution cycles of the sequential circuit in Fig. 13.7 as a combinatorial circuit. We use ILAs and their corresponding formulas interchangeably in this chapter.

Contemporary decision procedures for propositional logic (discussed in our survey [2] or [5, 12], for instance) can determine a satisfying assignment of Formula (13.6), i.e., an assignment to the variables $\bigcup_{0 \le i \le k} V_i$ which makes Formula (13.6) true. Each such assignment directly corresponds to a *correct* execution $s_0, \ldots, s_k$, effectively allowing us to "execute" the circuit symbolically.[2]

Example 1 Consider an execution in which the flip-flop is initialized to 0, kept at that value for two cycles, and is set to 1 in the third cycle and kept at 1 in the last cycle. The corresponding satisfying assignment for the iterative logic array in Fig. 13.8 is

$$\begin{array}{ll} \underbrace{Q_0 \mapsto 0, d_0 \mapsto 0, set_0 \mapsto 0,}_{\text{cycle 1}} & \underbrace{Q_1 \mapsto 0, d_1 \mapsto 1, set_1 \mapsto 0,}_{\text{cycle 2}} \\ \underbrace{Q_2 \mapsto 0, d_2 \mapsto 1, set_2 \mapsto 1,}_{\text{cycle 3}} & \underbrace{Q_3 \mapsto 1, d_3 \mapsto 1, set_3 \mapsto 0, Q_4 \mapsto 1}_{\text{cycle 4}}\ . \end{array} \tag{13.7}$$

(Note that the values of d_0, d_1, and d_3 are irrelevant.)

[2]The approach of unwinding transition relations into a propositional formula which is then passed to a SAT solver was popularized by the success of Bounded Model Checking (BMC) [13].

13.3.2 Consistency-Based Diagnosis of Faults

An unfolded formula or ILA (as in Formula (13.6) and Fig. 13.8) potentially has a large number of satisfying assignments and is a compact representation of the reachable sets of states in Fig. 13.4. As explained in Sect. 13.2, the observations obtained from the execution of an erroneous circuit do not correspond to a correct execution of the golden model and can therefore not be consistent with any satisfying assignment of our unfolding.

Example 2 Consider an execution of a faulty manufactured prototype of the circuit in Fig. 13.7 in which the lower AND gate is "stuck-at-0", and assume that the inputs are the same as in Example 1. After four execution cycles, the flip-flop stores the value 0 despite the fact that it was set to 1 in cycle 3 (cf. Eq. (13.7)). The observation

$$d_2 \mapsto 1,\ set_2 \mapsto 1,\ set_3 \mapsto 0,\ Q_3 \mapsto 1,\ Q_4 \mapsto 0 \tag{13.8}$$

is not consistent with any satisfying assignment of the ILA in Fig. 13.8, i.e., there is no reachable state of the circuit in Fig. 13.7 with the inputs as above such that $Q_4 \mapsto 0$.

The general situation is depicted in Fig. 13.9, where s_k is an erroneous state observed when executing the manufactured prototype, and T is the transition relation of the golden model. The observations are *inconsistent* with the golden model. By performing an analysis of the golden model in combination with the observations, we can determine the potential causes of this inconsistency; this approach is known as consistency-based diagnosis [14] and demonstrated in the following example.

Example 3 We continue working in the setting of Examples 1 and 2 and assume that the manufactured circuit contains a single faulty gate.

In the first step, we narrow down the cycles which contribute to the inconsistency: note that the observations in Eq. (13.8) are restricted to cycles 3 and 4. Since the value of the flip-flop is overwritten in cycle 3 by the value of d_2 (1), the prior execution history has no impact on subsequent cycles. Indeed, dropping the first two cycles from Fig. 13.8 still leaves us with a (sub-)ILA that is inconsistent with the inputs d_2, set_2, and set_3 and the observation $Q_4 \mapsto 0$ in Eq. (13.8). In fact, using the observation $Q_3 \mapsto 1$ we can further narrow down the fault to the final cycle, since the following formula (derived from Formula (13.5) and our observations) has no satisfying assignment:

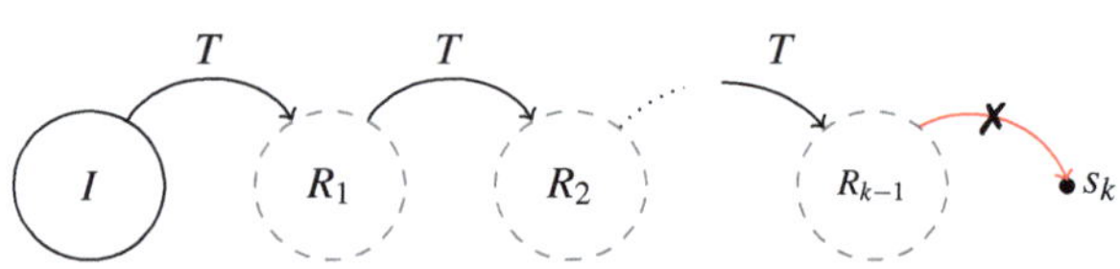

Fig. 13.9 Discrepancy between model and reality

$$\underbrace{Q_4 \Leftrightarrow ((\neg set_3 \wedge Q_3) \vee (set_3 \wedge d_3))}_{T(V_3,V_4)} \quad \wedge \quad \underbrace{(\neg set_3 \wedge Q_3 \wedge \neg Q_4)}_{\text{observations}} \tag{13.9}$$

In the following, we perform a case analysis of potential faults. Assume that the fault is caused by the upper AND gate in the final cycle. Since we assume a single-fault hypothesis, all other gates must function correctly. Consequently, the output of the lower AND gate in cycle 4 must be 1, since set_3 is 0 and Q_3 is 1. Since the OR gate is also fault-free, Q_4 cannot be 0 even if the upper AND gate malfunctioned.

After we excluded the upper AND gate as a possible culprit, we need to consider the remaining gates. It is obvious that a faulty OR gate (which outputs 0 in the final cycle despite one of its inputs being true) can result in the observed erroneous behavior. Moreover, a faulty lower AND gate, which outputs 0 despite Q_3 being 1 and set_3 being 0, is also a viable fault candidate.

Observe that the set of fault candidates in Example 3 does indeed contain the "stuck-at-0" fault described in Example 2. It does, however, also suggest that the OR gate might be at fault. This example demonstrates that in general, it is not always possible to determine the exact underlying cause of an error.

There is a number of possibilities to increase the accuracy of our diagnosis. Firstly, note that we did not make any reference to fault models in Example 3. Additional assumptions about *how* a gate may fail would allow us to reduce the set of fault candidates further. If we assume, for instance, that the malfunction we are looking for is a permanent stuck-at fault, then we can rule out the OR gate as a fault candidate: if the OR gate were indeed stuck at a constant value, it could not output 1 in cycle 3 and 0 in cycle 4. This approach is known as model-based diagnosis with *strong* fault models [15]. In absence of a fault hypothesis and if we allow arbitrary transient faults, however, we cannot make specific assumptions about the behavior of faulty gates. In this case, we must base our diagnosis on the description of the system's normal behavior only (i.e., use *weak* fault models [16]). Strong fault models include a description of some aspects of abnormal behavior but at the cost of increased computational complexity. In the following, we restrict our discussion to weak fault models.

Secondly, the accuracy of the diagnosis can be improved by increasing the number of observations and executions we take into account. As mentioned in Sect. 13.2, the accuracy of model-based diagnosis is contingent on the quantity and quality of the observations. If, for instance, the value of Q_3 is not known in Example 3 because the value of the flip-flop cannot be observed at runtime, we are unable to narrow down the fault to cycle 4 and must take into the account that it could have been triggered in cycle 3. A further consequence is that we cannot rule out that the *upper* AND gate failed in cycle 3, since Q_3 could then be 0 in cycle 4.

If we base our diagnosis on several different test scenarios and erroneous executions, we can combine the results of the individual analyzes to reduce the set of potential fault candidates. This approach, however, is complicated by the fact that transient faults do not necessarily occur consistently across repeated executions.

Consistent reproduction of erroneous behavior over multiple runs can be challenging [17, 18]. Therefore, we focus on the analysis of single executions.

13.3.3 Applying Satisfiability Solvers to Diagnose Faults

Diagnosis as described in Sect. 13.3.2 requires a tedious and error-prone case split. In the following, we describe how contemporary satisfiability (SAT) solvers [5], which are extremely good at performing case splits automatically, can be deployed for this task.

Most modern SAT solvers expect their input to be a propositional formula in Conjunctive Normal Form (CNF), i.e., a conjunction of *clauses*, where each clause is a disjunction of *literals* and each literal is a propositional variable or its negation. While a naive transformation of an encoding such as the one in Formula (13.5) into CNF might result in an exponential increase of the size of the formula, it is possible to derive a formula in CNF directly from the circuit, since each logic gate of a circuit has a corresponding representation in CNF. To this end, we introduce a fresh propositional variable representing the output of each gate. For instance, the output w_1 of the AND gate with inputs d and $\neg set$ in Formula (13.5) and Fig. 13.7 is defined by the formula $w_1 \Leftrightarrow (\neg set_i \wedge Q_i)$, which can be transformed into CNF in a few simple steps:

$$
\begin{aligned}
(w_1 \Rightarrow (\neg set_i \wedge Q_i)) \wedge ((\neg set_i \wedge Q_i) \Rightarrow w_1) &\equiv \\
(\neg w_1 \vee (\neg set_i \wedge Q_i)) \wedge (\neg(\neg set_i \wedge Q_i) \vee w_1) &\equiv \\
(\neg w_1 \vee \neg set_i) \wedge (\neg w_1 \vee Q_i) \wedge (set_i \vee \neg Q_i \vee w_1) &
\end{aligned}
\tag{13.10}
$$

Similarly, we define $w_2 \Leftrightarrow (set_i \wedge Q_i)$ for the second AND gate and $Q_{i+1} \Leftrightarrow (w_1 \vee w_2)$ for the OR gate in Fig. 13.7. Using similar transformations as in Eq. (13.10), we obtain a CNF representation of Formula (13.5). Any assignment that satisfies the resulting formula in CNF corresponds to a satisfying assignment of the original formula (in our case, Formula (13.5)). Moreover, the resulting formula in CNF is only a constant factor larger than the original formula [19].

Consequently, each gate of an ILA is encoded by one or more clauses, and each clause can be attributed to exactly one of the gates. If we additionally constrain the CNF encoding of the ILA with the variable values observed during the execution of the manufactured prototype (similar to Formula (13.9) in Example 3), we obtain an *unsatisfiable* formula.

Example 4 We revisit Example 3. The following formula is a CNF encoding of Formula (13.9):

$$T(V_3, V_4) \begin{cases} (\neg w_1 \vee \neg set_3) \wedge (\neg w_1 \vee Q_3) \wedge (set_3 \vee \neg Q_3 \vee w_1) \wedge \\ (\neg w_2 \vee set_3) \wedge (\neg w_2 \vee d_3) \wedge (\neg set_3 \vee \neg d_3 \vee w_2) \wedge \\ (\neg Q_4 \vee w_1 \vee w_2) \wedge (\neg w_1 \vee Q_4) \wedge (\neg w_2 \vee Q_4) \wedge \\ \underbrace{(\neg set_3) \wedge (Q_3) \wedge (\neg Q_4)}_{\text{observations}} \end{cases} \tag{13.11}$$

The auxiliary variable w_1 represents the output of the lower AND gate and w_2 represents the output of the upper AND gate in the fourth execution cycle of Fig. 13.8. There is no satisfying assignment that makes all clauses of Formula (13.11) true.

Note that, augmenting the encoding with additional execution cycles does not change its unsatisfiability: Formula (13.11) in Example 4 is a so-called *unsatisfiable core* of the CNF encoding of Fig. 13.8 and the observations from Example 3. The only means of restoring the satisfiability of Formula (13.11) is by *dropping* one (or more) of its clauses. In particular, we are interested in a *minimal* set of clauses that need to be dropped to make the formula satisfiable.

Such a minimal set of clauses that need to be dropped to make a CNF formula satisfiable is called a *Minimal Correction Set* (MCS) [20]. Note that, MCSes are not unique; a CNF encoding potentially has many different MCSes.

Dropping an MCS that contains clauses encoding a gate of the transition relation $T(V_3, V_4)$ corresponds to allowing the behavior of that gate to deviate from the golden model. The remaining formula corresponds to that part of the ILA that is consistent with the observations. Consequently, the "dropped" gate is a viable fault candidate.

As discussed in Example 3, model-based diagnosis requires us to consider each gate in the ILA; each viable fault candidate corresponds to an MCS.

Example 5 Consider the encoding in Formula (13.11) of Example 4. Dropping the clause $(set_3 \vee \neg Q_3 \vee w_1)$ from the first line of Formula (13.11) allows w_1 to be 0, making the formula satisfiable. Consequently, the set containing only the clause $(set_3 \vee \neg Q_3 \vee w_1)$ is an MCS, which represents the case that the lower AND gate malfunctions in cycle 4 and outputs 0 despite the fact that both its inputs are 1.

Similarly, dropping the clause $(\neg w_1 \vee Q_4)$ from the third line of Formula (13.11) allows Q_4 to be 0 despite w_1 being 1 (note that w_2 is false already). This corresponds to the case that the OR gate malfunctions in cycle 4.

MCSes that span more than a single gate do not correspond to fault candidates as long as we maintain our single-fault hypothesis. (SAT-based techniques to analyze multiple faults are discussed in [21], for instance.) Moreover, MCSes that contain clauses encoding the observations do not correspond to fault candidates, since the observations represent hard constraints that cannot be changed.

Extensions of SAT solvers (see for instance, [22–24]) enable us to enumerate all possible MCSes of a CNF encoding. Grouping clauses that correspond to a single gate and making observations "hard" constraints that cannot be dropped avoids the enumeration of MCSes that do not represent fault candidates.

The scalability of the SAT-based fault localization approach described above is limited by the size of the ILAs for which MCSes need to be enumerated. Given that erroneous executions in post-silicon validation can comprise thousands of cycles, the resulting ILAs can become unwieldy and quickly exceed the scalability limits of contemporary SAT solvers. The following section discusses a divide-and-conquer strategy that improves the scalability (potentially at the cost of accuracy).

13.3.4 Window-Based Fault Localization

Because of long error detection latencies (the time elapsed from a fault being triggered to the detection of an observable failure) in post-silicon validation the length of executions that need to be analyzed can reach millions of cycles [25, 26]. The scalability of the SAT-based approach described in Sect. 13.3.3 is restricted to a much smaller number of cycles (in the range of thousands). Consequently, the CNF encoding of a full-length execution (represented by an ILA) is typically prohibitively large. Luckily, as Example 3 demonstrates, only a fraction of the execution may be necessary to localize faults. Let $S_i(V_i)$ be a CNF formula encoding the observations recorded in cycle i (cf. Sect. 13.2). It is sufficient to find an *execution window* of length n starting after cycle m such that the following formula becomes unsatisfiable:

$$\left(\bigwedge_{i=m+1}^{m+n} T(V_{i-1}, V_i)\right) \wedge \left(\bigwedge_{i=m}^{m+n} S_i(V_i)\right) \tag{13.12}$$

If Formula (13.12) is satisfiable, then either the window of length n starting after cycle m does not include the point at which the fault was not triggered, or the observations are insufficient to locate the fault within that window.

Example 6 Consider the ILA from Example 3, but assume that the value of the flip-flop is only observable in the last cycle (i.e., the values of Q_2 and Q_3 are unknown).

Figure 13.10 illustrates an execution window for this scenario that is sufficiently long to enable the localization of the fault.

Note that, if the value of the flip-flop after the third cycle is known, it suffices to analyze a window comprising only the last (fourth) cycle. In absence of information about Q_3, though, the transition relation $T(V_3, V_4)$ of the final cycle constrained with the observations $d_3 \wedge set_3 \wedge Q_4$ is satisfiable.

The formalism introduced in Sect. 13.2 enables us to characterize the information which is given in a fixed execution window, is required to enable fault localization. Consider an execution of length k and an execution window of length n starting after cycle m (where $0 \leq m < (m+n) \leq k$) in which the fault was triggered. The sets of states $S_m, \ldots, S_{m+n}$ derived from the observations of the execution need to satisfy the following conditions:

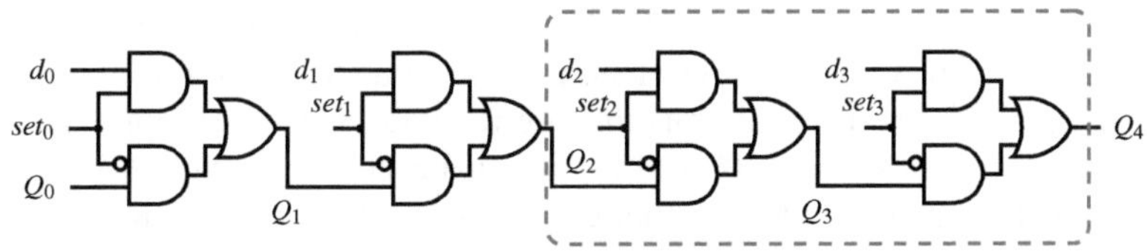

Fig. 13.10 Localization window in ILA from Fig. 13.8

$$
\begin{array}{rclcrcl}
C_m & \stackrel{\text{def}}{=} & S_m & \text{and} & C_i & \stackrel{\text{def}}{=} & T(C_{i-1}) \cap S_i \\
B_{m+n} & \stackrel{\text{def}}{=} & S_{m+n} & \text{and} & B_{i-1} & \stackrel{\text{def}}{=} & T^{-1}(B_i) \cap S_{i-1} \\
 & & & C_i \cap B_i = \emptyset & & &
\end{array}
\qquad (13.13)
$$

Intuitively, C_m represents a set of "correct" states and B_m represents "bad" states that are "infected" with an error that will propagate and eventually lead to an observable failure (as in Sect. 13.2). At any point in the window, we need sufficiently good observations to make sure that the correct states do not intersect with the bad states (the last condition in Eq. (13.13)). This condition may be violated if the constraints $S_m, \ldots, S_{m+n}$ derived from the observations are not strong enough or if the execution window is not long enough for the fault to be triggered and propagate to an observable latch or signal.

In practice, the quantity and quality of the observations $S_m, \ldots, S_{m+n}$ is contingent on logging mechanisms (such as trace buffers and scan chains). The length of execution windows, on the other hand, is bounded from above by the scalability of satisfiability solvers (the maximum size of encodings the solver can deal with) and from below by the error detection latency (the number of cycles from the occurrence of a fault to its propagation to an observable signal or latch). A number of techniques to tweak these parameters have been proposed:

- *BackSpace* [17, 18] accumulates observations over several executions by repeatedly rerunning failing tests and setting predecessor states (which are computed symbolically) of erroneous states as new breakpoints. At least in theory, BackSpace allows us to record arbitrarily long traces. In practice, the consistent reproduction of erroneous behavior over multiple runs can be challenging [17, 18].
- Identifying registers and flip-flops to which errors are likely to propagate can guide and aid the selection of signals recorded in trace buffers [27], decrease the error detection latency, and increase the chance that the recorded information contributes to fault localization. The technique is based on the injection of faults to identify test patterns [28] which reveal the registers to which faults propagate.
- The *Quick Error Detection* (QED) approach systematically generates families of post-silicon validation tests with a short error detection latency [25], thus reducing the need for lengthy window sizes and making SAT-based fault localization practical [26].

- E-QED [26] as well as BackSpace [17, 18] record *signatures*, a lossy representation of the sequence of values of a set of signals over time. E-QED, for instance, uses shift-registers in which certain bits that are fed back as well as input and output signals of design blocks are XORed with the values being shifted [26]. The shift-register is then also encoded symbolically as part of the execution window and constrained with the recorded information.
- The IFRA (Instruction Footprint Recording and Analysis) approach [29] relies on information about the chip architecture to record footprints. Unlike an architecture-agnostic approach like E-QED, IFRA requires considerable insight into the design of the circuit.

The techniques listed above aim at improving the recorded information and shortening the required window length. Techniques relying on sliding windows (such as [30]) relax the requirement that Formula (13.12) must be unsatisfiable by splitting windows into smaller satisfiable formulas. To explain how this works, we express the constraint in Eq. (13.13) using the notation used in Formula (13.12). At any point i within the execution window (where $m \leq i \leq (m+n)$), there must exist formulas $C_i(V_i)$ and $B_i(V_i)$ such that the following implications hold:

$$
\begin{aligned}
\overbrace{\left(\bigwedge_{j=m+1}^{i} T(V_{j-1}, V_j) \wedge \bigwedge_{j=m}^{i} S_j(V_j)\right)}^{\text{part } A} &\Rightarrow \qquad C_i(V_i) \\
\underbrace{\left(\bigwedge_{j=i+1}^{m+n} T(V_{j-1}, V_j) \wedge \bigwedge_{j=i}^{m+n} S_j(V_j)\right)}_{\text{part } B} &\Rightarrow \qquad B_i(V_i) \qquad\qquad (13.14) \\
C_i(V_i) &\Rightarrow \neg B_i(V_i)
\end{aligned}
$$

Intuitively, $\neg B_i(V_i)$ can be seen as a lower bound and $C_i(V_i)$ as an upper bound for the information that needs to be propagated to guarantee that the encoding of the execution window is unsatisfiable. Formally, any formula $F(V_i)$ which is implied by $C_i(V_i)$ (part A in (13.14), respectively) and inconsistent with $B_i(V_i)$ (part B in (13.14), respectively) is an *Craig interpolant* [31] for the two partitions of the execution window. Interpolants are used extensively in model checking [2, 32], but also have applications in design debugging [33].

Given such an interpolant, the window can be split into two parts which can be analyzed separately (since $C_i(V_i)$ in conjunction with part B is unsatisfiable, and $B_i(V_i)$ in conjunction with part A is unsatisfiable).

Example 7 We revisit the setting from Example 6. Consider the following encoding of the final cycle of the ILA in Fig. 13.8:

$$T(V_3, V_4)\begin{cases}(\neg w_1 \vee \neg set_3) \wedge (\neg w_1 \vee Q_3) \wedge (set_3 \vee \neg Q_3 \vee w_1) \wedge \\ (\neg w_2 \vee set_3) \wedge (\neg w_2 \vee d_3) \wedge (\neg set_3 \vee \neg d_3 \vee w_2) \wedge \\ (\neg Q_4 \vee w_1 \vee w_2) \wedge (\neg w_1 \vee Q_4) \wedge (\neg w_2 \vee Q_4) \wedge \\ \underbrace{(\neg set_3) \wedge (\neg Q_4)}_{\text{observations}}\end{cases} \quad (13.15)$$

Unlike in Example 3, we lack information about the value of Q_3; consequently, Formula (13.15) is satisfiable. Note that in all satisfying assignments of Formula (13.15) the variable Q_3 takes the value 0, i.e., Formula (13.15) implies $\neg Q_3$. Conversely, we can construct a similar formula $T(V_2, V_3)$ for cycle 3 constrained with the observations $(d_2) \wedge (set_2)$ (as in Example 2). Again that formula is satisfiable, and Q_3 takes the value 1 in all its satisfying assignments. It follows that the formula Q_3 is an interpolant for the formulas encoding the third and fourth cycle of the execution, respectively. If we constrain Formula (13.15) with the interpolant Q_3, it becomes unsatisfiable and we can proceed with our diagnosis as in Example 5.

Intuitively, the underlying assumption in Example 7 is that cycle 3 is fault-free, and that the information observed in this cycle can, therefore, be safely propagated, leading to a contradiction in cycle 4. Conversely, if we assume that cycle 4 is fault-free and backpropagates the information derived there (i.e., $\neg Q_3$), then this leads to a contradiction in cycle 3, allowing us to diagnose a fault there.

In Example 7, we were able to derive an interpolant Q_3 that satisfies the constraints in Eq. (13.14) by arguing that Q_3 takes the *same value* in *all* satisfying assignments. A literal with this property is said to be part of the *backbone* of the satisfiable formula. Backbones can be efficiently computed using a SAT solver [34, 35] and have been used for fault localization [30, 36].

The information obtained via backbones, however, is not necessarily sufficient to construct an interpolant. In general, it might be necessary to derive stronger consequences or to improve the information recorded during execution using the strategies discussed earlier (e.g., a strategic selection of the signals of the circuit tracked by trace buffers significantly increases the chances to succeed [27]).

13.4 Conclusion

Satisfiability solvers can be deployed to successfully localize faults in post-silicon validation. Recently, published work shows that the approach is effective and efficient [26] reports that symbolic reasoning combined with low-overhead logging mechanisms and state-of-the-art testing techniques [25] can narrow down electrical bugs to a few dozen flip-flops for designs with one million flip-flops completely automatically within hours. This success is owed in part to the impressive advances of satisfiability solvers [2, 5]. Recent improvements in computing unsatisfiable cores [37] and minimal correction sets [23] are bound to push the scalability of consistency-based diagnoses even further and increase the practicality of SAT-based fault localization.

Acknowledgements Parts of this chapter are based on the habilitation of the first author [38] and describe joint work with Charlie Shucheng Zhu [27, 30, 35, 36]. The first author is supported by the Austrian National Research Network S11403-N23 (RiSE) and by the Vienna Science and Technology Fund (WWTF) through grant VRG11-005.

References

1. E. Clarke, O. Grumberg, D. Peled, *Model Checking* (MIT Press, Cambridge, 1999)
2. Y. Vizel, G. Weissenbacher, S. Malik, Boolean satisfiability solvers and their applications in model checking. Proc. IEEE **103**(11), 2021–2035 (2015)
3. M. Abramovici, P. Bradley, K. Dwarakanath, P. Levin, G. Memmi, D. Miller, A reconfigurable design-for-debug infrastructure for SoCs, in *Design Automation Conference (DAC)* (ACM, 2006), pp. 7–12
4. M. Abramovici, M.A. Breuer, A.D. Friedman, *Digital Systems Testing and Testable Design* (Computer Science Press, USA, 1990)
5. A. Biere, M.J.H. Heule, H. van Maaren, T. Walsh (eds.), *Handbook of Satisfiability*. Frontiers in Artificial Intelligence and Applications, vol. 185 (IOS Press, Amsterdam, 2009)
6. J.R. Burch, E.M. Clarke, K.L. McMillan, D.L. Dill, L.J. Hwang, Symbolic model checking: 10^{20} states and beyond, *Logic in Computer Science (LICS)* (IEEE, New York, 1990), pp. 428–439
7. K.L. McMillan, *Symbolic Model Checking* (Kluwer, Dordrecht, 1993)
8. R.E. Bryant, Graph-based algorithms for Boolean function manipulation. IEEE Trans. Comput. **35**(8), 677–691 (1986)
9. J.C. King, A program verifier, Ph.D. thesis, Carnegie Mellon University, Pittsburgh, PA, USA (1970)
10. T. Ball, B. Cook, V. Levin, S.K. Rajamani, Slam and static driver verifier: technology transfer of formal methods inside microsoft, *Integrated Formal Verification (IFM)*. LNCS, vol. 2999 (Springer, Berlin, 2004)
11. S. Lee, K.A. Sakallah, Unbounded scalable verification based on approximate property-directed reachability and datapath abstraction, in *Computer Aided Verification*. LNCS, vol. 8559, ed. by A. Biere, R. Bloem (Springer, Berlin, 2014), pp. 849–865
12. D. Kroening, O. Strichman, *Decision Procedures: An Algorithmic Point of View*, Texts in Theoretical Computer Science (EATCS) (Springer, Berlin, 2008)
13. A. Biere, A. Cimatti, E.M. Clarke, Y. Zhu, Symbolic model checking without BDDs, *Tools and Algorithms for the Construction and Analysis of Systems*. LNCS, vol. 1579 (Springer, Berlin, 1999), pp. 193–207
14. R. Reiter, A theory of diagnosis from first principles. Artif. Intell. **32**(1), 57–95 (1987)
15. P. Struss, O. Dressler, Physical negation – integrating fault models into the general diagnostic engine, in *International Joint Conference on Artificial Intelligence (IJCAI)*, ed. by N.S. Sridharan (Morgan Kaufmann, USA, 1989), pp. 1318–1323
16. J. de Kleer, A.K. Mackworth, R. Reiter, Characterizing diagnoses and systems. Artif. Intell. **56**(2–3), 197–222 (1992)
17. F.M. de Paula, M. Gort, A.J. Hu, S.J.E. Wilton, J. Yang, Backspace: formal analysis for post-silicon debug, *Formal Methods in Computer-Aided Design (FMCAD)* (IEEE, New York, 2008), pp. 1–10

18. F.M. de Paula, A. Nahir, Z. Nevo, A. Orni, A.J. Hu, TAB-backspace: unlimited-length trace buffers with zero additional on-chip overhead, in *Design Automation Conference (DAC)* (ACM, 2011), pp. 411–416
19. G. Tseitin, On the complexity of proofs in poropositional logics, in *Automation of Reasoning: Classical Papers in Computational Logic 1967–1970*, vol. 2, ed. by J. Siekmann, G. Wrightson (Springer, Berlin, 1983). Originally published 1970
20. M.H. Liffiton, K.A. Sakallah, Algorithms for computing minimal unsatisfiable subsets of constraints. J. Autom. Reason. **40**(1), 1–33 (2008)
21. A. Sülflow, G. Fey, R. Bloem, R. Drechsler, Using unsatisfiable cores to debug multiple design errors, in *Great Lakes Symposium on VLSI*, ed. by V. Narayanan, Z. Yan, E. Macii, S. Bhanja (ACM, New York, 2008), pp. 77–82
22. A. Ignatiev, A. Previti, M.H. Liffiton, J. Marques-Silva, Smallest MUS extraction with minimal hitting set dualization, in *Principles and Practice of Constraint Programming (CP)*. LNCS, vol. 9255, ed. by G. Pesant (Springer, Berlin, 2015), pp. 173–182
23. C. Mencía, A. Ignatiev, A. Previti, J. Marques-Silva, MCS extraction with sublinear oracle queries, in *Theory and Applications of Satisfiability Testing (SAT)*. LNCS, vol. 9710, ed. by N. Creignou, D. Le Berre (Springer, Berlin, 2016), pp. 342–360
24. C. Mencía, A. Previti, J. Marques-Silva, Literal-based MCS extraction, in *International Joint Conference on Artificial Intelligence (IJCAI)*, ed. by Q. Yang, M. Wooldridge (AAAI Press, London, 2015), pp. 1973–1979
25. D. Lin, T. Hong, Y. Li, S. Eswaran, S. Kumar, F. Fallah, N. Hakim, D.S. Gardner, S. Mitra, Effective post-silicon validation of system-on-chips using quick error detection. IEEE Trans. Comput. Aided Des. Integr. Circuits Syst. (TCAD) **33**(10), 1573–1590 (2014)
26. E. Singh, C.W. Barrett, S. Mitra, E-QED: electrical bug localization during post-silicon validation enabled by quick error detection and formal methods, in *Computer Aided Verification*. LNCS, vol. 10427, ed. by R. Majumdar, V. Kuncak (Springer, Berlin, 2017), pp. 104–125
27. C.S. Zhu, G. Weissenbacher, S. Malik, Coverage-based trace signal selection for fault localisation in post-silicon validation, *Haifa Verification Conference (HVC)*. LNCS, vol. 7857 (Springer, Berlin, 2012), pp. 132–147
28. K.-T. Cheng, L.-C. Wang, Automatic test pattern generation, *EDA for IC System Design, Verification, and Testing* (CRC Press, Boca Raton, 2006)
29. S.-B. Park, T. Hong, S. Mitra, Post-silicon bug localization in processors using instruction footprint recording and analysis (IFRA). IEEE Trans. Comput. Aided Des. Integr. Circuits Syst. (TCAD) **28**(10), 1545–1558 (2009)
30. C.S. Zhu, G. Weissenbacher, S. Malik, Silicon fault diagnosis using sequence interpolation with backbones, *Computer-Aided Design (ICCAD)* (IEEE, New York, 2014), pp. 348–355
31. W. Craig, Linear reasoning. A new form of the Herbrand-Gentzen theorem. J. Symb. Log. **22**(3), 250–268 (1957)
32. K.L. McMillan, Interpolation and SAT-based model checking, *Computer Aided Verification*. LNCS, vol. 2725 (Springer, Berlin, 2003), pp. 1–13
33. B. Keng, A.G. Veneris, Scaling VLSI design debugging with interpolation, *Formal Methods in Computer-Aided Design (FMCAD)* (IEEE, New York, 2009), pp. 144–151
34. J. Marques-Silva, M. Janota, I. Lynce, On computing backbones of propositional theories, in *European Conference on Artificial Intelligence (ECAI)* (IOS Press, 2010), pp. 15–20
35. C.S. Zhu, G. Weissenbacher, D. Sethi, S. Malik, SAT-based techniques for determining backbones for post-silicon fault localisation, *High Level Design Validation and Test Workshop (HLDVT)* (IEEE, New York, 2011), pp. 84–91
36. C.S. Zhu, G. Weissenbacher, S. Malik, Post-silicon fault localisation using maximum satisfiability and backbones, *Formal Methods in Computer-Aided Design (FMCAD)* (IEEE, New York, 2011), pp. 63–66
37. M.H. Liffiton, A. Previti, A. Malik, J. Marques-Silva, Fast, flexible MUS enumeration. Constraints **21**(2), 223–250 (2016)
38. G. Weissenbacher, Logical methods in automated hardware and software verification, Habilitation thesis at TU Wien (2016)

Chapter 14
Coverage Evaluation and Analysis of Post-Silicon Tests with Virtual Prototypes

Kai Cong and Fei Xie

14.1 Introduction

14.1.1 Motivation and Problem Statement

14.1.1.1 Motivation

New computer systems, smartphones, wearable devices, tablets, laptops, servers, etc., are entering the marketplace at an ever-accelerating pace. This brings enormous pressures on the product development teams to shorten the time-to-market. A recent study by International Business Strategies indicates that a 3-month delay to market reduces revenue by about 30% for chip manufacturers in general, and the penalty is even more severe for fast-evolving markets such as mobile devices [1]. To exacerbate the pressures, the complexities of these systems, both their hardware and software, have been increasing significantly. Quoting a SoC architect for a mobile platform, "a state-of-the-art mobile platform is considered more complex than a server due to the many types of technologies it integrates while the product cycle is often as short as two years." A crucial stage in the product development cycle is post-silicon validation, i.e., validation conducted on actual devices or silicon prototypes with corresponding drivers. Post-silicon validation is a significant, fastest-growing component of validation cost. According to recent industry reports [2], post-silicon validation effort often consumes more than 50% of a 65 nm SoC's overall design effort. This demands innovative approaches to speed-up post-silicon validation and reduces its cost.

Though post-silicon validation covers many aspects ranging from electronics properties of hardware to performance and power consumption of whole systems,

K. Cong (✉) · F. Xie
CS Department, Portland State University, Portland, OR 97201, USA
e-mail: congkai@cs.pdx.edu

F. Xie
e-mail: xie@cs.pdx.edu

P. Mishra and F. Farahmandi (eds.), *Post-Silicon Validation and Debug*,
https://doi.org/10.1007/978-3-319-98116-1_14

a central task remains validating functional correctness of both hardware and its integration with software. Recently, virtual prototypes are increasingly used in hardware/software development to enable driver development and validation at an early stage even before silicon prototypes become available [3]. An example is how Intel used virtual prototypes to enable driver development for their 40G Ethernet adapter (E40G) before the silicon prototype became available [4]. An E40G virtual prototype was created and used to test and validate the E40G driver being developed. Bugs were found in the driver using the E40G virtual device, even before the real E40G device became available. Since virtual prototypes are utilized as a transaction-level replacement for silicon devices to support driver development and validation, it is greatly desired to extend the effectiveness of virtual prototypes into post-silicon functional validation so that the major efforts invested can be fully utilized. We see major potentials of virtual prototypes in post-silicon functional validation of both hardware and its integration with software.

14.1.1.2 Problem Statement

This work is concerned with how to speed-up post-silicon functional validation with virtual prototypes for both hardware and software developments. We observe two major challenges in achieving our goal:

- *Limited Silicon Observability and Traceability.* The silicon device is typically a black box. The amount of runtime information that can be retrieved from the device internal with build-in test circuitries and advanced logic analyzers is still quite limited. Such limited observability and traceability make post-silicon validation difficult.
- *Lack of Good Test Coverage Estimation.* There lacks good test coverage metrics over a silicon device. Therefore, it is difficult to assess the effectiveness of test cases and prioritize their application. In addition, coverage metrics rooted in hardware design are not well suited for testing the integration with software.

14.1.2 Proposed Solution

We propose an approach to accelerating post-silicon functional validation and reducing its cost with virtual prototypes. Our approach mainly supports two components:

- *Coverage evaluation*: While a silicon device is often a black box, its corresponding virtual prototype is a white box, i.e., its internal structures and workings are visible. The virtual prototype often models transaction-level behaviors of the silicon device. Therefore, the virtual prototype can be utilized to estimate the coverage of post-silicon validation tests on the functionalities of the silicon device.
- *Runtime analysis*: Since virtual prototypes provide all device functionalities, post-silicon tests can be issued to virtual prototypes to verify desired functionalities.

While a test is issued to a virtual prototype, developers can observe the corresponding device transactions and state changes. Therefore, the virtual prototype can be utilized to analyze and debug how post-silicon validation tests control the devices.

For coverage evaluation and runtime analysis, we employ symbolic execution of virtual prototypes as the foundation. More details about these components are elaborated below:

Coverage Evaluation of Post-silicon Tests. Test coverage is an important metric for evaluating the quality and readiness of post-silicon tests. We propose an online-capture offline-replay approach to coverage analysis of post-silicon validation tests with virtual prototypes for estimating silicon device test coverage [5]. We first capture necessary data from a concrete execution of the virtual prototype within a virtual platform under a given test, and then compute the test coverage by efficiently replaying this execution offline on the virtual prototype itself. Our approach provides early feedback on quality of post-silicon validation tests before silicon is ready. To ensure fidelity of early coverage evaluation, our approach has been further extended to support coverage evaluation and conformance checking in the post-silicon stage.

Runtime Analysis of Post-silicon Tests. We developed a shadow execution framework to run a virtual prototype in a shadow environment with the normal execution. Such shadow execution can be used for observing device behaviors triggered by the post-silicon tests and debugging incorrect and undesired device state changes. More, under certain state triggered by the tests, we run symbolic execution on virtual prototypes to generate test cases. Generated test cases enable developers to better observe and trace any variable change in the virtual device along this path. Developers can execute a device model based on a concrete test case back and forth, step by step and observe variable changes at each step. Developers can also inspect values of all virtual device variables easily at any time and check whether the state conforms to the specification. With such observability and traceability, developers can understand what paths exist and get a better understanding of the device behaviors.

14.1.3 Chapter Outline

The remainder of this chapter is organized as follows. Section 14.2 introduces a brief overview of background including virtual prototypes and symbolic execution. Section 14.3 elaborates coverage evaluation of post-silicon tests. Section 14.4 presents runtime analysis of virtual prototypes which allows developers to analyze and debug post-silicon tests. Section 14.5 discusses related work. Section 14.6 concludes our work.

14.2 Background

14.2.1 Virtual Prototypes and QEMU Virtual Devices

Virtual prototypes are fast, fully functional software models of hardware systems, which enable unmodified execution of software code. QEMU is a generic, open-source machine emulator and virtualizer [6, 7]. We adopt QEMU virtual devices as the virtual prototypes for our study due to the open-source nature of QEMU and its wide varieties of virtual devices. Technology developed on QEMU virtual devices can be readily generalized to other open-source or commercial virtual prototyping environments due to their similarity in virtualization concepts, despite their different levels of modeling details.

To better understand the concept of virtual prototype, we illustrate it with a QEMU virtual device for the Intel E1000 Gigabit network adapter. The E1000 adapter is a Peripheral Component Interconnect (PCI) device which communicates with its control software through interface registers and interrupts. The E1000 virtual device has corresponding functions to support such communication, for instance, interface register functions and interrupt functions. In order to realize the functionalities of silicon devices, the E1000 virtual device also needs to maintain the device state and implement functions that virtualize device transactions and environment inputs. As shown in Fig. 14.1, the E1000 virtual device has the following components:

- The device state, *E1000State*, which keeps track of the state of the E1000 device and the device configuration;
- The interface register functions such as *write_reg* which are invoked by QEMU to access interface registers and trigger transaction functions;
- The device transaction functions such as *start_xmit* which are invoked by the interface register functions to realize the functionality;
- The environment functions such as *receive* which are invoked by QEMU to pass environment inputs such as a packet received to the virtual device.

Both the device transaction functions and environment functions may access DMA data by calling DMA functions *pci_dma_write* and *pci_dma_read*, as well as fire interrupts by calling interrupt function *set_irq*. Both PCI interface functions and environment input functions are device entry functions which are invoked by QEMU to trigger device functionalities.

14.2.2 Symbolic Execution

Symbolic execution executes a program with symbolic values as inputs instead of concrete ones and represents the values of program variables as symbolic expressions.

```
// 1. Device state
typedef struct E1000State_st {
    PCIDevice dev; //PCI configuration
    uint32_t mac_reg[0x8000]; //Interface registers
    ......
    uint32_t rxbuf_size; //Internal variables
    ......
} E1000State;

// 2. Interface register function: write register
static void write_reg(void *opaque, uint64_t index, uint32_t value) {
    E1000State *s = (E1000State *)opaque;
    ......
    if(index == TRANSMIT) {
        s->mac_reg[index] = value;
        start_xmit(s); //Invoking transaction function
    }
    ......
}

// 3. Device transaction function: transmit packets
static void start_xmit(E1000State *s) {
    ......
    pci_dma_read(&s->dev, base, &desc, sizeof(desc)); //Invoking DMA functions
    ......
    set_irq(s->dev.irq[0],1); //Invoking interrupt function
}

// 4. Environment function: receive packets
static ssize_t receive(NetClientState *nc, const uint8_t *buf, size_t size) {
    ......
    pci_dma_write(&s->dev, base, &desc, sizeof(desc)); //Invoking DMA functions
    ....
    set_irq(s->dev.irq[0],1); //Invoking interrupt function
}
```

Fig. 14.1 Excerpt of QEMU E1000 virtual device

Consequently, the outputs computed by the program are expressed as a function of input symbolic values. The symbolic state of a program includes the symbolic values of program variables, a path condition, and a program counter. The path condition is a Boolean formula over the symbolic inputs; it accumulates constraints which the inputs must satisfy for the symbolic execution to follow the particular path. The program counter points to the next statement to execute. A symbolic execution tree captures the paths explored by the symbolic execution of a program: the nodes represent the symbolic program states and the arcs represent the state transitions.

We use the program in Fig. 14.2 to illustrate how symbolic execution is conducted. At the entry, x has a symbolic value, i.e., any value allowed by its type (in this case, integer). At each branching point, the path condition is updated with conditions on the inputs to select between the two alternative paths. For this example, we can get three paths based on symbolic execution. Each path will have its own path condition, for example, $x < 0$ for the leftmost path.

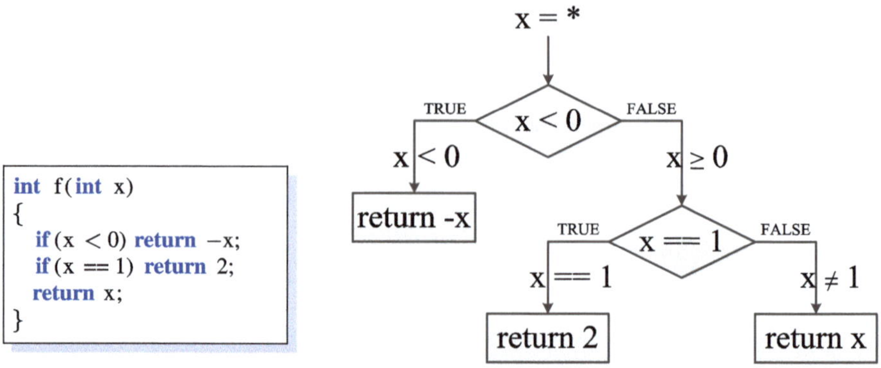

Fig. 14.2 An example of symbolic execution

14.2.3 Post-silicon Conformance Checking

In [8–10], we have developed an approach to post-silicon conformance checking of a silicon device with its virtual device. The conformance between the silicon and virtual devices is defined over their interface states. The request sequence issued to the device is first captured on the silicon device, and then replayed on the virtual device to check if the interface states of the silicon and virtual devices are consistent.

In coverage evaluation of post-silicon validation tests, we utilize conformance checking to ensure fidelity of coverage evaluation.

14.3 Coverage Evaluation of Post-silicon Tests

14.3.1 Motivation and Overview

Post-silicon validation has become a bottleneck in system development cycle and is a significant, growing part of overall validation cost [11]. To speed-up post-silicon validation, some tasks should be conducted early in the pre-silicon stage, e.g., development and evaluation of post-silicon validation tests. Test coverage is an important metric for evaluating the quality and readiness of post-silicon validation tests. Precise coverage results are necessary for engineers to judge whether existing test suites can achieve sufficient coverage and cover desired functionalities on the device.

Before the first silicon prototype is ready, it is very challenging to quantify coverage of post-silicon validation tests since we do not have a silicon device to run these tests on. Even if a silicon prototype is ready, the black box nature of the silicon prototype only supports limited observability and traceability that makes post-silicon validation difficult.

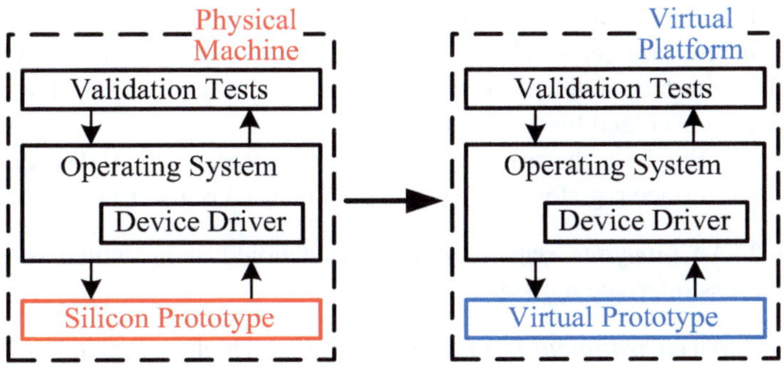

Fig. 14.3 From physical to virtual

As shown in Fig. 14.3, virtual prototypes and silicon devices are running, respectively, in virtual platforms and physical machines. Virtual prototypes can provide the same transaction-level functionalities as silicon devices to support driver development and validation. Virtual prototypes have major potential to play a crucial role in estimating silicon device functional coverage of post-silicon validation tests. The white box nature of virtual prototypes brings complete observability and traceability that evades silicon devices. It is possible to have thorough test coverage evaluation over virtual prototypes.

This section presents an online-capture offline-replay approach to coverage evaluation of post-silicon validation tests with virtual prototypes. We first capture necessary runtime data, including the initial device state and device requests from a concrete execution of the virtual prototype within a virtual platform under a given test. We then compute the test coverage by efficiently replaying captured data offline on the virtual prototype itself. To evaluate the coverage, we have adopted four typical software coverage metrics and developed two hardware-specific coverage metrics: register and transaction coverage. To ensure fidelity of coverage estimation on the silicon device, we further extend our approach to compute coverage after the silicon device becomes ready and check conformance with coverage estimate on the virtual prototype.

We have implemented this approach in Device Coverage Analyzer (DCA), a coverage analysis tool using virtual prototypes. We have applied our approach to evaluate a suite of common tests with virtual prototypes of five network adapters. Our approach was able to reliably estimate that this suite achieves high functional coverage on all five silicon devices.

14.3.2 Preliminary Definitions for Virtual Devices

In order to help better understand Sects. 14.3 and 14.4, we introduce several definitions and define a formal model for a virtual device.

Definition 1 A ***device state*** is denoted as $\boldsymbol{s} = \langle \boldsymbol{s}_I, \boldsymbol{s}_N \rangle$ where $\boldsymbol{s}_I$ is the interface state including all interface registers and $\boldsymbol{s}_N$ is the internal state including all internal registers. The interface state $\boldsymbol{s}_I$ can be accessed by a high-level software (e.g., driver) while $\boldsymbol{s}_N$ is only accessed by the device itself.

As shown in Fig. 14.1, the structure *E*1000*State* represents the E1000 device state and includes interface registers *mac_reg* and an internal register *rxbuf_size*.

Definition 2 An ***interface register request*** is denoted as $\boldsymbol{r}_{ir}$ which is issued by drivers to access interface registers.

Definition 3 An ***environment input*** is denoted as $\boldsymbol{r}_{ei}$ which is received by the device from the environment.

Definition 4 A ***device request*** is denoted as $\boldsymbol{r}$ which is issued by high-level software to control and operate the device.

As shown in Fig. 14.1, the parameters *index* and *value* of interface register function *write_reg* can be treated as a request $\boldsymbol{r}$, which is issued by the driver to modify the interface register and trigger the transaction function.

Direct Memory Access (DMA) is a feature of modern computers that allows certain devices to access system memory independent of CPU. In order to process a device request $\boldsymbol{r}$, a device might read/write data using DMA.

Definition 5 A ***DMA sequence*** is denoted as $\boldsymbol{d} = \boldsymbol{d}_1, \boldsymbol{d}_2, \ldots, \boldsymbol{d}_n$ where $\boldsymbol{d}_i$ is the *i*th DMA data accessed for processing one request.

Definition 6 A ***device event*** is denoted as $\boldsymbol{e} = \langle \boldsymbol{r}, \boldsymbol{d} \rangle$ where $\boldsymbol{r}$ is a device request and $\boldsymbol{d}$ is a sequence of DMA data. For some event $\boldsymbol{e}$, $\boldsymbol{d}$ might be *null* since no DMA data is needed for processing $\boldsymbol{r}$.

Definition 7 A ***sequence of device events*** is denoted as $\boldsymbol{seq} = \boldsymbol{e}_1, \boldsymbol{e}_2, \ldots, \boldsymbol{e}_n$. A subsequence $\boldsymbol{seq}_k$ of $\boldsymbol{seq}$ contains the first k events of $\boldsymbol{seq}$ where $\boldsymbol{seq}_k = \boldsymbol{e}_1, \boldsymbol{e}_2, \ldots, \boldsymbol{e}_k$. After processing a sequence of device events, the device can be transitioned to a new state from the initial state.

Definition 8 A ***test case*** is denoted as $\boldsymbol{tc} = \langle \boldsymbol{seq}, \boldsymbol{e} \rangle$, where $\boldsymbol{seq}$ is a sequence of device events and $\boldsymbol{e}$ is an additional device event. The device is transitioned to a desired state from the initial state after processing $\boldsymbol{seq}$. Then, the device event *e* is issued to the device to trigger the desired device functionality.

Definition 9 A ***state under test*** is denoted as $\boldsymbol{s}_{ut}$ where $\boldsymbol{s}_{ut}$ is the device state on which test cases are generated.

Devices are transactional in nature: device requests are processed by device transactions. For a virtual device (which is a program), given a state *s* and a device request *r*, a program path of the virtual device is executed and the device is transitioned into a new state. Each distinct program path of the virtual device represents a distinct device transaction.

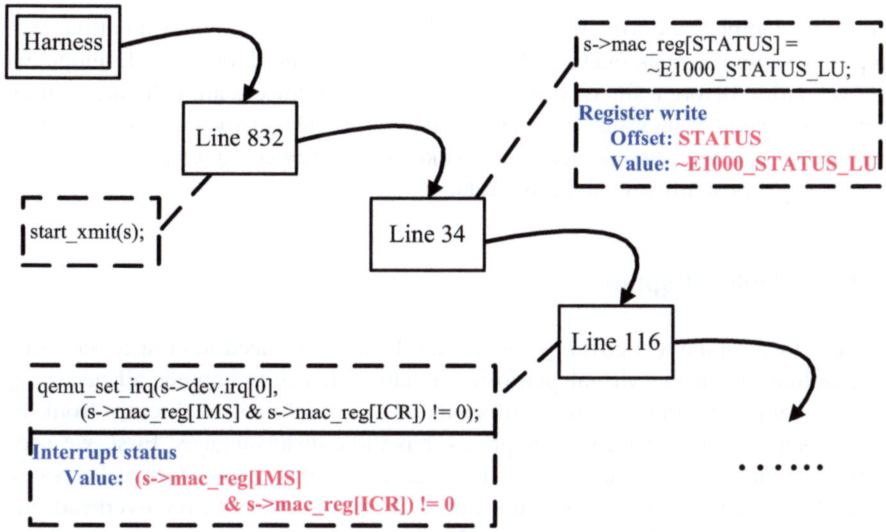

Fig. 14.4 A transaction example

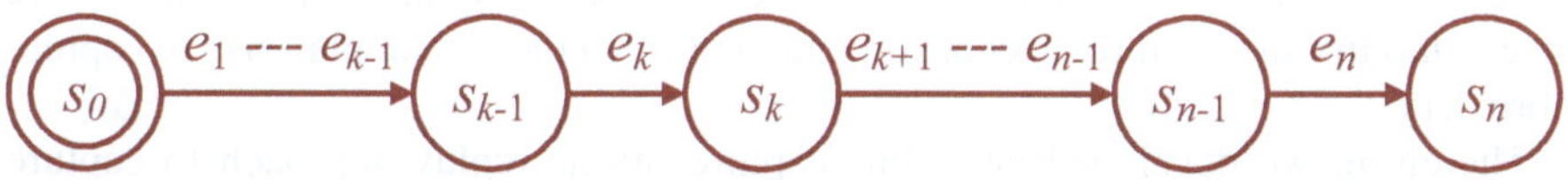

Fig. 14.5 A graph representation of state transitions

Definition 10 A ***device transaction***, denoted as $\boldsymbol{t} = \boldsymbol{l}_1, \boldsymbol{l}_2, \ldots, \boldsymbol{l}_n$., is a program path of a virtual device. Each step $\boldsymbol{l}$ in the path is a tuple ($\boldsymbol{\lambda}$, $\boldsymbol{\gamma}$, $\boldsymbol{\xi}$), where $\boldsymbol{\lambda}$ is the code statement executed, $\boldsymbol{\gamma}$ is the registers accessed, and $\boldsymbol{\xi}$ is the interrupt status.

Figure 14.4 gives an example of a transaction. Besides the basic code statement sequence, the transaction $\boldsymbol{t}$ also contains hardware-related information, such as the registers accessed and the interrupt status.

A virtual device is a transaction-level model of hardware design which can be represented as an event-driven state transition graph. As shown in Fig. 14.5, given a device state s_{k-1} and a device event e_k, the device will transit to a new device state s_k. We use $s \xrightarrow{e} s'$ to denote a transaction.

14.3.3 Online-Capture Offline-Replay Coverage Evaluation

Before a silicon device is ready, post-silicon validation tests can be evaluated using RTL emulation. However, emulating hardware design has certain limitations. First, RTL emulators can be very expensive. Second, RTL emulation is often slow. Third, it requires a complete working RTL design [4] to evaluate post-silicon validation tests. Recently, virtual devices and virtual platforms have been used for driver development and validation before a silicon device is ready. Virtual devices are software components. Compared to their hardware counterparts, it is easier to achieve observability and traceability on virtual devices. This makes virtual devices amenable to coverage evaluation of post-silicon validation tests.

14.3.3.1 Online-Capture

In order to compute test coverage on virtual devices, we need to collect necessary runtime data from the virtual platform. A naïve idea is to capture all necessary runtime data including execution information of virtual devices directly from the virtual platform. However, such approach has three disadvantages. First, we need to instrument virtual devices to capture execution information of virtual devices. Second, capturing detailed execution information introduces heavy overhead into the virtual platform. Third, we need to decide what kinds of information should be captured before runtime execution of the virtual platform. It is hard to guarantee that captured information is sufficient. Once a new metric is added, it is possible that we have to modify the capture mechanism and then rerun the virtual platform to capture more data.

Therefore, we developed an online-capture offline-replay approach to capture minimum necessary data at runtime, and then replay the runtime data on the virtual device itself offline to collect necessary execution information.

A device can be treated as a state transition system. As shown in Fig. 14.5, given a device state $\boldsymbol{s}_{k-1}$ and a device event $\boldsymbol{e}_k$, the device will transit to a new device state $\boldsymbol{s}_k$. Therefore, with the initial state $\boldsymbol{s}_0$ and the whole event sequence *seq*, we can infer all states and reproduce all state transitions. In other words, capturing $\boldsymbol{s}_0$ and *seq* from the concrete execution of a virtual device within the virtual platform should introduce the lowest overhead and deliver the most effective data.

14.3.3.2 Offline-Replay

Our offline-replay mechanism reproduces runtime execution on virtual devices with $\boldsymbol{s}_0$ and *seq*, which provides flexible analysis mechanism and powerful debug capability.

(1) Flexible analysis mechanism: The replay process is independent of the virtual platform/physical machine. Once runtime data is captured, users can replay the event sequence and reproduce the execution at any time. Based on different user requirements, users can generate different coverage reports from the replay process with different metrics. *(2) Powerful debug capability*: The replay mechanism provides capability for debugging interesting execution traces on virtual devices statement by statement, backward, and forward.

Algorithm 1 REPLAY_EVENTS (s_0, seq)

```
i ← 0; //loop iteration
s ← s_0; //Set initial device state
while i < seq.size() do
    e ← seq[i];
    ⟨t, s_next⟩ ←Execute_Virtual_Device (s, e);
    T.save(t);
    s ← s_next; //Set next device state
    i ← i + 1;
end while
Generate_Report (T);
```

Algorithm 1 illustrates how to replay all events with s_0 and seq to collect necessary execution information. In Algorithm 1, T is a temporary vector for saving execution information for all events. The algorithm takes the initial device state s_0 and the event sequence seq as inputs. Before replaying the event sequence, we set s_0 as the device state s. We run the virtual device with each event e in the event sequence seq and the corresponding state s to compute the execution information t and the next state s_{next}. Then t is saved in T and s_{next} is assigned to s. After replaying all events, we generate coverage reports based on T and user configuration.

14.3.3.3 Coverage Computation and Conformance Checking in the Post-silicon Stage

In our approach, we use coverage evaluation of virtual prototypes to estimate functional coverage on silicon devices. In order to make our approach practical and reliable, we need to address the following two key challenges:

(1) Accuracy: In our approach, we capture runtime data from the concrete execution of virtual devices within a virtual platform. Events ($\boldsymbol{E}_v$) issued to virtual devices within a virtual platform can be different from events ($\boldsymbol{E}_s$) issued to silicon devices within a physical machine for the same tests. The concern is whether the coverage ($\boldsymbol{C}_v$) computed on ($\boldsymbol{E}_v$) is a good approximation of the coverage ($\boldsymbol{C}_s$) computed on ($\boldsymbol{E}_s$).

(2) Conformance: Another challenge is whether coverage estimation on virtual devices can really reflect silicon device functional coverage. Although both virtual devices and silicon devices are developed according to the same specification, whether they conform to each other is still a major concern.

To address the above two challenges, we have extended our approach to support coverage computation and conformance checking after the silicon device is ready. We first reset the silicon device, and then capture runtime data, including all silicon device states $SS = \{ss_0, ss_1, \ldots, ss_n\}$ and the device event sequence $seq = e_1, e_2, ldots, e_n$, from the concrete execution of a silicon device within a physical machine. For a silicon device, interface registers are observable while the internal registers are not observable in general. Therefore, it is only possible to record all silicon device interface states $SS_I = \{ss_{I0}, ss_{I1}, \ldots, ss_{In}\}$ due to the limited observability. Algorithm 2 shows the extended algorithm for replaying SS_I and seq on the virtual device.

Algorithm 2 EXTENDED_REPLAY_EVENTS (SS_I, seq)

$k \leftarrow 0$; //loop iteration
$s \leftarrow$ *Reset_Virtual_Device ()*; //$s = \langle s_I, s_N \rangle$
while $k < seq.size()$ **do**
  $s_I \leftarrow ss_{Ik}$; //Load captured silicon device interface state
  $e \leftarrow seq[k+1]$;
  $\langle t, s' \rangle \leftarrow$ *Execute_Virtual_Device (s, e)*; //$s' = \langle s'_I, s'_N \rangle$
  T.save(t);
  Check_Conformance (s'_I, $ss_{I(k+1)}$);
  $s_N \leftarrow s'_N$;
  $k \leftarrow k + 1$;
end while
Generate_Report (T);

In Algorithm 2, we first reset the virtual device to get the initial device state s. We assume that the internal states of the silicon device and its virtual device are the same after resetting devices. Even if both internal states are not exactly the same, a few differences should not cause a large number of functional differences according to device specifications. We take the captured device state ss_{Ik} and e_{k+1} as inputs to replay one event. The virtual device is executed with s and e_{k+1} to compute the execution information and the state s' after processing e_{k+1}. Then, conformance checking is conducted between the computed interface state s'_I on the virtual device and the captured interface state $ss_{I(k+1)}$ on the silicon device to detect inconsistencies. After replaying one event, we keep the internal state and load next interface state captured to compose the device state. After replaying all events, we can get coverage reports and inconsistency report.

We utilize the coverage evaluation and conformance checking results in three aspects to assure the coverage estimation accuracy. First, we compare $\boldsymbol{C}_s$ and $\boldsymbol{C}_v$ to detect differences. If we can verify that there is no difference or few differences between $\boldsymbol{C}_v$ and $\boldsymbol{C}_s$, we can better trust that $\boldsymbol{C}_v$ can be a good approximation of $\boldsymbol{C}_s$. Second, the number of inconsistencies provides basic measurement how many differences there are between the silicon device and the virtual prototype. After analyzing the inconsistencies, we further evaluate whether these inconsistencies cause different device behaviors. If there are few inconsistencies found and there is no significant effect on the device, it can increase our confidence on coverage estimation. Third, it is easy to fix the detected inconsistencies on the virtual device so that the fixed virtual device conforms with the silicon device. Then, we compute coverage again on the fixed virtual device using the same test cases. By comparing the coverage report on the fixed virtual device with that on the silicon device, we further verify that the differences in coverage caused by the inconsistencies are removed.

14.3.4 Coverage Metrics

Computing test coverage requires appropriate coverage metrics. In our approach, we use virtual prototype coverage to estimate silicon device functional coverage. A virtual prototype is not only a software program but also models the characteristics of the silicon device. Therefore, we have employed two kinds of coverage metrics: we have adopted the typical software coverage metrics and developed two hardware-specific coverage metrics: register coverage and transaction coverage.

14.3.4.1 Code Coverage

Code coverage is a typical measure used in software testing. Virtual devices are software models. We can apply all code coverage metrics to virtual devices. We select four common coverage metrics: function coverage, statement coverage, block coverage, and branch coverage.

14.3.4.2 Register Coverage

A hardware register stores bits of information in such a way that systems can write to or read out from it all the bits simultaneously. High-level software can determine the state of the device by reading registers, and control and operate the device by writing registers. It is critical for engineers to know what registers have been accessed so they can check whether the device is accessed correctly according to the specification. Virtual devices provide complete observability, and therefore we can capture accesses on both interface and internal registers. Actually, in our approach, we capture all register accesses and deliver different kinds of register coverage reports according to user configuration.

14.3.4.3 Transaction Coverage

Devices and, therefore, virtual devices are transactional in nature: they receive interface register requests and environment inputs, and process them concurrently without interference. Thus, an interesting and useful metric is transaction coverage. For a virtual device (which is a C program), given a state s and a device request r, a program path of the virtual device is executed and the device is transitioned into a new state. Each distinct program path of the virtual device represents a distinct device transaction. When computing coverage, the impact of a test case on the virtual device in terms of what transactions it hits and how often they are hit is recorded. The impact of a test suite can be recorded the same way. The coverage statistics can be visualized

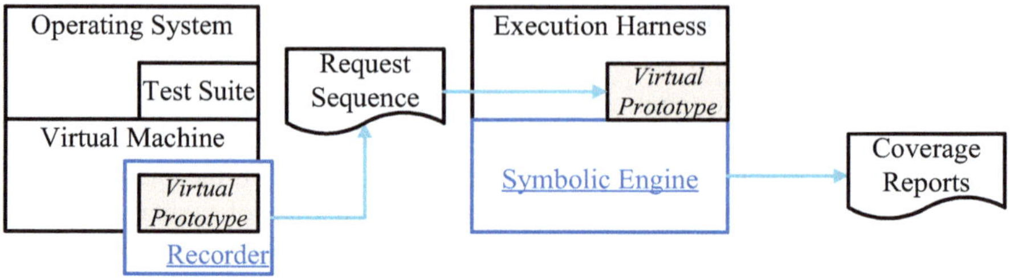

Fig. 14.6 Workflow for coverage evaluation

using pie or bar charts in terms of what and how many requests were made, what and how many transactions were hit, and what percentages they account for among all requests. Moreover, the details of a transaction are recorded, such as registers accessed and interrupt status.

14.3.5 Implementation

As shown in Fig. 14.6, we first capture necessary data from a concrete execution of the virtual prototype within a virtual platform under a given test, and then compute the test coverage by efficiently replaying this execution offline on the virtual prototype itself. Our approach provides early feedback on quality of post-silicon validation tests before silicon is ready.

14.3.5.1 Coverage on Different Levels

To generate coverage reports, we first analyze virtual devices statically to get program information, such as the position of branches and the number of functions, and then generate all kinds of coverage reports based on the execution traces computed by the replay engine. Our approach provides flexibility to generate reports on two different levels:

(1) Event Level: Given an event, a user can check what transaction is explored, what registers are accessed, and whether any interrupt is fired. Moreover, the user can debug the execution trace step by step using the replay engine.

(2) Test Case/Suite Level: A test case/suite issues a sequence of requests to a device. Simultaneously, the device may receive environment inputs and read DMA data. Given a test case/suite, all device events are captured. The replay engine replays all captured events and generates the code coverage, the register coverage, and the transaction coverage for the test case/suite.

14.3.5.2 Implementation Details

We implement our approach on the QEMU virtual platform. The event capture mechanism is implemented as a QEMU module which can be used for hooking QEMU virtual devices. Device interface functions are invoked by the QEMU framework. For instance, a driver issues a read register request, and the QEMU invokes the corresponding read register function defined in the virtual device. Our module hooks all the interface functions when the virtual device registers these functions to QEMU. In this way, the module captures the device events when there is an interface register request, an environment input, or a DMA access. This module provides capability to hook different virtual devices without modifying virtual devices. For capturing events on silicon devices in physical machines, we modified device drivers to achieve it.

We construct our replay engine using the symbolic execution engine KLEE [12]. We modify KLEE in three aspects. First, we implement some special function handler for loading events and DMA data. Second, we capture execution trace during execution of virtual devices. Third, we realize our own module for coverage generation.

14.3.6 Experimental Results

We have implemented and applied Device Coverage Analyzer (DCA) to QEMU-based virtual devices for five popular network adapters: Intel E1000, Broadcom Tigon3, Intel EEPro100, AMD RTL8139, and Realtek PCNet. While our tool currently focuses on QEMU-based virtual devices, the principles also apply to other virtual prototypes. The experiments were performed on a desktop with an 8-core Intel(R) Xeon(R) X3470 CPU, 8 GB of RAM, 250 GB and 7200RPM IDE disk drive and running the Ubuntu Linux OS with 64-bit kernel version 3.0.61.

14.3.6.1 Online-Capture and Offline-Replay Overhead

In order to evaluate our approach, we capture a request sequence triggered by a test suite. The test suite includes most common network testing programs, such as ifconfig and ethtool [13]. DCA needs to capture the initial device state and device events at runtime, which brings overhead to runtime QEMU environment. With the capture mechanism, both QEMU and virtual devices work normally.

To evaluate the overhead of online-capture mechanism, we illustrate the time usage for the whole test suite under the capture configuration and no-capture configuration in Fig. 14.7. Between the capture and no-capture configurations, there is low running time overhead introduced. For example, the overhead for E1000 is about $(570-550)/550 = 3.6\%$.

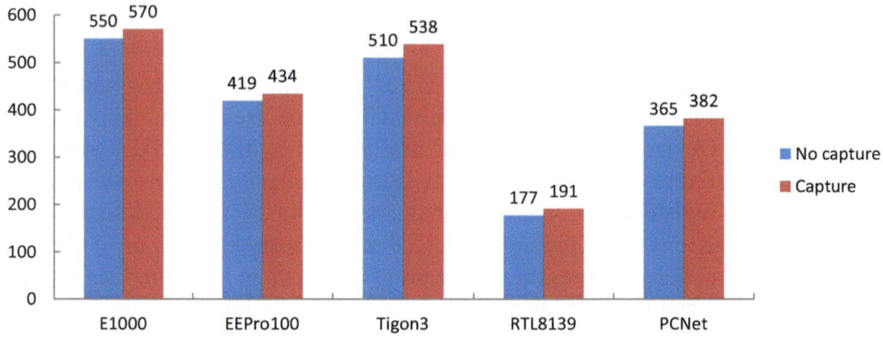

Fig. 14.7 Time usage (seconds) for online-capture

Table 14.1 Time and memory usages for offline-replay

	Events (#)	Time (Min)	Memory (Mb)
E1000	65530	10.5	268.24
Tigon3	89032	12.0	336.35
EEPro100	30112	6.0	213.18
RTL8139	43228	7.0	225.26
PCNet	54016	8.5	254.60

We further evaluated time and memory usages for the offline-replay process. As shown in Table 14.1, time and memory usages of the offline-replay are modest. It only takes a few minutes to process tens of thousands of events.

14.3.6.2 Coverage Results

We demonstrate our coverage results in three aspects: code coverage (statement/block/branch/function coverage), register coverage, and transaction coverage. Due to space limitation, we only illustrate coverage results for E1000 below although we have finished coverage evaluation on all five devices.

Figure 14.8 uses a stack to show incremental coverage of different test programs on E1000 under different code coverage metrics. We evaluate the coverage for both a test case, such as sending a ping packet, and a test suite including most common testing programs. These coverage results can give engineers basic measurement of the quality of test cases.

Figure 14.9 shows partial register coverage results for E1000. Each register is identified using the register offset, such as 0x0 and 0x8. The figure shows that how many times and how much percentage top ten registers are accessed. For instance, the most accessed register is register 0x8 (status register), which is accessed 21927

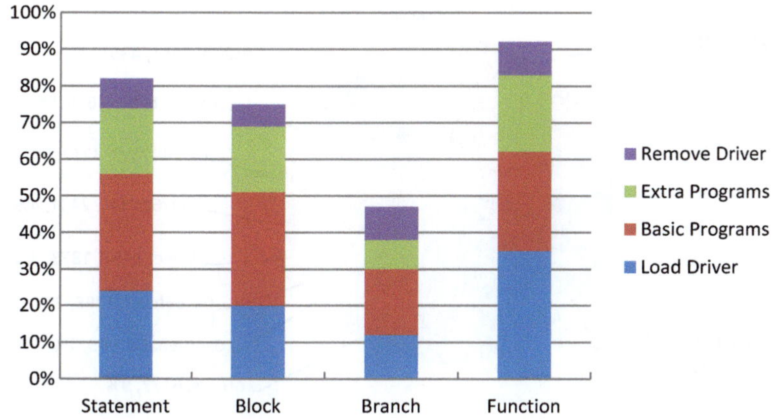

Fig. 14.8 Code coverage results for E1000

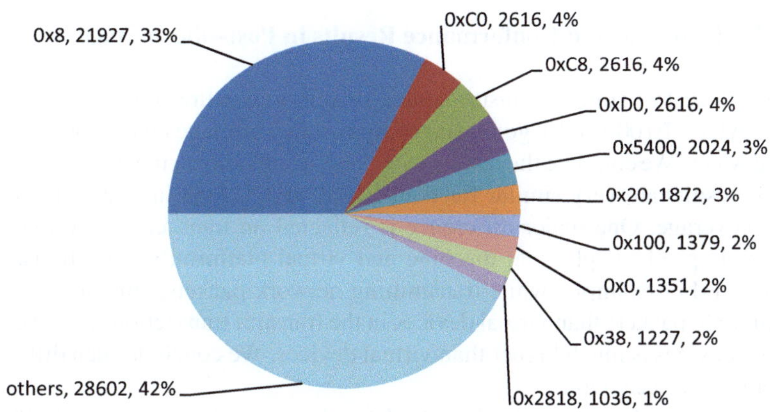

Fig. 14.9 Top ten accessed registers for E1000

times. The system software reads this register very frequently to query the device state.

Figure 14.10 shows partial transaction coverage results for E1000. Each transaction is identified using a hash value, such as 0xd4e4d3ed. It shows that how many times and how much percentage top ten transactions are accessed. By analyzing transaction coverage, engineers can know what functionalities have been tested. By analyzing execution information of each transaction, engineers can further observe register accesses.

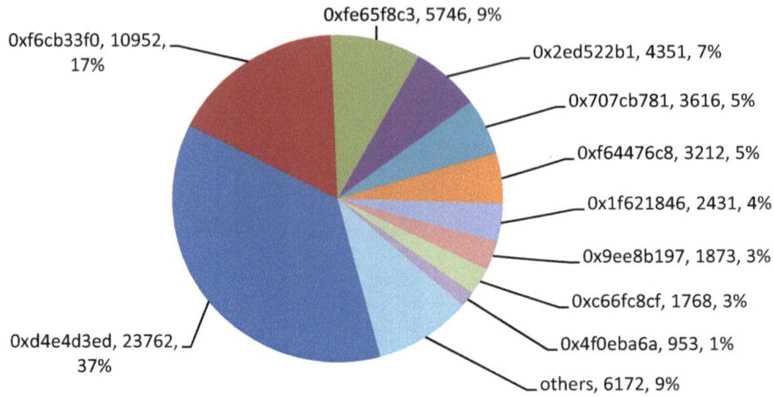

Fig. 14.10 Top ten transactions for E1000

14.3.6.3 Coverage and Conformance Results in Post-silicon Stage

With the same test suite, we instrumented drivers to capture runtime data on two silicon devices, E1000 and Tigon3, and computed the coverage on the corresponding virtual devices. We compare the results with these results shown in Sect. 14.3.6.2. The coverage results are very similar for both E1000 and Tigon3 in terms of code and register coverage. One major difference is reflected on transaction coverage. Due to different speeds of physical machine and virtual platform, several transactions are affected. For example, while transmitting network packets, silicon devices can transmit more packets than virtual devices in the transmit transaction since the speed of silicon devices is much higher than virtual devices. We conclude such differences in coverage are acceptable.

We applied conformance checking to detect inconsistencies between E1000 and Tigon3 and their corresponding virtual devices. There are 13 inconsistencies discovered between the two network adapters and their virtual devices under the given tests: 7 in Intel E1000 and 6 in Broadcom BCM5751. We modified 21 lines of code in virtual devices to fix all 13 inconsistencies. Then, we rerun coverage tools on fixed virtual devices to generate new coverage reports. After comparing the new reports with the post-silicon coverage reports, we found no differences except the known transaction differences.

Remarks: Coverage evaluation in the post-silicon stage often requires instrumenting the device driver and comes too late. Coverage evaluation on virtual prototypes can be available much earlier; therefore, it can guide improvement of post-silicon tests. From conformance checking results and coverage report comparison, it is clear the more conforming the virtual and silicon devices are, the more accurate the coverage evaluation on the virtual device. Even if there exist inconsistencies, conforming checking facilitates quick correction of coverage estimate in the post-silicon stage by conveniently detecting these inconsistencies.

14.3.7 Summary

Quantifying coverage of post-silicon validation tests is very challenging due to limited hardware observability [14]. In this section, we have presented an approach to early coverage evaluation of post-silicon validation tests with virtual prototypes, which fully leverages the observability and traceability of virtual prototypes. We have applied our approach to evaluate a suite of common tests on virtual prototypes of five network adapters. We have also established high confidence in fidelity of coverage evaluation by further conducting coverage evaluation and conformance checking on silicon devices.

14.4 Runtime Analysis of Post-silicon Tests

Runtime shadow execution allows the developer to monitor or diagnose a virtual device's behavior at runtime, ideal for the driver development and testing environment. We can follow the device sequence to trace all state transitions from the initial state. The detailed information of each state transition can be observed. In this section, we describe how to utilize runtime shadow execution to conduct runtime analysis of post-silicon tests.

14.4.1 Symbolic Execution of Virtual Prototypes

14.4.1.1 Motivation

Symbolic execution of virtual prototypes is the foundation for our approach to runtime shadow execution. In order to symbolically execute virtual prototypes, we must address the following technical challenges:

- *Environment modeling.* A virtual device is not a stand-alone program. There are two issues with this incompleteness. First, the virtual device needs to be properly initialized and its entry functions properly exercised. Second, the virtual device may invoke libraries in its environment. Therefore, we need a solution to enclose the virtual device so that the symbolic execution engine can consume it and perform accurate and efficient analysis.
- *Symbolic execution engine adaptation.* We symbolically execute virtual devices using the KLEE symbolic execution engine. KLEE is not specially designed for executing virtual devices while virtual devices have specific characteristics. Hence, we need to adapt KLEE to execute virtual devices efficiently and provide more hardware-specific information.

Sections 14.4.1.2 and 14.4.1.3 show the solutions for solving the above two challenges. Furthermore, we demonstrate that our approach employs symbolic execution engine to support both symbolic execution and analysis of virtual prototypes in Sect. 14.4.2.

14.4.1.2 Harness Generation

For symbolic execution of QEMU virtual devices, we adapt KLEE to handle the non-deterministic entry function calls and symbolic inputs to device models. Since the virtual device by itself is not a stand-alone program, in order for the symbolic engine to execute a virtual device, a harness must be provided for the virtual device. A key challenge here is how to create such a harness. This harness has to be faithful so that the symbolic execution of the virtual device will not generate too many paths infeasible in the real device. On the other hand, it has to be simple enough so that symbolic engine can handle the symbolic execution efficiently. To an extreme, the complete QEMU with the guest OS can serve as the harness which, however, is impractical for the symbolic engine to handle.

Currently, we generate harnesses manually for major device categories. Since devices fall into device categories depending on interface types such as PCI and USB and on functionalities such as network adapters and massive storage devices, we started with creating harnesses for major device categories, e.g., PCI network adapters, and improved such a harness as we experiment on devices in this category. Manual harness generation involves examining how QEMU invokes the virtual device, what QEMU APIs that a virtual device invokes, and what these APIs invoke recursively, and deciding what to include. At times, it may be necessary to make an API produce non-deterministic outputs by throwing away its implementation. The harness includes the following parts as shown in Fig. 14.11:

- Declarations of state variables and parameters of entry functions. A virtual device is not a stand-alone program. If a virtual device is running in a virtual machine, it will register its entry functions with the virtual machine. Moreover, the virtual machine will help the virtual device manage its state variables. Every time an entry function is invoked, the state variables and necessary parameters of the function will be made available to the function by the virtual machine. In order to exercise a virtual device symbolically, we need to handle the state variables and function parameters. Hence, we add declarations of state variables and inputs of entry functions to the harness.
- Code for loading the concrete state and making parameters of entry functions symbolic. In order to cover as many paths as possible in an entry function, we need to make certain inputs of the entry function symbolic. The inputs of an entry function contain state variables and necessary parameters. We implement two utility functions that are specially handled by the engine. Function "load_state" is used for loading the concrete state. Function "make_symbolic" is used for initializing the inputs symbolically.

```
//Declarations of necessary variables
E1000State state; //Device state
target_phys_addr_t address; //Address
......

int main() {
    //Load the concrete state
    load_state(&state, sizeof(state), "state");

    //Make parameters symbolic
    make_symbolic(&address, sizeof(address), "address");
    ......

    //Non-deterministic calls to entry functions
    switch(svd_deviceEntry) {
    case MMIO_WRITE:
        write_reg((void *)&state, address, value);
        break;
    case MMIO_READ:
        read_reg((void *)&state, address);
        break;
        ......
    }
}

//Stub functions
uint16_t net_checksum_finish(uint32_t sum) {
    ......
}
```

Fig. 14.11 Excerpt of E1000 virtual device harness

- Non-deterministic calls to virtual device entry functions. For a real device, there are many ways for the OS and the environment to communicate with it. Similarly, virtual devices provide many types of entry functions for communicating with the OS and the environment. To analyze a virtual device, we go through all entry functions with symbolic inputs. We define a symbolic variable in the harness. With this symbolic variable, we make non-deterministic calls to all entry functions.
- Stub functions for virtual machine API functions invoked by virtual devices. Virtual devices often invoke API functions of virtual machines to achieve certain functionalities. Stubs for these functions need to be provided to complete the harness and are created manually as discussed above.

14.4.1.3 Symbolic Execution Engine Adaptation

To improve efficiency of symbolic execution, we modify KLEE to address four key technical challenges for symbolic execution of virtual devices.

Path Explosion Problem

Path explosion is a major limitation for symbolic execution to thoroughly test software programs. The number of paths through a program is roughly exponential in program size. The problem also exists in executing virtual devices symbolically.

We apply two constraints when executing the virtual device to combat the path explosion problem. First, we add a loop bound to each loop whose loop condition is a symbolic expression. With the loop bound, the user controls the depth of each loop explored. Currently, we add the loop bounds manually in virtual devices. This is practical since there are only a few loops in our analysis of three virtual devices. Second, we can add a time bound to ensure that symbolic execution will terminate in a given amount of time. If the symbolic execution does not complete within the given time bound, there may be unfinished paths. For such paths, we still generate test cases with path constraints obtained so far.

Environment Interaction Problem

A virtual device is a software component and may invoke outside API functions to interact with its environment. We divide such interactions into two categories based on whether a function call affects the values of variables in virtual devices. We detect whether the function has any pointer argument, accesses global variables, or returns a value. If so, this function potentially affects the values of variables in virtual devices. We then use two different mechanisms to handle functions in these two categories.

- If the function call does not affect the values of variables in virtual devices, we instruct KLEE to ignore it and issue a warning.
- If the function call may affect values of variables in virtual devices, we implement this function in our stubs. As there are not many such function calls for a category of virtual devices, such manual effort is acceptable.

Handling DMA

When a virtual device is processing a request, DMA data may be needed. QEMU provides two functions "pci_dma_read" and "pci_dma_write" for reading and writing DMA data separately. We ignore "pci_dma_write" function because it does not affect the device state. We instruct the symbolic execution engine to specially handle "pci_dma_read" function.

We hook "pci_dma_read" function to capture all runtime DMA read data in the concrete execution of the virtual device within the virtual machine. Then, we utilize the captured data in both replay process and test generation process. In the replay process, every time "pci_dma_read" function is invoked, and the corresponding data is loaded into the virtual device by the symbolic execution engine. In the test generation process, we compose a symbolic DMA sequence using the captured DMA data to guide test case generation.

```
//Declarations of a sparse function pointer array
static uint32_t (*macreg_readops[])(E1000State *, int) = {
    [RCTL] = mac_readreg, [TCTL] = mac_readreg, [ICS] =
        mac_readreg,
    [GPTC] = mac_read_clr4, [TPR] = mac_read_clr4, [TPT] =
        mac_read_clr4,
    [ICR] = mac_icr_read, [EECD] = get_eecd, [EERD] =
        flash_eerd_read,
    ......
}
enum { NREADOPS = ARRAY_SIZE(macreg_readops) };

//Invoke the function using the function pointer
static uint64_t e1000_mmio_read(void *opaque, target_phys_addr_t addr,
     unsigned size)
{
    E1000State *s = opaque;
    unsigned int index = (addr & 0x1ffff) >> 2;

    if (index < NREADOPS && macreg_readops[index])
    {
        return macreg_readops[index](s, index);
    }
    ......
}
```

Fig. 14.12 An example of a sparse function pointer array

Sparse Function Pointer Array Problem

A virtual device provides many different functions for realizing different device behaviors. For example, if a write register operation is issued to the virtual device, different functions can be triggered depending on different register offsets. Therefore, it is common for virtual devices to utilize a sparse function pointer array for accessing different functions, which makes the code concise. A sparse function pointer array is shown in Fig. 14.12 which is used by QEMU E1000 virtual device.

If a symbolic execution engine invokes a function defined in the sparse function pointer array with a symbolic offset, the engine tries to explore all possible array offsets in order to cover all functions in the array. In this example, the symbolic engine needs to fork 5845 branches when the "macreg_readops" array is accessed. It takes much time to explore all 5845 branches. Actually, only seven functions are included in this function array. We summarize this information by static analysis of the virtual device. We modify the symbolic execution engine to specially handle sparse function pointer arrays. Every time a sparse function pointer array is accessed, we only fork branches according to the number of valid functions. In this example, we only fork seven branches.

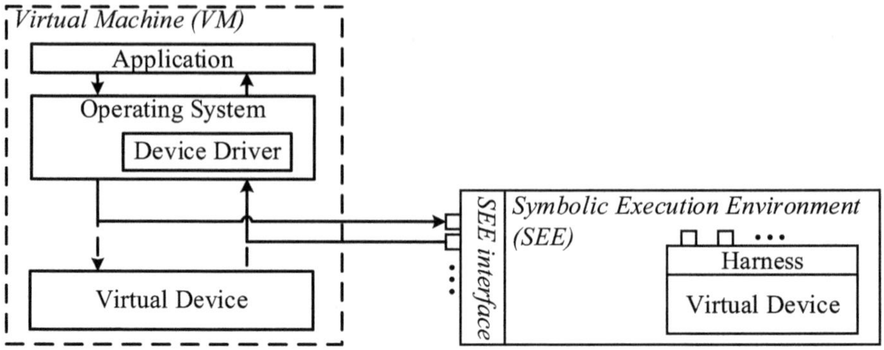

Fig. 14.13 Framework for runtime analysis

14.4.2 Runtime Shadow Execution

14.4.2.1 Runtime Shadow Execution Framework

To better understand device state transitions, we integrate symbolic execution of virtual devices into the virtual machine at runtime. The framework for runtime shadow execution is shown in Fig. 14.13. The SEE interface has been implemented as the bridge between virtual machine and SEE. Our framework can support two modes: monitor and analysis modes.

Our runtime framework does not change the normal working process of the virtual machine. The framework only intercepts the communications between the virtual machine and the virtual device. Furthermore, our framework implements the SEE interface to act as the bridge transferring data from the virtual machine to SEE. The SEE interface intercepts three types of data:

(1) Device states: It captures concrete device states when runtime shadow execution is enabled.

(2) I/O requests and packets: It captures I/O requests and packets when there is a device request from either the driver or the environment.

(3) DMA data: It captures DMA data when DMA data is accessed for processing a device request.

With the captured data, the virtual device can be executed concretely in monitor mode or symbolically in analysis mode.

14.4.2.2 Runtime Monitor Mode

In the monitor mode, the virtual device is executed concretely in both the virtual machine and SEE simultaneously. With the captured data through SEE interface, concrete execution can be conducted to enable runtime step-by-step analysis, which

```
//Declarations of necessary variables
E1000State state; //Device state
target_phys_addr_t address; //Address
......

int main() {
    //Load the concrete device state
    load_state(&state, sizeof(E1000State), "state");

    //Load the concrete device request and request type
    load_request(&address, sizeof(address), "address");
    ......

    //Calls to interface functions
    switch(svd_deviceEntry) {
    case MMIO_WRITE:
        e1000_mmio_write((void *)&state, address, value);
        break;
    case MMIO_READ:
        e1000_mmio_read((void *)&state, address);
        break;
        ......
    }
}
```

Fig. 14.14 Complete harness for runtime monitor mode

helps developers thoroughly understand the concrete state transition for processing a device request.

The harness for runtime monitor is slightly different from the harness for static analysis. An example of such harness is shown in Fig. 14.14. Two special functions "*load_state*" and "*load_request*" are employed to load captured concrete device states and request information within the SEE. Then, the corresponding entry function is invoked according to the request type.

Usually, developers would like to analyze some desired state transitions. We provide two mechanisms to help developers select desired state transitions.

First, we provide a special user-level program to issue special I/O requests to label the start and finish points of a test case. The SEE interface parses all I/O requests. Once the special I/O requests are found, the SEE checks what kind of flag the request stands for. If it is a start flag, the SEE starts analyzing the upcoming requests. If it is a finish flag, the SEE stops analyzing the requests that follow.

Second, we provide two kinds of breakpoints to help developers select desired state transitions. *(1) Statement breakpoint*: The user can select one or more statements as the breakpoints. *(2) Path breakpoint*: Since we collect path information for each test case generated by static analysis, the user can select one path as the breakpoint. Once a breakpoint is hit, the virtual machine is paused and the corresponding path is analyzed within the SEE.

To evaluate a test case, one important mechanism is coverage analysis. Our framework provides an approach to conduct runtime coverage analysis. The VD is executed concretely both in the VM and SEE simultaneously. We collect a concrete execution path for processing a device request. After processing a test case, we can get a set of concrete execution paths. To improve the runtime profiling efficiency, we calculate the coverage report for the test case with the code information pre-generated by static analysis. If we can estimate the coverage of a test case at runtime, it helps developers to understand the quality of the test case. Although an online-capture offline-replay approach has been implemented to deliver detailed coverage report as described in Sect. 14.3, runtime coverage analysis is still very important for developers to get a quick estimation.

14.4.2.3 Runtime Analysis Mode

The virtual device is executed concretely in the virtual machine and symbolically in the SEE simultaneously. The SEE executes the virtual device with the concrete device state and symbolic requests. It computes all the feasible execution paths under the current device state and generates runtime analysis test cases for the covered paths. The feasible paths are provided to the user. This mode enables step-by-step debugging and error injection, by observing device behaviors with generated test cases under the current device state.

The harness for runtime analysis is the same as the one shown in Fig. 14.11. Our approach assists developers in analyzing a virtual device symbolically at runtime. Once a request is selected or a breakpoint is hit in monitor mode, the virtual device can be executed symbolically with a symbolic request under the concrete state. All possible paths are explored. For each possible path, a runtime analysis test case is generated which contains the concrete device state and inputs that can be used to replay the corresponding path symbolically explored. Replaying the test case enables developers better observe and trace any variable change in the virtual device along this path. Furthermore, all paths explored at runtime are reachable. The developers can confirm what paths covered by static analysis can be covered at runtime. The developers can also alter the virtual device execution by injecting a device request identified by the SEE. Developers can execute a device model based on a concrete test case back and forth, step by step, and observe variable changes at each step. Developers can also inspect values of all VD variables easily at any time and check whether the state conforms to the specification. With such observability and traceability, developers can understand what paths exist and get a better understanding of the device behaviors.

14.4.2.4 Further Potentials

This section illustrates how to employ symbolic execution of virtual prototypes to support runtime monitoring and analysis. It can thoroughly analyze each state transition and collect related information. It can further support coverage evaluation which is demonstrated in Sect. 14.3 and test generation [13].

14.4.3 Experimental Results

In this section, we evaluate our approach from the following two key aspects:

1. *Feasibility.* Can virtual devices from popular virtual machines and for real-world devices be analyzed using our approach? How much extra effort is needed to apply our approach to analyze virtual devices?
2. *Usefulness.* Can our approach help developers achieve better observability and traceability? Can we provide a user-friendly tool for our approach?

Based on analyzing the QEMU virtual devices of five popular network adapters, our results demonstrate that our approach can

1. Be readily applied to the device models of all five virtual devices. To execute the device models, we only need to create a small harness for each virtual device and implement a common stub for all virtual devices in the network adapter category.
2. Provide a user-friendly interface to assist developers in achieving better understanding of virtual device behaviors.

14.4.3.1 Feasibility

QEMU includes many virtual devices, which provides a broad range of test cases for our approach. We applied our approach to five virtual devices for popular network adapters, which are released with QEMU, as shown in Table 14.2.

Table 14.2 Five virtual devices for network adapters analyzed

Device	Vendor	Description
E1000	Intel	Pro/1000 gigabit Ethernet adapter
EEPro100	Intel	Pro/100 Ethernet adapter
PCNet	AMD	PCNet32 10/100 Ethernet adapter
RTL8139	Realtek	PCI fast Ethernet adapter
Tigon3	Broadcom	BCM57xx-based gigabit Ethernet adapter

Table 14.3 Summary of five device models

Device	Virtual device		Harness	
	Lines of code	Number of functions	Lines of code	Number of entry functions
E1000	2099	53	74	4
EEPro100	2178	70	85	7
RTL8139	3528	110	111	13
PCNet	2139	50	112	13
Tigon3	4648	34	80	4

To execute virtual devices symbolically, we manually created a simple harness for each virtual device. We also created a common library of stub functions for all five virtual devices. The stub library has 481 lines of C code. More details about the device models and their harnesses are given in Table 14.3. All device models are non-trivial in size ranging from 2099 lines to 4648 lines of C code. All harnesses are relatively easy to create, having about 100 lines only. Only several hours are needed to create and fine-tune each harness and the stub library.

The experiments were performed on a laptop with an 8-core Intel(R) Core(TM)2 i7 CPU, 8 GB of RAM, 320 GB and 7200RPM IDE disk drive and running the Ubuntu Linux OS with 64-bit kernel version 2.6.38.

14.4.3.2 Usefulness

Our approach assists developers in analyzing a virtual device and generating test cases. A test case contains the concrete values for the device state and inputs that can be used to replay the corresponding path symbolically explored. Replaying the test case enables developers better observe and trace any variable change in the virtual device along this path. Developers can execute a device model based on a concrete test case back and forth, step by step, and observe variable changes at each step. Developers can also inspect values of all virtual device variables easily at any time and check whether the state conforms to the specification. With such observability and traceability, developers can understand what paths exist and get a better understanding of the device behaviors.

We have implemented an RCP-based (Eclipse Rich Client Platform) Graphical User Interface (GUI) to assist developers in replaying test cases, which is illustrated in Fig. 14.15. We have demonstrated our tool to industry developers. The feedbacks from these developers are that our tool is useful, can provide in-depth knowledge about virtual devices, and they are willing to use our tool.

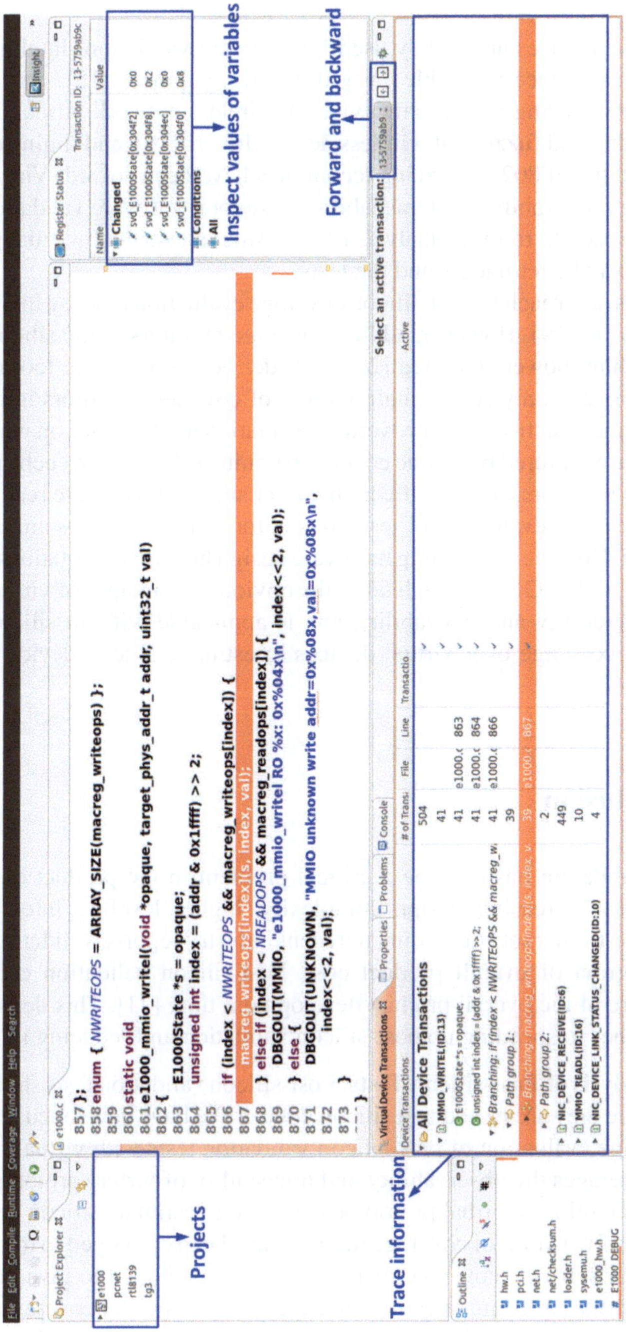

Fig. 14.15 Test case replay GUI

14.5 Related Work

Recently, virtual devices are widely used for software validation. Intel has utilized a network virtual device to enable early driver development [15], and bugs were found in the driver using the virtual device. A wireless network virtual device was created for testing and fuzzing of wireless device drivers [16], and timing difficulties inherent to traditional 802.11 fuzzing techniques have been solved. Virtual devices bring complete observability and traceability to support software validation, and we present an approach to runtime analysis of post-silicon tests with virtual prototypes to help developers better understand the tests.

One common approach to post-silicon coverage evaluation is to use in-silicon coverage monitors [17–19]. However, adding coverage monitors to the silicon is costly in terms of timing, power, and area [20]. In order not to introduce too much overhead, developers can only add a small number of coverage monitors in the design. Consequently, the effectiveness of coverage evaluation highly relies on what kinds of device signals are captured by in-line coverage monitors. Moreover, such approach of using coverage monitors can take effect only after silicon devices are ready. Another approach to coverage evaluation of test cases before silicon devices are available is RTL emulation. However, emulating hardware design has some limitations as we discussed in Sect. 14.3.1. Our approach takes the obvious advantages of virtual devices: complete observability and traceability, and is applicable without silicon devices. We utilize test coverage over virtual devices to estimate silicon device functional coverage.

14.6 Conclusion

Post-silicon validation has become a critical problem in the product development cycle, driven by increasing design complexity, higher level of integration, and decreasing time-to-market. According to recent industry reports, validation accounts for a large portion of overall product cost. Post-silicon validation consumes an increasing share of the overall product development time [21]. This demands innovative approaches to speeding up post-silicon validation and reducing its cost.

Coverage Analysis of Post-silicon Tests. Post-silicon validation tests should be well evaluated before they are issued to a silicon device. We have developed an approach to early coverage evaluation of post-silicon validation tests with virtual prototypes, which fully leverages the observability and traceability of virtual prototypes.

The approach utilizes virtual prototype coverage to estimate silicon device functional coverage. Two kinds of coverage metrics have been employed to the evaluation process. Typical software coverage metrics have been adopted to give basic coverage indication. Two hardware-specific coverage metrics, register coverage and transaction coverage, have been developed to deliver more accurate hardware-oriented coverage results.

The approach has been used for evaluating a suite of common tests on virtual prototypes of five network adapters. High confidence has been established in fidelity of coverage evaluation by further conducting coverage evaluation and conformance checking on silicon devices. With this early coverage estimation, it can guide further test generation.

Runtime Analysis of Post-silicon Tests. Early analysis of post-silicon tests is critical for developers to better understand device behaviors and improve the quality of the tests. We have developed an approach to early runtime analysis of post-silicon validation tests with virtual prototypes, which fully take advantage of the whitebox nature of virtual prototypes.

The approach employs a shadow execution environment to execute a virtual prototype in a controlled environment simultaneously along the normal execution. Such shadow execution allows developers to execute any device transaction back and forth on a virtual prototype and observe any device state change. Furthermore, the framework provides a thorough analysis on each device transaction to allow developers analyze a different transaction under same state with a different request.

An eclipse-based user interface has been developed to provide a user-friendly GUI to help developers to do runtime analysis. The approach has been evaluated on virtual prototypes of five network adapters. It shows powerful analysis and debugging capabilities.

References

1. International Business Strategies, Inc., Global systemIC industry service monthly reports (2014), http://www.ibs-inc.net
2. A. Nahir, A. Ziv, M. Abramovici, A. Camilleri, R. Galivanche, B. Bentley, H. Foster, A. Hu, V. Bertacco, S. Kapoor, Bridging pre-silicon verification and post-silicon validation, in *DAC* (2010)
3. P. Sampath, B. Rachana Rao, Efficient embedded software development using QEMU, in *13th Real Time Linux Workshop* (2011)
4. S. Nelson, P. Waskiewicz, Virtualization: writing (and testing) device drivers without hardware, in *Linux Plumbers Conference* (2011)
5. K. Cong, L. Lei, Z. Yang, F. Xie, Coverage evaluation of post-silicon validation tests with virtual prototypes, in *DATE* (2014)
6. F. Bellard, QEMU, a fast and portable dynamic translator, in *USENIX ATEC* (2005)
7. B. Fabrice, QEMU (2013), http://wiki.qemu.org/Main_Page
8. L. Lei, F. Xie, K. Cong, Post-silicon conformance checking with virtual prototypes, in *DAC* (2013)
9. L. Lei, K. Cong, F. Xie, Optimizing post-silicon conformance checking, in *ICCD* (2013)
10. L. Lei, K. Cong, Z. Yang, F. Xie, Validating direct memory access interfaces with conformance checking, in *Proceedings of the 2014 IEEE/ACM International Conference on Computer-Aided Design* (2014)
11. J. Keshava, N. Hakim, C. Prudvi, Post-silicon validation challenges: how EDA and academia can help, in *DAC* (2010)
12. C. Cadar, D. Dunbar, D. Engler, KLEE: unassisted and automatic generation of high-coverage tests for complex systems programs, in *OSDI* (2008)

13. K. Cong, F. Xie, L. Lei, Automatic concolic test generation with virtual prototypes for post-silicon validation, in *ICCAD* (2013)
14. S Mitra, S.A. Seshia, N. Nicolici, Post-silicon validation opportunities, challenges and recent advances, in *DAC* (2010)
15. S. Nelson, P. Waskiewicz, Virtualization: writing (and testing) device drivers without hardware (2011), http://www.linuxplumbersconf.org/2011/ocw/sessions/243
16. S. Keil, C. Kolbitsch, Stateful fuzzing of wireless device drivers in an emulated environment, in *Black Hat Japan* (2007)
17. K. Balston, M. Karimibiuki, A.J. Hu, A. Ivanov, S.J.E. Wilton, Post-silicon code coverage for multiprocessor system-on-chip designs. IEEE Trans. Comput. (2011)
18. T. Bojan, M.A. Arreola, E. Shlomo, T. Shachar, Functional coverage measurements and results in post-silicon validation of Core^{TM} 2 duo family, in *HLVDT* (2007)
19. X. Liu, Q. Xu, Trace signal selection for visibility enhancement in post-silicon validation, in *DATE* (2009)
20. A. Adir, A. Nahir, A. Ziv, C. Meissner, J. Schumann, Reaching coverage closure in post-silicon validation, in *HVC* (2010)
21. E. Singerman, Y. Abarbanel, S. Baartmans, Transaction based pre-to-post silicon validation, in *DAC* (2011)

Chapter 15
Utilization of Debug Infrastructure for Post-Silicon Coverage Analysis

Farimah Farahmandi and Prabhat Mishra

15.1 Introduction

The exponential growth of System-on-Chip (SoC) complexity, time-to-market reduction, and the huge gap between simulation speed and hardware emulation speed force the verification engineers to shorten the pre-silicon validation phase. There is a high chance that many bugs escape from pre-silicon analysis and it affects the functionality of the manufactured circuit. To ensure the correct operation of the design, post-silicon validation is necessary. However, post-silicon validation is a bottleneck due to limited observability, controllability, and technologies to cope with future systems [20]. There is a critical need to develop efficient post-silicon validation techniques.

Currently, there is no effective way to collect coverage of certain events directly and independently on silicon. Engineers need to assume that they will cover at least the same set of post-silicon coverage events as they cover with pre-silicon exercisers using accelerators/emulators [2]. However, we cannot be sure about the accuracy of these coverage metrics since silicon behaves differently than the simulated/emulated design mainly because of asynchronous interfaces. Moreover, because of time constraints, validation engineers are not able to hit all of the desired coverage events during pre-silicon validation or some coverage events are not activated enough. Therefore, they are seeking for an accurate and efficient way to know the coverage of desired events on silicon (Fig. 15.1).

Post-silicon validation techniques consider many important aspects such as effective use of hardware verification techniques [21] and stimuli generation [1]. Several approaches are also focused on test generation techniques [9, 11] that can address various challenges associated with post-silicon debug. Assertions and associated checkers are widely used for design coverage analysis in pre-silicon validation to reduce debugging time. They can also be synthesized and used in the form of cov-

F. Farahmandi (✉) · P. Mishra
Department of Computer and Information Science and Engineering,
University of Florida, Gainesville, USA
e-mail: ffarahmandi@ufl.edu

P. Mishra
e-mail: prabhat@ufl.edu

P. Mishra and F. Farahmandi (eds.), *Post-Silicon Validation and Debug*,
https://doi.org/10.1007/978-3-319-98116-1_15

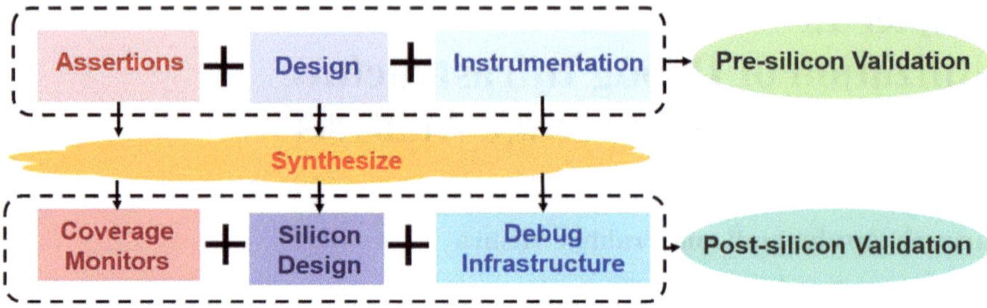

Fig. 15.1 Coverage analysis using synthesized monitors

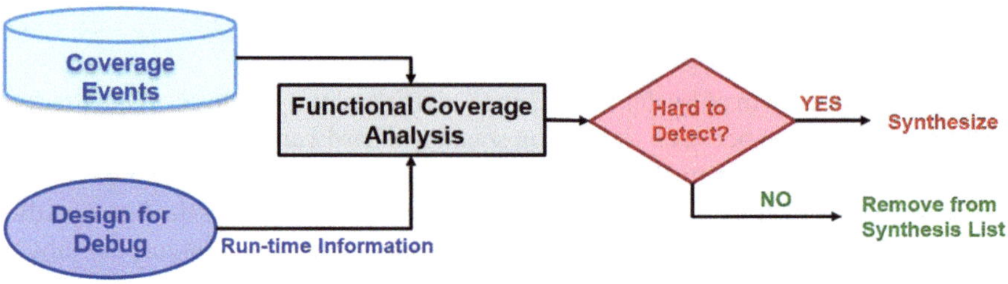

Fig. 15.2 Overview of coverage analysis using design for debug infrastructure

erage monitors to address controllability and observability issues in post-silicon [4]. Figure 15.2 shows an overview of the coverage analysis using synthesized coverage monitors. The pre-silicon assertions are converted to automatons and gate-level hardware to monitor certain events during post-silicon validation [4, 6]. Assertion multiplexing and assertion grouping can be utilized to reduce the number of synthesized coverage monitors. The process of generating coverage monitors from pre-silicon assertions can be automated [8]. However, there are other challenges that need to be addressed effectively in order to use coverage monitors in post-silicon validation. Finding minimum set of assertions to synthesize in order to improve the design observability, efficiently gathering status (pass/fail) information of hardware assertions, and distributing economical set of enable/disable signals without using many gates are the issues that should be considered [3]. Coverage monitors can be reconfigured during run-time to change the focus of the observability [13].

Unfortunately, synthesizing coverage monitors introduce additional area, power and energy overhead that may violate the design constraints. Adir et al. proposed a method to utilize post-silicon exerciser on a pre-silicon acceleration platform in order to collect coverage information from pre-silicon [3]. However, the collected pre-silicon coverage may not accurately reflect post-silicon coverage in many scenarios. To address these limitations, different frameworks are introduced to reduce the number of coverage monitors in post-silicon while utilizing the existing debug infrastructure to enable functional coverage analysis [10, 12, 15]. In this chapter, we describe the main idea of these approaches.

Knowledge of internal signal states during post-silicon execution helps to trace the failure propagation to debug the circuit. There are several built-in debug-for-debug (DfD) mechanisms such as trace buffers and performance monitors in order to enhance the design observability during post-silicon validation and reduce debugging effort. Trace buffers record the values of a limited number of selected signals (typically less than 1% of all signals in the design) during silicon execution for specified number of clock cycles. The trace buffer values can be analyzed off-chip to restore the values of untraced signals. There are different techniques to select trace signals such as structure/metric-based selection [5, 18], simulation-based selection, as well as hybrid of both approaches [16]. Recently, Ma et al. have proposed a metric that models behavioral coverage [19]. However, none of these approaches consider functional coverage analysis as a constraint for signal selection.

This chapter presents an approach to utilize the information that can be extracted from on-chip trace buffer in order to determine easy-to-detect functional coverage events. The overview of the approach is shown in Fig. 15.2. The trace-based coverage analysis enables the trade-off between observability and hardware overhead. Although the proposed method can provide coverage data only for the recorded cycles (instead of the complete execution), it can significantly reduce the post-silicon validation effort for various reasons. First, the traced data for specific cycles are able to restore untraced signals for additional cycles. Moreover, when we know that certain events have been hit (covered) by trace analysis, the validation effort can be focused on the remaining set of events. From a practical perspective, it is not valuable to collect coverage data when generating test case for an exerciser [14] or during checking a test case since the exerciser code is relatively simple, repetitive, and not expected to hit bugs on silicon.

The remainder of the chapter is organized as follows. Section 15.2 provides an overview about assertion-based validation. Section 15.3 describes the post-silicon functional coverage analysis framework. Section 15.4 presents some experimental results. Finally, Sect. 15.5 concludes this chapter.

15.2 Background

Based on the functional coverage goal, a design is instrumented to check specific conditions of few internal signals. For example, assertions are inserted in a design to monitor any deviation from the specification. These days, designers mostly use one of the powerful assertion languages such as PSL (Property Specification Language) to describe interesting behavioral events. First, we give an overview of PSL assertions and then we describe our method using them. However, the presented method is not dependent on any assertion language. There are two general types of assertions: *assert* and *cover* statements. There is a single bit associated with *assert* which indicates the pass or fail status of the assertion. The *cover* assertion triggers at the end of the execution when the assertion is not covered during the runtime. PSL assertions contain several layers such as Boolean and temporal layers and it can be used on

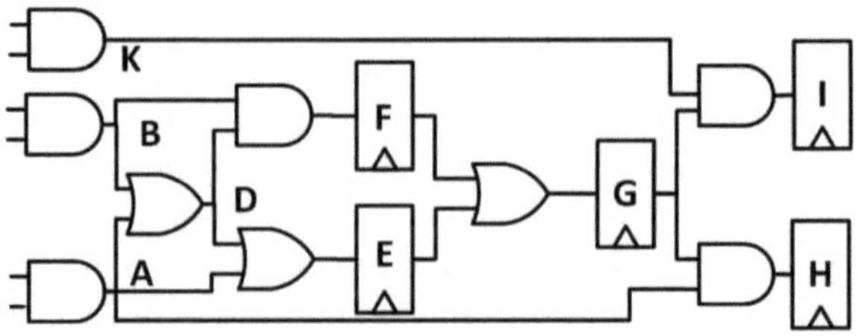

Fig. 15.3 A simple circuit to illustrate design properties

top of different HDL languages including Verilog and VHDL. It denotes temporal sequence using different operators such as ";" (notation of one clock cycle step), ":" operator for concatenation and operators $[low : high]$, $[*]$ or $[+]$ present the notation of bounded or unbounded repetition. Operator $[*]$ shows the repetition of zero or more instances; however, $[+]$ denotes one or more repetition. Operators "&" and "|" show the logical AND and OR between sequences. Different operators such as *always*, *eventually*! and implication and nonoverlapped implication, which means right-hand side property can hold one cycle after occurrence of left-hand side sequence can be combined with other arguments and operators. Assertions are usually accompanied by an active edge of the clock (usually the rising edge of the clock is considered as default). PSL assertion can include Boolean expressions (b_i), sequences of events of Boolean primitives ($s_i = b_l; ...; b_r$) and properties on Boolean expressions and sequences. Assertions can be classified into two groups: conditional and obligation [7]. The goal of a conditional assertions is to detect a failure. Therefore, it is activated every time all of its events are observed. For example, "$assert\ never\ b_1$" is called a conditional assertion as every time b_1 is evaluated true, a failure will happen and assertion should be triggered. On the other hand, an assertion is in obligation mode when a failure in its sequence triggers it. For example, "$assert\ always\ b_2$" is in obligation mode as b_2 should be always true and every time it is evaluated to false, the assertion is activated.

Example 1 Consider a part of a circuit shown in Fig. 15.3. Suppose that we have two design properties: first, whenever signal E asserted, signal H is supposed to be asserted within one–three cycles. The following assertion describes this property $A_1 : assert\ always(E \rightarrow \{[1:3]; H\})\ @rising\_edge(clk)$.

Consider a second property where we would like to cover functional scenarios such that D and I signals are not true at the same time. This property can be formulated as $A_2 : assert\ never(D\&I)$.

15.3 Post-silicon Functional Coverage Analysis

To have full observability in post-silicon, one option is to synthesize all of the functional scenarios (typically thousands of assertions, coverage events, etc.) to coverage monitors and track their status during post-silicon execution. However, this option is not practical due to unacceptable design overhead. Therefore, designers would like to remove all or some of the coverage monitors to meet area and energy budgets. It creates a fundamental challenge that how we can decide which coverage monitors can be removed. In this chapter, we present an approach to evaluate the assertion activation efforts by on-chip trace buffer and rank them based on the difficulty in covering/detecting them. Clearly, the hard-to-detect ones should be synthesized, whereas the easy-to-detect ones can be ignored (trace analysis can cover them). The presented approach consists of four major steps: decomposition of coverage scenarios, signal restoration, coverage analysis, and signal selection. The remainder of this section describes these steps.

15.3.1 Decomposition of Coverage Scenarios

Suppose that a gate-level design D as well as a set of pre-silicon RTL assertions $\mathbb{A}$ are given and the goal is to use trace buffer information to determine activation of $\mathbb{A}$ in model D during silicon execution. The decomposition can be done in pre-silicon. First, RTL assertions are scanned to extract their signals and their corresponding gate-level signals based on name mapping methods. Next, each RTL assertion from set $\mathbb{A}$ is mapped to a set of clauses such that each clause contains assignments to a set of signals in specific cycles. Figure 15.4 shows the overview of the assertion decomposition approach.

Formally, each pre-silicon assertion $A_i \in \mathbb{A}$ is scanned and its signals and its corresponding gate-level signals are defined. Then, A_i is decomposed to a set of clauses $A \equiv \mathbb{C} = \{C_1, C_2, ..., C_n\}$ based on its mode (conditional or obligation). Each C_j can be formalized as $C_j = \{\alpha_1 \Delta_1 \alpha_2 \Delta_2 ... \Delta_{m-1} \alpha_m\}$. Each α_k presents a Boolean assignment on gate-level signal $n \subset \mathbb{N}$ ($\mathbb{N}$ shows all corresponding gate-level signals of set $\mathbb{A}$) on cycle c_t where $1 \leq c_t \leq CC$ (Suppose that we know that the manufactured design will be simulated for maximum CC clock cycles) such as $a_i : \{n = val\ in\ cycle[c_t]\}$ where $val \in \{0, 1\}$. Operators Δ_k can be one of the logical operations such as AND, OR or NOT. As a result, the original assertion is translated as a set of clauses $\mathbb{C}$ in a way that activation of one of them triggers the original assertion.

From now on, we assume that signals of A_i are mapped to corresponding assertion on gate-level signals. The following rules are used to generate the set of clauses ($\mathbb{C}$):

- If an assertion is in obligation mode and it contains an AND operator ($p \wedge q$), the operands are negated and AND will be changed to a OR operator ($p \vee q$). For example, in assertion

$$assert\ \ always\ \ p\ \ \&\ \ q,$$

  whenever conditions p or q are false, the assertion is activated. Therefore, the assertion is translated to a set of clauses as $\mathbb{C} = \bigcup_{t=1}^{CC} \{p = 0[t] \vee q = 0[t]\}$ (clauses are also expanded over time).

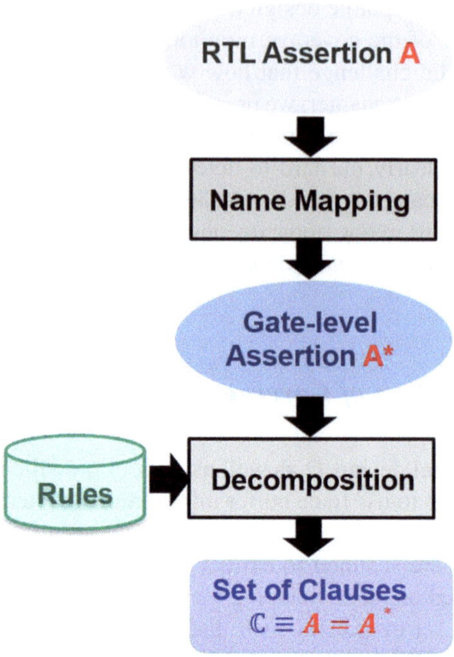

Fig. 15.4 Decomposition of coverage scenario "A" to a set of clauses

- If an assertion is in obligation and it contains OR operator such as

$$assert \quad always \quad (p \;\mid\; q),$$

  the conditions will be negated and the set of clauses is extracted as: $\mathbb{C} = \bigcup_{t=1}^{CC}\{p = 0[t] \wedge q = 0[t]\}$.
- If an assertion is in obligation mode and it contains an implication operator, antecedent conditions are not modified. However, consequent conditions are negated. For example, if we have

$$assert \quad always \quad (p \rightarrow next \quad q),$$

  it is converted to $\mathbb{C} = \bigcup_{t=1}^{CC}\{p = 1[t] \wedge q = 0[t+1]\}$. Next operation shows its effect in condition $q = 0[t+1]$.
- If an assertion is in conditional and it contains OR operator such as

$$assert \quad never \quad (p \;\mid\; q),$$

  it is translated to $\mathbb{C} = \bigcup_{t=1}^{CC}\{p = 1[t] \vee q = 1[t]\}$.

- If an assertion is in conditional mode and it contains an AND operator $(p \wedge q)$, the operation is kept as it is. For example, in assertion

$$assert \quad never \quad (p \quad \& \quad q),$$

if p and q are true at the same time, the assertion will be activated. Therefore, the assertion is translated as $\mathbb{C} = \bigcup_{t=1}^{CC}\{p = 1[t] \wedge q = 1[t]\}$.
- If an assertion is in conditional mode and it contains implication operator, antecedent, and consequent conditions remain the same as the original assertion. For example, if we have

$$assert \quad never \quad (p \rightarrow next \quad q),$$

it is converted to $\mathbb{C} = \bigcup_{t=1}^{CC}\{p = 1[t] \wedge q = 1[t+1]\}$.
- If there are *eventually!* or *until* operators in an assertion, based on the mode of assertion it shows its effect in generating repeating conditions in different clock cycles. For example,

$$assert \quad always \quad (p \rightarrow \quad eventually \quad q)$$

is translated to $\mathbb{C} = \bigcup_{t=1}^{CC}\{(p = 1[t]) \wedge (q = 0[t] \wedge q = 0[t+1] \wedge ... \wedge q = 0[CC])\}$. On the other hand, if we have an assertion as

$$assert \quad always \quad (p \quad until \quad q),$$

the conditions are found as: $\mathbb{C} = \bigcup_{t=1}^{t+n=CC}\{(p = 1[t]) \wedge (p = 0[t+1] \vee p = 0[t+2] \vee ... \vee p = 0[t+n-1]) \wedge (q = 1[t+n])\}$.

Example 2 Consider the assertions in Example 1 form Sect. 15.2. We assume that the circuit will be executed for 10 clock cycles for post-silicon validation. The first property (A_1) will be decomposed to equivalent conditions as follows:
$\mathbb{C}_{A_1} : \{\{E = 1[1] \wedge H = 0[1] \wedge H = 0[2] \wedge H = 0[3]\}, \{E = 1[2] \wedge H = 0[3] \wedge H = 0[4] \wedge H = 0[5]\}, ..., \{E = 1[7] \wedge H = 0[7] \wedge H = 0[9] \wedge H = 0[10]\}\}$

The assertion is activated if signal E is asserted and signal H remains false for next three cycles. The second property is decomposed as shown below since it will be activated if both D and I are true at the same time.
$\mathbb{C}_{A_2} : \bigcup_{t=1}^{10}\{D = 1[t] \wedge I = 1[t]\}$

The computed conditions are used to detect activation of assertions during post-silicon validation as described in Sect. 15.3.3.

15.3.2 Restoration of Signal States

Suppose that we have a gate-level design with $\mathbb{G}$ internal signals and the design has been executed for CC clock cycles during post-silicon validation. A set of signals ($\mathbb{S}\ where\ \mathbb{S} \subset \mathbb{G}$) are sampled and their values are stored in trace buffer T during post-silicon execution for CC_t clock cycles ($CC_t \leq CC$). The information of the trace buffer (with $|\mathbb{S}|$ and CC_t dimensions) can be used to find the values of other signals ($\mathbb{G} - \mathbb{S}$). The restoration starts from the stored values of $\mathbb{S}$ signals over CC_t cycles and go forward and backward to fill the values of matrix $\mathbb{M}_{\mathbb{G}xCC}$. Matrix $\mathbb{M}$ is used to present states of the design during CC clock cycles. Each cell of matrix $\mathbb{M}$ can have value 0, 1, or X. Value X in $m_{i,j} \in \mathbb{M}$ presents the fact that the value of signal i in clock cycle j cannot be restored based on traced values of $\mathbb{S}$ sampled signals. Matrix $\mathbb{M}$ information will be utilized to determine if any of assertions is definitely covered during runtime.

15.3.3 Coverage Analysis

The goal is to use both traced and restored values to check the clauses that are found in Sect. 15.3.1 to define easy-to-detect assertions. In order to find coverage for assertions in set $\mathbb{A}$, each assertion $A_i \in \mathbb{A}$ is decomposed to clauses, set $\mathbb{C}$, as described in Sect. 15.3.1. Set $\mathbb{C}$ is designed in a way that if one of the $C_i \in \mathbb{C}$ can be evaluated to true on matrix $\mathbb{M}$, assertion A_i is triggered. Using the proposed method, each C_i contains a set of Boolean functions (α_j) and each $\alpha_j : n = val$ in cycle t where $1 \leq t \leq CC$ is mapped to one cell of matrix $\mathbb{M}$ ($m_{n,t}$). If the value of $m_{n,t}$ is equal to $val \in \{0, 1\}$, the condition α_j is evaluated true. Condition C_i is evaluated true when the expression consisting of all α_j and Δs evaluated to be true. An assertion is called covered during post-silicon validation if one of its C_is is evaluated to be true.

For assertions that originally contains implication operator ($A : assert\ p \rightarrow q$), the information that which Boolean α_j belongs to the precondition (p) is kept and which conditions belongs to the fulfilling condition (q) when their conditions over $\mathbb{M}$ are checked. Checking assertion A starts from rows which belongs to signals existing in antecedent and check every cycle to find the desired value. Then, the search is continued for consequent from those cycles when antecedent is true to find values that make whole A true. In other words, to be able to find out the activation of assertion A, we need to minimize the number of X values in cells of matrix $\mathbb{M}$. The approach can count how many times assertion A is activated for sure. Note that for checking conditions, 3-valued (ternary) logic is used. In other words, condition $p \vee q$ is evaluated as true if signal p is true and q has X value and vice versa. The proposed framework counts the number of times an assertion is activated during execution. If count is equal to zero, it means that the activation of the assertion cannot be determined based on trace buffer values. The coverage percentage is computed by counting the assertions that have been activated at least once and dividing it by the number of total assertions.

Signal/ Cycle	1	2	3	4	5	6	7	8	9	10
A	0	1	0	1	1	0	0	0	1	0
B	1	0	0	1	0	0	0	1	1	1
K	X	X	X	X	X	X	X	X	X	X
D	1	1	0	1	1	0	0	1	1	1
E	X	1	1	0	1	1	0	0	1	1
F	X	1	0	0	1	0	0	0	1	1
G	X	X	1	1	0	1	1	0	0	1
I	X	X	X	X	X	0	X	X	0	0
H	X	0	X	0	1	0	0	0	0	0

Fig. 15.5 Restored signals for circuit shown in Fig. 15.3 when A and B are trace signals

Example 3 Consider the circuit shown in Fig. 15.3 and the associated assertions shown in Example 1. Suppose that the only two signals (A and B) can be traced during post-silicon validation (the width of trace buffer is two). Note that the signal selection is not limited to flip-flops and every internal signal can be considered as potential sampled signal. Figure 15.5 shows the states of the design based on the stored values of A and B signals. In fact, Fig. 15.5 shows matrix $\mathbb{M}$. Suppose that we take the clauses shown in Example 2 and the matrix shown in Fig. 15.5 as inputs to compute the coverage of these assertions during runtime. Based on information shown in Table 5, assertion A_1 is activated since signal E is asserted in cycle 6 and signal H remains zero in the next three cycles, 7, 8, and 9. However, we cannot comment on A_2 since the respective conditions cannot be evaluated.

Until now, we identified which assertions are activated during runtime for sure. The assertions are ranked based on the required efforts to detect them using the proposed method of Sect. 15.3.3 to decide which assertions are better to be kept as coverage monitors in post-silicon to improve the design observability and increase the assertion coverage. For example, if we have the budget to synthesize 10% of assertions, we can choose these assertions from hard-to-detect assertions. In other words, the assertions that can be easily detected using the proposed method can have a lower priority. The assertions that are hard-to-detect (for example, cannot be detected even one time using the proposed method) or represent critical functional scenarios are best candidates to be kept as coverage monitors in silicon to improve the design observability.

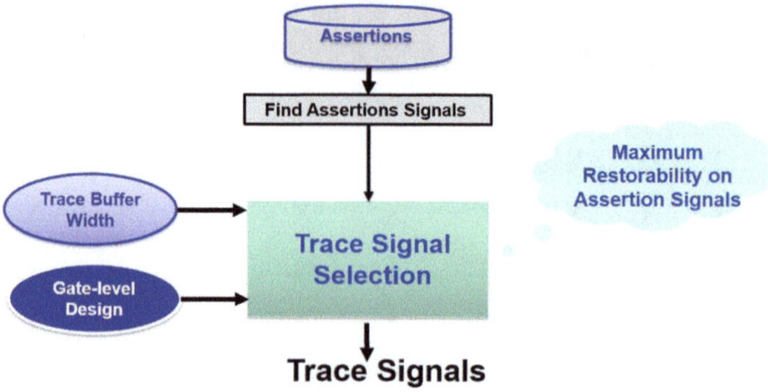

Fig. 15.6 Coverage-aware trace signal selection

Signal/ Cycle	1	2	3	4	5	6	7	8	9	10
A	0	1	0	1	1	0	0	0	1	0
B	X	X	X	X	X	X	X	X	1	X
K	X	1	1	1	X	0	1	X	1	X
D	X	1	X	1	1	X	X	X	1	X
E	X	X	1	X	1	1	X	X	X	1
F	X	X	X	X	X	X	X	X	X	1
G	X	1	1	1	X	1	1	X	0	1
I	0	0	1	1	1	0	0	1	0	0
H	X	0	1	0	1	X	0	0	0	0

Fig. 15.7 Restored signals for circuit shown in Fig. 15.3 when A and I are trace signals

15.3.4 Coverage-Aware Signal Selection

Traditional signal selection methods select signals that have priority over other design signals as they may have a better restorability and more internal signals might be restored during the off-chip analysis. However, if we select trace signals that have better restorability on signals appear on assertions, we can increase the chance of finding the activation of assertions. In order to select trace signals that have a better restorability on assertion signals, pre-silicon assertions $\mathbb{A}$ are scanned to find their signals and their importance based on how many times a specific signal is repeated. Existing simulation-based trace signal selection algorithms can be modified to select signals which have maximum restoration ratio on assertion signals based on the simulated values of random test vectors for several cycles. Figure 15.6 shows the overall flow of the signal selection approach.

Example 4 As it can be seen from Example 3, the activation status of assertion A_2 cannot be detected based on information of Fig. 15.5. Using coverage-aware trace signal selection approach, A and I are selected as trace signals based on their good restorability on assertion signals (E, H, D and I). Restoration and coverage analysis (using the traced values of new signals) would be able to detect activation of both assertions in Example 1 as shown in Fig. 15.7.

15.4 Experiments

Table 15.1 presents results for assertion coverage of total 12,000 (4000 for each trace buffer configuration) assertions for each benchmark. The trace buffers were chosen with a widths of 8, 16, and 32 and depth of 1024.[1] Assertions are in obligation and conditional modes based on the presented method of [17]. The benchmarks are simulated using different trace buffers for 1024 clock cycles with random test vector to model post-silicon validation. The first three columns show the type of the benchmark, the number of its gates and the width of its trace buffer, respectively. The fourth column shows the restoration ratio based on existing trace signal selection. The fifth column shows the coverage of assertions using trace buffer (without introducing any overhead). Note that, *Observability-aware SS* represents our assertion coverage analysis framework on top of existing signal selection techniques. We improved the assertion coverage using our proposed signal selection algorithm with negligible effect on restorability of whole design (sixth and seventh columns). Note that, *Coverage-aware SS* represents our assertion coverage analysis framework on top of our coverage-aware signal selection method. Since signal selection algorithms are based on heuristic methods, in some of the cases, our coverage-aware signal selection algorithms improves the restorability of the design (such as s15850).

The results shown in Tables 15.1 show the extent of functional coverage analysis without introducing any hardware penalty for synthesized coverage monitors. Figure 15.8 shows coverage improvement if we randomly choose 10–90% of the remaining assertions that we cannot be sure about their activation using trace buffer information. The straight line shows the coverage when our method is not used and observability is provided only by using synthesized coverage monitors (the percentage of observability is equal to the percentage of synthesized assertions). On the other hand, the approach presented in Sect. 15.3.3 is used to select hard to detect coverage monitors.

As it can be seen in Fig. 15.8, 100% observability is achieved with significant reduction in overhead (40–50% coverage monitors with observability-aware signal selection can provide 100% functional coverage). Figure 15.9 shows the result

[1]A trace buffer with width 32 and depth 1024 represents that it can trace the values of 32 signals over 1024 clock cycles.

Table 15.1 Assertion coverage when the total number of assertions for each row is 4000 (12,000 per benchmark)

Benchmark			Signal selection			
Type	#Gates	#Traces	Observability-aware SS		Coverage-aware SS	
			Restoration %	Asser. Cov. %	Restoration %	Asser. Cov. %
S5378	2995	8	60.97	46.97	58.57	49.6
		16	79.27	63.6	76.95	64.23
		32	93.10	88.5	92.26	90.57
S9234	5844	8	84.85	65.4	76.03	65.8
		16	90.19	75.27	83.70	83.12
		32	94.54	90.67	93.8	93.45
s15850	10383	8	72.03	59.7	76.03	65.8
		16	80.97	61.6	75.55	68.7
		32	84.14	72.9	82.66	74.9
s35932	17828	8	41.09	26.825	41.63	27.02
		16	41.35	26.825	41.88	27.05
		32	41.79	26.825	42.22	27.25
s38417	23843	8	36.53	23.025	36.97	23.23
		16	43.76	28.575	46.91	32.58
		32	49.77	35.075	55.82	42.33
s38584	20717	8	72.97	47.625	67.78	59.8
		16	79.15	63.65	76.53	69.3
		32	88.85	73.65	87.27	82.78

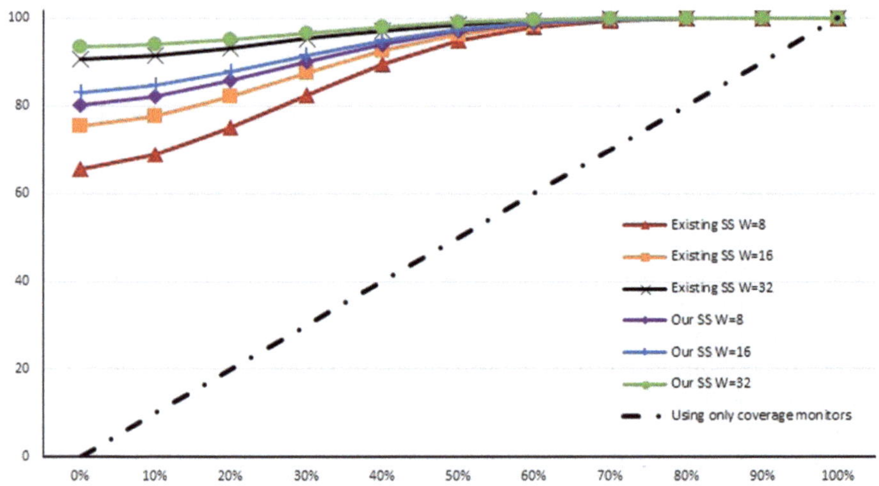

Fig. 15.8 Coverage analysis for s9234: coverage monitors are selected randomly

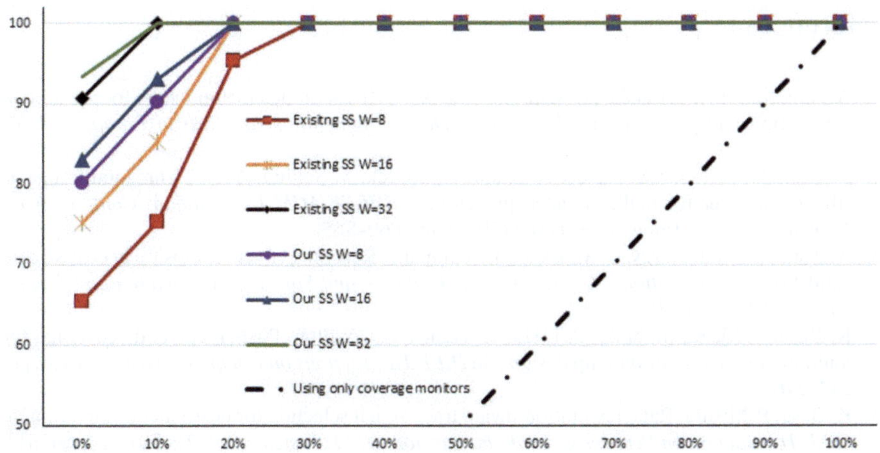

Fig. 15.9 Coverage analysis for s9234: coverage monitors are selected from hard-to-detect events

of observability of s9234 with different trace buffer widths and different coverage monitor selection strategies (the straight line is cut on 50% for improved illustration). As it can be seen, when our signal selection algorithm was used to choose 32 trace signals and our coverage monitor algorithm was used, synthesizing only 10% assertion leads to 100% observability. Although, we presented the result for s9234, we obtained similar results for other ISCAS89 benchmarks.

It can be argued that it takes little effort to cover the first 90%, but significantly more to cover the remaining 10%. Based on our proposed method, if the remaining 10% of the assertions are synthesized, 100% coverage is achieved. However, if it is not possible to synthesize those assertions due to design constraints, increasing the width or depth of trace signals can be considered. Dynamic signal selection capability (if available) can be utilized to focus in tracing of the remaining 10% assertions.

15.5 Summary

This chapter presented an efficient technique to find functional coverage on silicon without introducing any overhead. The proposed method utilizes the existing debug infrastructure in modern designs to rank coverage monitors in terms of required efforts to detect them. A signal selection algorithm was also introduced to improve the coverage analysis with negligible impact on restoration ratio.

References

1. A. Adir, S. Copty, S. Landa, A. Nahir, G. Shurek, A. Ziv, A unified methodology for pre-silicon verification and post-silicon validation, in *Design Automation and Test in Europe (DATE)* (2011), pp. 1590–1595
2. A. Adir, M. Golubev, S. Landa, A. Nahir, G. Shurek, V. Sokhin, A. Ziv, Threadmill: a post-silicon exerciser for multi-threaded processors, in *IEEE/ACM International Conference on Computer Design Automation (DAC)* (2011), pp. 860–865
3. A. Adir, A. Nahir, A. Ziv, C. Meissner, J. Schumann, Reaching coverage closure in post-silicon validation, in *International Conference on Hardware and Software: Verification and Testing (HVC)* (2010), p. 6075
4. K. Balston, M. Karimibiuki, A.J. Hu, A. Ivanov, S.J. Wilton, Post-silicon code coverage for multiprocessor system-on-chip designs, in *IEEE Transactions on Computers* (IEEE, 2013), pp. 242–246
5. K. Basu, P. Mishra, Rats: restoration-aware trace signal selection for post-silicon validation, in *IEEE Transaction on Very Large Scale Integration (VLSI) Systems*, vol. 21 (2013), p. 605613
6. M. Boule, J. Chenard, Z. Zilic, Adding debug enhancements to assertion checkers for hardware emulation and silicon debug, in *International Conference on Computer Design (ICCD)* (2006), pp. 294–299
7. M. Boulè, Z. Zilic, Automata-based assertion-checker synthesis of psl properties. ACM Trans. Des. Autom. Electr. Syst. **13**(1) (2008)
8. M. Boulè, Z. Zilic, *Generating Hardware Assertion Checkers for Hardware Verification, Emulation*, (Springer Publishing Company Incorporated, New York, 2008)
9. M. Chen, X. Qin, H. Koo, P. Mishra, *System-Level Validation - High-level Modeling and Directed Test Generation Techniques* (Springer, Berlin, 2012)
10. E. El Mandouh, A. Gamal, A. Khaled, T. Ibrahim, A.G. Wassal, E. Hemayed, Construction of coverage data for post-silicon validation using big data techniques, in *24th IEEE International Conference on Electronics, Circuits and Systems (ICECS)* (IEEE, 2017), pp. 46–49
11. F. Farahmandi, P. Mishra, S. Ray, Exploiting transaction level models for observability-aware post-silicon test generation, in *Design Automation and Test in Europe (DATE)* (2016), pp. 1477–1480
12. F. Farahmandi, R. Morad, A. Ziv, Z. Nevo, P. Mishra, Cost-effective analysis of post-silicon functional coverage events, in *2017 Design, Automation & Test in Europe Conference & Exhibition (DATE)* (IEEE, 2017), pp. 392–397
13. M. Gao, K.-T. Cheng, A reconfigurable design-for-debug infrastructure for socs, in *High Level Design Validation and Test Workshop (HLDVT)* (2010), pp. 90–96
14. W. Kadry, D. Krestyashyn, A. Morgenshtein, A. Nahir, V. Sokhin, J.S. Park, S.-B. Park, W. Jeong, J.C. Son, Comparative study of test generation methods for simulation accelerators, in *2015 Design, Automation and Test in Europe Conference and Exhibition (DATE)* (IEEE, 2015), pp. 321–324
15. B. Kumar, K. Basu, M. Fujita, V. Singh, Rtl level trace signal selection and coverage estimation during post-silicon validation, in *2017 IEEE International High Level Design Validation and Test Workshop (HLDVT)* (IEEE, 2017), pp. 59–66
16. M. Li, A. Davoodi, A hybrid approach for fast and accurate trace signal selection for post-silicon debug. Des. Autom. Test Eur. (DATE), 485–490 (2013)
17. L. Liu, D. Sheridan, W. Tuohy, S. Vasudevan, Towards coverage closure: Using goldmine assertions for generating design validation stimulus. Des. Autom. Test Eur. (DATE), 173–178 (2011)
18. X. Liu, Q. Xu, Trace signal selection for visibility enhancement in post-silicon validation. Des. Autom. Test Eur. (DATE), 1338–1343 (2009)
19. S. Ma, D. Pal, R. Jiang, S. Ray, S. Vasudevan, Can't see the forest for the trees: State restoration's limitations in post-silicon trace signal selection, in *International Conference On Computer Aided Design (ICCAD)* (2015), pp. 146:1–146:6

20. P. Mishra, R. Morad, A. Ziv, S. Ray, Post-silicon validation in the soc era: a tutorial introduction. IEEE Des. Test **34**(3), 68–92 (2017)
21. S. Mitra, S.A. Seshia, N. Nicolici, Post-silicon validation opportunities, challenges and recent advances, in *IEEE/ACM International Conference on Computer Design Automation (DAC)* (2010), pp. 12–17

Part V
Case Studies

Chapter 16
Network-on-Chip Validation and Debug

Subodha Charles and Prabhat Mishra

16.1 Introduction

System-on-Chip (SoC) designs consists of a wide variety of components including processor cores, co-processors, memories, controllers, and so on. Network-on-Chip (NoC) is utilized to enable communication between components. Continuous advances in chip manufacturing technologies enabled computer architects to integrate increasing number of general-purpose as well as specialized processors on the same SoC. For example, Intel Xeon Phi processors, code named "Knight's Landing" (KNL), features 64–72 Atom cores and 144 vector processing units [1]. Furthermore, larger number of cores inevitably increase the memory requirements. To satisfy this need, a suite of integrated Memory Controllers (MCs) provide multiple interfaces. This causes the NoC to connect and communicate between cores, off-chip memory, and other components in the SoC. As a result, the NoC that interconnects over hundred on-chip resources plays a critical role in both performance and power consumption [2, 3].

As complexity increases, getting the full system to work properly becomes increasingly difficult. It requires better pre-silicon and post-silicon verification and debug solutions. Pre-silicon verification uses formal and simulation-based methods to detect and eliminate bugs before it is manufactured. However, this alone is not sufficient to guarantee bug-free first silicon. It is inevitable that functional and non-functional bugs will exist in the manufactured chip, and thus, identifying the escaped bugs as soon as possible is of utmost importance. Compared with pre-silicon verification, the challenge of post-silicon debug is the observability of the internal signals and implementation of solutions with acceptable overhead.

SoC verification effort can be broadly categorized in verification of processor cores, memory, and interconnection network. With the shift from single-core to mul-

S. Charles (✉) · P. Mishra
Department of Computer and Information Science and Engineering, University of Florida, Gainesville, USA
e-mail: charles@cise.ufl.edu

P. Mishra and F. Farahmandi (eds.), *Post-Silicon Validation and Debug*,
https://doi.org/10.1007/978-3-319-98116-1_16

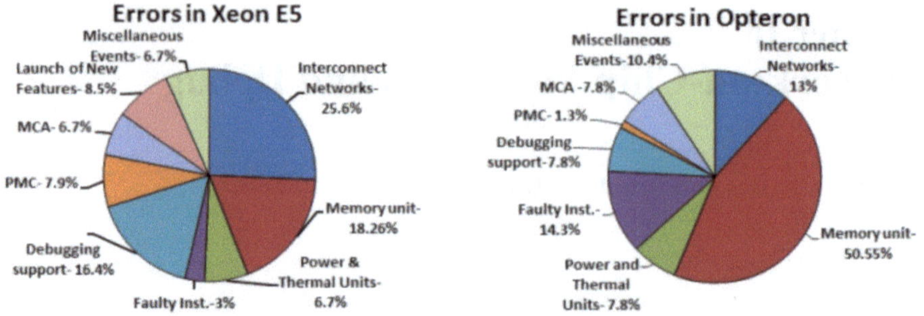

Fig. 16.1 Design error analysis for two product families—Intel Xeon Processor E5 v2 (Revision date: April 2015) [4] and AMD Opteron (Revision date: June 2009) [5, 6]

Table 16.1 Summary of errors found in NoC as a percentage of total errors in multicore architectures [6]

Processor	# Errors in NoC (%)
Intel Xeon E5	25.6
AMD Opteron	13
Intel Xeon Phi	16.07 [7]
ARM MX6	48

ticore, there is an increasing focus toward the validation of interconnection network since analysis has shown NoC to be a significant source of errors. Intel Xeon E5 v2 processor family error report shows that Peripheral Component Interconnect express (PCIe) and Quick Path Interconnect (QPI) contribute to nearly 26% of total errors [4]. Similarly, in AMD Opteron, hypertransport (Lightning data transfer) errors account for 13% [5] as shown in Fig. 16.1. A summary of NoC error percentages compared to the total errors reported in several multicore architectures is shown in Table 16.1. These statistics have urged the verification efforts to consider NoC as one of the major sources of errors.

In this chapter, we survey the NoC verification challenges and associated solutions. Section 16.2 gives an overview of NoC considering the evolution of communication architectures from traditional bus-based to the state-of-the-art NoCs. Unique verification challenges of NoC which sets NoC verification apart from core/memory verification are outlined in Sect. 16.3. NoC verification solutions given by researchers overcoming these challenges are elaborated in Sect. 16.4. Finally, Sect. 16.5 concludes the chapter.

16.2 NoC Overview

Consider a designer who is responsible for designing the road network of a large city. Roads should be laid out giving easy access to all the offices, schools, houses, parks, etc. If all the most common places are situated close to each other, it is inevitable that the roads in that area will get congested and other areas will be idle. The designer should make sure that such instances do not occur and the traffic is uniformly distributed as much as possible. Alternatively, the roads should have more lanes and parking lots in such congested areas to cater to the requirement. In addition to accessibility and traffic distribution, the architect should also consider intersections, traffic lights, priority lanes, and potential detours due to occasional road maintenance. Moreover, self-driving cars and drones that deliver mail and other packages might come into picture in the future as well. Analogous to this, the designer of an SoC faces a similar set of challenges when designing the communication infrastructure connecting all the cores.

The early SoCs employed bus- and crossbar-based architectures. Traditional bus architecture had dedicated point-to-point connections, with one wire dedicated to each signal. With the number of core count in an SoC being low, buses were most cost-effective and simple to implement. In fact, buses are also successfully implemented in many complex architectures. ARM's AMBA (Advanced Microcontroller Bus Architecture) bus [8] and IBM's CoreConnect [9] are two popular examples. Buses do not classify activities depending on their characteristics. For example, the general classifications as transaction, transport, and physical layer behavior are not distinguished by buses. This is one of the main reasons why they cannot adapt to changes in architecture or make use of advances in silicon process technology. Due to increasing SoC complexity coupled with increasing number of cores, buses often become the performance bottleneck in complex SoCs. This together with several other major drawbacks such as non-scalability, increased power consumption, non-reusability, variable wire delay, increased verification costs, etc. motivated researchers to search for alternative solutions.

The inspiration came from traditional networking solutions, more specifically, the Internet. The network-on-chip, a miniature version of the wide area network with routers, packets, and links was proposed as the solution [10, 11]. This used a packet-based communication approach. A request or response that goes to a core/cache or to off-chip memory is divided into packets, and subsequently, to flits and injected on the network. Flits are routed on the interconnection network through links and routers using multiple hops and arrive at the destination. A generic NoC architecture is shown in Fig. 16.2. The new paradigm described a way of communicating between cores including features such as routing protocols, flow control, switching, arbitration, and buffering. With increased scalability, resource reuse performance and reduced costs, design risk, time-to-market, etc., NoC became the solution for the emerging SoCs.

With power and performance requirements becoming more and more demanding, multicore architectures seem to be the most feasible solution to satisfy them. The International Technology Roadmap for Semiconductors (ITRS) roadmap projected

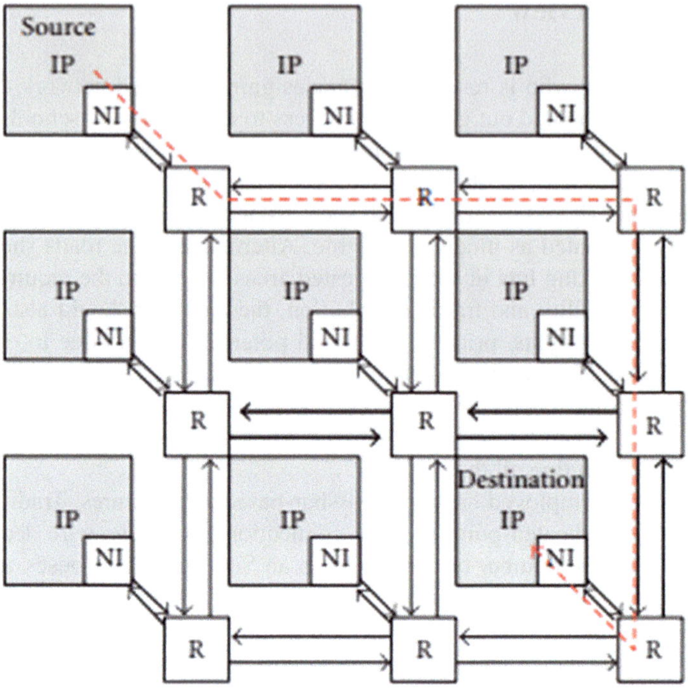

Fig. 16.2 A general NoC architecture with Intellectual Property (IP) cores arranged in a mesh topology. Communication happens through routers (R) and Network Interface (NI) [12]

that the performance demand would continue to grow to 300x by 2022 [13], which in turn would require chips with 100x more cores than the current state-of-the-art design. Out of the main contributors to power in an SoC, the NoC can consume around 30–40% of the total chip power, and thus, design of high-performance and low-power interconnects is essential for sustaining the multicore growth in coming years.

16.3 Unique Challenges of NoC Verification and Debug

While the verification problem of SoC components is already challenging, NoC introduces another layer of complexity, mainly due to its parallel communication characteristics. The increase in NoC complexity rapidly increases its verification effort and as a result, the need for efficient and scalable solutions becomes inevitable. In the following subsections, we outline some of the important challenges in NoC verification [14].

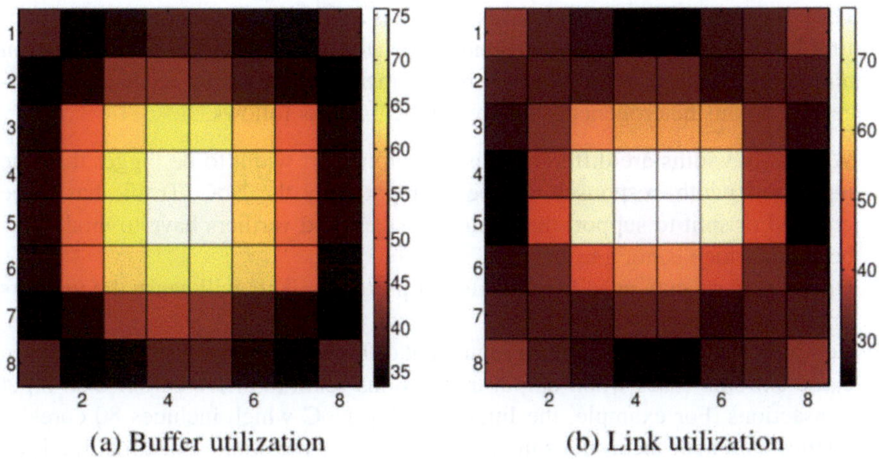

Fig. 16.3 Buffer and link utilization (in percentage) across all routers in an 8x8 mesh on heat map scale [15]

16.3.1 Handling Heterogeneous Behavior

With different topologies being employed for NoCs, it is observed that the NxN mesh topology is the most widely employed topology due to its scalability, regularity, and ease of implementation in silicon. While variations such as Torus, Butterfly, Ring, Crossbar, etc. have been proposed, all of them are structured around the same homogeneous router design. The mesh topology together with the most widely used X-Y/Y-X deterministic routing protocol is known to show non-uniform traffic behavior across routers where the center routers tend to carry most traffic while edge routers being less congested. This can lead to hotspots causing performance degradation. A study conducted by Mishra et al. [15] shows that an 8x8 mesh with uniform random traffic pattern causes the center routers to be highly utilized (~75%), whereas the edge routers have low (~25%) utilization. The heat map for average buffer and link utilization is shown in Fig. 16.3. This scenario is common across most traffic patterns and benchmarks.

Several such drawbacks in the homogeneous setup of NoCs have caused the designers to re-think the uniform resource allocation across the interconnect. As a solution for the previous example of central routers being highly utilized, one can argue that center routers should be upgraded with more Virtual Channels (VCs), wider link width, and more buffers, while the edge routers employ less of those resources. This changes the homogeneous NoC setup to a heterogeneous NoC, which is becoming the state-of-the-art in NoC design. The need for a heterogeneous NoC is further aggravated by the fact that future SoCs will have many cores and components that are heterogeneous in nature. Inherently, these components will exhibit very different resource requirements that should be supported by the interconnection network.

Thus, it is evident that, unlike buses, NoC has a fundamental requirement of supporting heterogeneous components and associated heterogeneous communication requirements. Some of the functional requirements which cause verification challenges due to this heterogeneous nature can be listed as follows:

1. When link widths are different causing output port width to be bigger than the input port width, responses will be aggregated by the NoC. Traffic has to be grouped or split to support these varying sizes, and verifiers have to model this behavior which is not a trivial task.
2. Different ports may need to use different protocols. This will cause the verifiers to handle protocol translations.
3. Components on the NoC can be running at different clock speeds. This may even cause extreme cases when output transactions seem to arrive earlier than input transactions. For example, the Intel TeraFlop NoC which includes 80 cores in 80 tiles uses a mesochronous interface for its inter-tile communication. It allows clock phase-insensitive communication across tiles and synchronous operation within each tile. This causes a synchronization latency penalty for distributing a single frequency clock across tiles. In such cases, verification includes additional tasks of handling clock domains and task synchronization across tiles [16].

16.3.2 Security and Trust

As SoCs grow in complexity and number of components, there is an increasing interest in manufacturers to use third-party IPs to minimize the cost and to meet tight time-to-market deadlines. It has come to the point that no single manufacturer puts together the whole thing, but rather, IPs are acquired from all over the globe to come up with the final SoC. This has lead to serious security concerns. One of the key assumptions in trustworthy computing is that software is running on top of secure hardware. This assumption no longer holds true with the use of third-party IPs and this calls for additional measures to ensure SoC security.

There is a growing interest in the industry to use NoC to secure the SoC as evident from NoC-Lock [17] and FlexNoC resilience package [18]. On the other hand, the NoC itself can be a threat when different IP blocks come from different vendors. A compromised IP in the NoC can corrupt data, degrade performance, or even steal sensitive data. The fact that NoC has access to all system data, NoC spans across the entire SoC and NoC elements are repetitive in a way that any modification can be easily replicated, gives implementation of security at the NoC level a strong motivation. Since the NoC is in close proximity to all the on-chip elements, a distributed security mechanism based on NoC will have lower area and power overhead. However, developing efficient and flexible solutions at lower cost with minimal impact on performance as well as how to certify these solutions remain as challenges to the industry. Several mechanisms including placing security measures in the router, network interface and also, hardware modules in between routers similar to firewalls

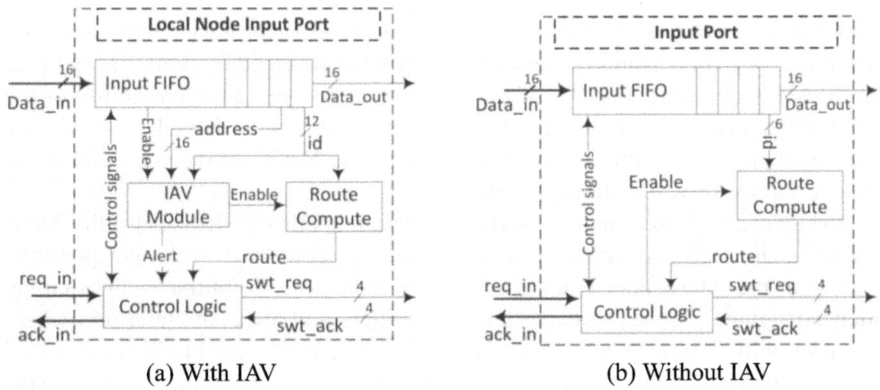

Fig. 16.4 Additional hardware complexity comparison of the router input port with and without IAV unit proposed in [19]

to secure access to sensitive data have been studied. These additional measures have caused more verification challenges.

Verification process needs to care for secure devices and be smart enough not to forward transactions to ports which are flagged as insecure. Use of additional hardware and check bits in the packet also increases the state space of verification. With varying placements and techniques of security measures implemented across the NoC depending on the requirement, coming up with generalized verification techniques is becoming almost obsolete. Additional hardware complexity of the proposed ID and address verification (IAV) unit to control access to secure nodes in an NoC-based many-core shared memory architecture by Ahmed et al. is shown in Fig. 16.4 [19].

16.3.3 Dynamic Behavior

Achieving better performance comes at the cost of power. Taking the simple proportionality between voltage and frequency, one can easily see that working at higher speed consumes more power. Thus, techniques that enable to reduce power while achieving the desired performance (e.g., meeting the task deadlines in real-time systems) are essential for SoC design. The varying characteristics of applications cause a static configuration to be suboptimal. Runtime/dynamic reconfiguration has become an avenue for research. Dynamic Voltage and Frequency Scaling (DVFS) which optimizes power and performance by reducing voltage and frequency depending on the nature of the task and architectural level optimizations such as Very Large Instruction Word (VLIW) processor are two well-known techniques.

When the IP blocks in the SoC become increasingly reconfigurable, the underlying interconnection network needs to keep up with the pace. Static NoC configuration decided at design time will either provide below average performance (when designed for average case) or consume excessive power (when designed for worst case). Since current NoCs consume 30–40% of the total power budget, reconfigurability of the NoC has become increasingly significant. As a result, studies have been done in a wide variety of reconfiguration techniques specific to NoC over the years which has added to the list of verification challenges.

Latest NoCs embed dynamic routing and shortcuts that prioritizes traffic. Thus, estimating the path depending on a single routing protocol is no longer possible. Routing in the interconnect may not be fully related to the initial address either. Sideband signals or other software level controllers will affect the final port where the packet will be transferred. This causes the verification model to be aware of all elements in order to predict and verify the traffic observed in the interconnect. NoC provides a major opportunity for reconfiguration due to the ability to play with order and disorder. *GOOLO* (Globally out of order, locally ordered) techniques are invented to avoid bottlenecks and hotspots in the interconnect. Yet, some applications may require strong or partial ordering to be satisfied. This adds to the verifier complexity when matching correctness of input and output transactions arriving out of order. In addition to this, increased complexity of hardware components (routers, network interfaces, links, etc.) adds to the possible verification state space.

16.3.4 Error Handling

As technology scales, NoCs will be more and more susceptible to various noise sources such as crosstalk, coupling noise, process variations, etc. This causes the wire delay as a fraction of the total delay to increase. The delay in crossing the chip diagonally for 45 nm technology is around 6–10 cycles [20]. Not only the delay but also the variation of delay caused due to process variations can lead to timing errors. As a solution to this, systems are designed to dynamically detect and correct circuit timing errors. A system is considered safe if there are no timing violations in the presence of noise.

In addition to timing errors, the errors that can occur in NoC links are broadly categorized as transient and permanent errors [21]. Architectural and design-level support is required to handle these errors. Flow control-based methods which combine the error control code with the retransmission mechanism to tolerate transient fault occurring while transmission is one potential solution. Feng et al. [22] proposed a hybrid automatic repeat request, and forward error correction scheme for this purpose. Fault-tolerant routing is another alternative which utilizes structural redundancies in an NoC to find alternative paths to route packets in case some links/routers fail. An ideal fault-tolerant mechanism should ensure that no packets are lost within the NoC. In such a setup, data transfers may happen in optimized kind of bursts with a large number of transfers in each (e.g., the latest AMBA protocols allow enlargement from 16 to 256 transfers). In case one of the early data bursts contain an error, the system will realize that sending more data is useless and stop transmission. This allows un-

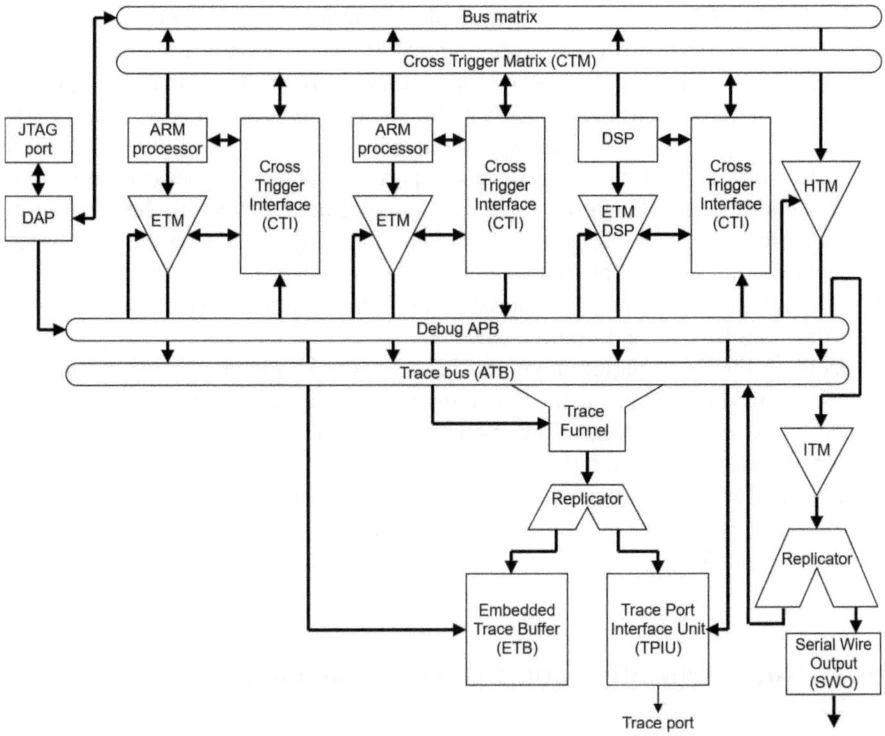

Fig. 16.5 ARM CoreSight™ debugging environment [24]

needed packets to be removed from the network. As a result, the verification debug techniques need to handle these error-related scenarios correctly.

16.3.5 Observability Constraints

As described in earlier sections, observability of the design is key for verification and debug. On the other hand, the higher the observability, the higher the risk of security breaches. Thus, it is important to strike a balance between the two. System-level debug solutions rely on observability and controllability. Even though controllability increases as the number of cores increase, observability decreases since the ratio between the number of cores and I/O pins increases. Observability and eventually, verification of computational units, and the traditional bus-based architecture are well-studied research problems and in fact, modules to observe those components are quite common [23]. ARM's CoreSight [24] debugging environment, shown as a block diagram in Fig. 16.5, offers a set of comprehensive debug and trace tools for ARM SoCs.

When the problem is extended to the state-of-the-art NoC, it becomes more complex as NoCs employ truly parallel communication paths. Having a centralized monitoring system similar to busses is no longer possible. With studies in NoC exploring more ways to optimize traffic transfers within the NoC in terms of power and performance, keeping up with efficient NoC monitoring mechanisms is a must. Multiple monitoring probes working in parallel are required for NoC verification to support its inherent parallelism. This poses a new challenge of interconnecting the monitoring probes themselves. Now, in addition to interconnecting the IPs on the SoC, interconnection of monitoring probes should also be considered. Any such interconnect should have the following characteristics [14]: minimum overhead, configurable, scalable, nonintrusive, and runtime usable.

Deciding the exact requirement and functionality of the distributed monitoring probes is almost impossible at the beginning of the design stage since NoC is a result of a complex synthesis design flow. Some requirements may pop up few steps into the design process, for example, verifying the packet traversal will depend on the topology and routing protocols. This will affect the placement of the monitoring probes. Thus, monitoring solutions should be found within or at least tightly coupled with the design flow.

16.4 NoC Verification and Debug Methodologies

The previous section outlined various challenges associated with NoC validation. This section presents several NoC verifications and debug methodologies used today.

Interconnects can be categorized into two types: off-die and on-chip. Off-Die Interconnect Network (ODIN) typically support transactions between last level cache and memory. In ODINs, several methodologies including trace buffer-based [25] and scan-based [26] have been proposed to capture internal states in the design. While trace buffer-based techniques face the limitation of size, boundary scan techniques lead to reduced test time and are mainly used to test ODIN [26]. The on-chip network, more commonly referred to as NoC, is the main focus of this chapter. Methods to verify and debug the NoC are described in subsequent subsections under several categories addressing their unique challenges. In general, the idea is to monitor the network, collect data, and perform on-chip or off-chip analysis for functional/nonfunctional verification.

16.4.1 Enhancing NoC Observability

NoC observability poses a unique set of challenges. There are several efforts to address these challenges. Hardware probes attached to NoC components, routers, and Network Interfaces (NIs) [27] containing data sniffers to gather data and an event generator to handle data processing. Gathered data is processed at an external

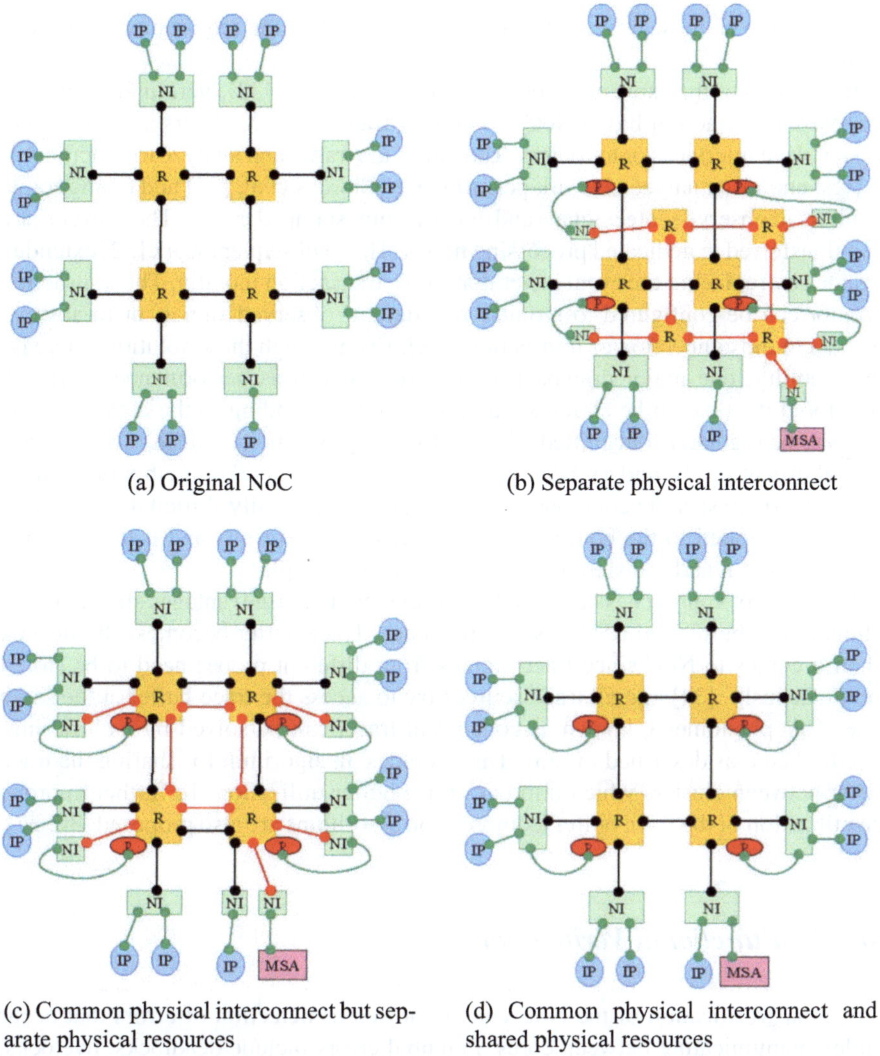

Fig. 16.6 Monitoring transport options [28]

processor connected to the NoC via a monitoring network interface. This setup can be used as a general-purpose monitoring platform where the number of probes is decided depending on the topology with off-chip centralized processing. The interconnect and resources required to connect the probes can either be separate from the original NoC or shared among both as shown in Fig. 16.6.

The three options (b, c, and d) shown in Fig. 16.6 are available in the Æthereal NoC [29, 30]. The selection of an option depends on impact on the overall NoC design flow, area cost, nonintrusiveness, and reuse potential of debug resources. The tradeoff between these aspects depending on the application requirements has been studied.

Deviations of the same concept can be seen in [27, 31, 32]. Vermulen et al. [31] proposed a transaction-based NoC monitoring framework. Monitors are attached to master/slave interfaces or to routers. They filter the traffic and analyze transactions of interest and also analyze network performance. Ciordas et al. [27] add configurable monitors to observe router signals and generate timestamped events. Those events are then transferred to dedicated processing nodes. Their subsequent work [32] extended this idea by replacing the event generator with a transaction monitor. The transaction monitor can be configured to monitor raw data of observed signals or to abstract the data to the connection or transaction level. Even though these solutions increase observability, one major drawback is the area as well as performance overhead. Monitored data has to be stored in large trace buffers adding to the area overhead or has to be regularly transmitted off-chip for analysis which perturbs the network's normal execution. Boundary Scan Registers (BSRs) are another methodology used to apply tests to specific components. Test data is then serially shifted through these BSRs and applied to the component being tested. Test results and traces are then serially read out and transferred off-chip for debugging [33].

More recent approaches have tried to circumvent the disadvantages shown by traditional trace buffer and BSR-based approaches. Trace buffer becomes a bottleneck when it comes to NoC since trace signals from different probes need to be stored simultaneously [34] and all trace sources try to access the trace buffer at the same time. This phenomena, known as concurrent trace, can be solved by a clustering-based scheme as described in [35]. This includes an algorithm for sharing the trace buffer between clusters which improves trace buffer utilization. To further improve the utilization of the trace buffer, compression algorithms are also proposed [36, 37].

16.4.2 Functional Verification

Functional verification of the NoC can be viewed as detection of errors that occur while communicating between cores. Potential errors include deadlocks, livelocks, starvation, packet data corruption, dropped and duplicated packets, and packet mis-routing. To ensure these errors do not occur, verification should guarantee non-dropping and non-corruption of packets, non-generation of new packets within the network, non-breaking of the protocol under stressed conditions, time-bounded packet delivery, ensuring secure transactions, and generation of errors for unsecure transactions.

In contrast to the trace buffer and boundary scan-based techniques used to enhance visibility, Abdel-Khalek et al. [33] proposed to observe the network by periodically taking snapshots, store the data in each core's cache, and analyze using a software algorithm for functional bugs. Routers take snapshots of packets traversing in the network and store them in a designated part of the L2 local cache corresponding to that node. These logs (samples of the observed traffic) are then analyzed using a

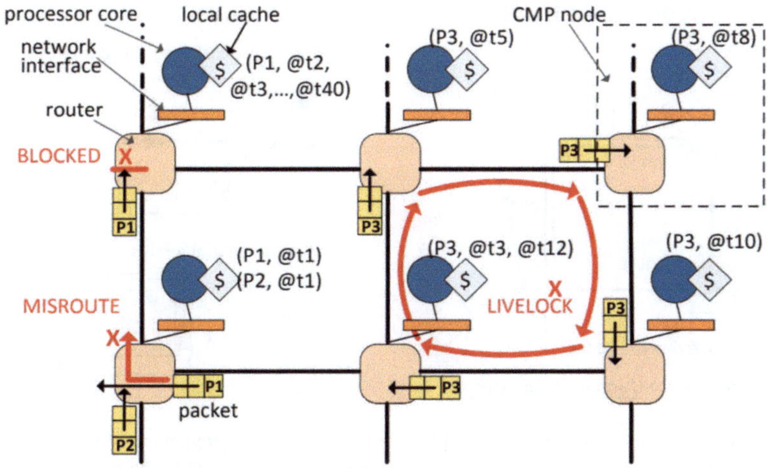

Fig. 16.7 Overview of NoC post-silicon validation platform proposed in [33]. Observability is increased by instrumenting the routers to periodically monitor network traffic. Figure shows the types of information collected and three kinds of bugs that can be discovered using this method

software checking algorithm that runs on each core for errors. If an error is to occur, at least one packet will be affected and will be captured by the software. Once an error is detected, logs of each router will be used to reconstruct the paths followed by packets. Aggregation of these paths will provide an overview of the network packet traversal which enables verification and debug of the network. The advantage of using L2 local cache is that space can be released after verification and no additional space is required for trace buffers. This leads to minimum area overhead. The solution is also independent of the NoC, routing algorithm, topology, and router architecture. An overview of this method is shown in Fig. 16.7.

The same authors proposed another approach for functional verification by instrumenting routers with checkers [38]. During the verification process, a packet's original data is replaced with debug data and routed through the network. Debug information contains a packet's current state during its network traversal. Packet, upon reaching the destination, is stored in the local cache of the destination node, and a checker in the router is used to check for errors. If an error is detected, a software algorithm, which can be run on the core or off-chip, analyzes the packet data. First, each core analyzes its local data and then sends off-chip for a global analysis. Additional hardware required to implement this idea are a register and a packet counter which are added to the router. An overview of the approach is shown in Fig. 16.8 [38].

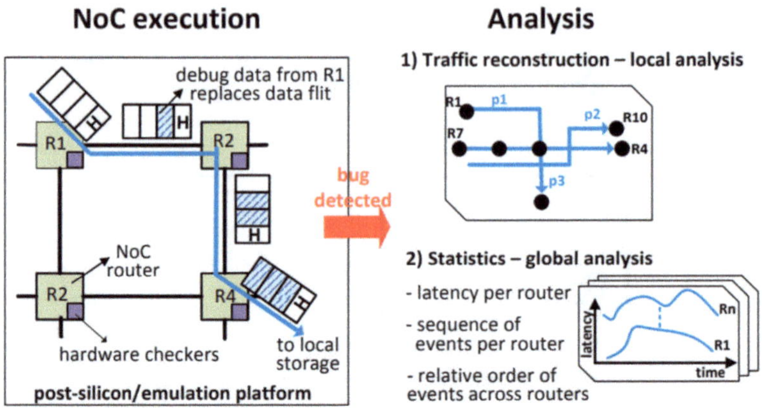

Fig. 16.8 Overview of solution given in [38]. Debug data is collected at every hop and stored in the packet during NoC execution. Hardware checkers are implemented in the routers that monitor the network and flag functional errors. If an error is detected, collected data is used to reconstruct traffic as well as provide other relevant statistics

16.4.3 Power and Performance Validation

An NoC can be functionally correct but might not meet power and performance criteria of the design. For example, packets may traverse among cores without dropping or causing any errors but might take longer than expected to arrive at the destination. DIAMOND solution proposed by Abdel-Khalek et al. [38] addresses the issue of performance validation as well. Since the packet contains debug data, it can record packet latencies at each hop. This data can then be aggregated across the path, from source to destination to see the packet traversal latency. Statistics derived from this can be used to identify performance bottlenecks of the network. For example, *dedup* benchmark of the PARSEC benchmark suite shows the highest average packet latency at router 49 when running on an 8x8 mesh network as shown in Fig. 16.9. Analysis has shown that this indeed is caused by the credit-based flow control that releases the output virtual channel only after the entire packet is transferred.

Observing key system parameters such as power and performance in real time gives an idea of the resource utilization as well. If resource utilization at any given time is higher than expected, it can lead to erroneous behaviors. On the other hand, if resources are underutilized, it means design was overestimated resulting in a higher bill of material than necessary. Being able to strike the perfect balance prevents over-design costs while maintaining the expected QoS standards. The NoC monitoring infrastructure proposed by Vermeulen et al. [31] includes measurements for bandwidth utilization and transaction latency to capture performance statistics. A user can specify the locations in the communication architecture where the monitors should be added. This allows to get more localized measurements of network parameters.

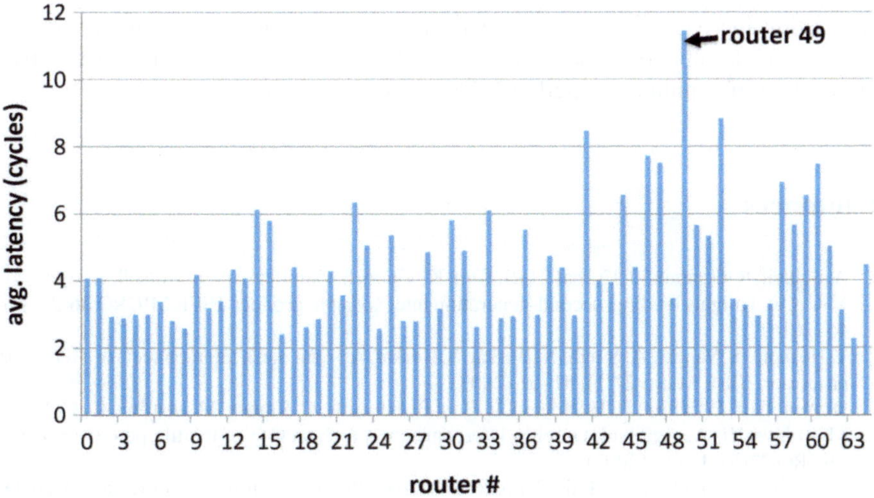

Fig. 16.9 Average packet latency observed at each router for *dedup* benchmark. Router 49 shows an unusually high latency

16.4.4 NoC Verification using Formal Methods

Simulation-based methods do not provide a guarantee on completeness, i.e., 100% coverage. In contrast, formal methods are effective in finding errors in a short period of time given a compatible design with less cost. Gharehbaghi et al. [39] utilized bounded model checking on transaction-level models to detect the origin of the error. The method mainly focuses on capturing errors in transactions between the cores. Critical information in each transaction—source, destination, and type (read or write)—is extracted using a transaction extractor. This information is then used to construct a finite state machine. Path analysis is performed based on constraints, transactions sent, and received for a state and assertions for each state in the state machine. This is done for all generated paths to collect all erroneous transactions.

Other model-based NoC verification approaches are described by Helmy et al. [40] and Zaman et al. [41]. The first solution describes a method based on a consequential enhancement of a *generic formal model* for the representation of the communications on a chip—GeNoC [42]—while the second focuses on formal verification of circuit-switched NoC.

16.5 Conclusion

This chapter provided an overview of challenges associated with NoC verification and debug. It presented several state-of-the-art NoC verification and debug techniques.

The existing approaches use a wide variety of trace collection and analysis techniques. The future NoC validation and debug approaches are expected to utilize an effective combination of simulation-based techniques and formal methods.

References

1. A. Sodani, R. Gramunt, J. Corbal, H.-S. Kim, K. Vinod, S. Chinthamani, S. Hutsell, R. Agarwal, Y.-C. Liu, Knights landing: second-generation intel xeon phi product. IEEE MICRO **36**(2), 34–46 (2016)
2. Y. Hoskote, S. Vangal, A. Singh, N. Borkar, S. Borkar, A 5-ghz mesh interconnect for a teraflops processor. IEEE Micro **27**(5), 51–61 (2007)
3. D. Wentzlaff, P. Griffin, H. Hoffmann, L. Bao, B. Edwards, C. Ramey, M. Mattina, C.-C. Miao, J.F. Brown III, A. Agarwal, On-chip interconnection architecture of the tile processor. IEEE MICRO **27**(5), 15–31 (2007)
4. Intel. 2015b. Intel Xeon Phi Coprocessor x100 Product Family Specification Update, www.intel.com/content/dam/www/public/us/en/documents/specification-updates/xeon-phi-coprocessor-specification-update.pdf
5. AMD. 2009. Revision Guide for AMD Athlon 64 and AMD Opteron Processors, http://support.amd.com/TechDocs/25759.pdf
6. P. Jayaraman, R. Parthasarathi, A survey on post-silicon functional validation for multicore architectures. ACM Comput. Surv. (CSUR) **50**(4), 61 (2017)
7. Intel. 2015a. Intel Xeon Processor E5 v2 Product Family Specification Update, www.intel.in/content/dam/www/public/us/en/documents/specification-updates/xeon-e5-v2-spec-update.pdf
8. ARM: AMBA specification, https://www.arm.com/products/system-ip/amba-specifications, Accessed August 2018.
9. IBM CoreConnect, https://www-03.ibm.com/press/us/en/pressrelease/2140.wss, Accessed August 2018.
10. L. Benini, G. De Micheli, Networks on chips: a new soc paradigm. Computer **35**(1), 70–78 (2002)
11. J. William, Dally and brian towles. route packets, not wires: on-chip interconnection networks, in *Proceedings of the Design Automation Conference* (IEEE, 2001), pp. 684–689
12. W.-C. Tsai, Y.-C. Lan, H. Yu-Hen, S.-J. Chen, Networks on chips: structure and design methodologies. J. Electr. Comput. Eng. **2012**, 2 (2012)
13. ITRS (International Technology Roadmap for Semiconductors), www.itrs2.net
14. P. Oury, N. Heaton, S. Penman, Methodology to verify, debug and evaluate performances of noc based interconnects, in *Proceedings of the 8th International Workshop on Network on Chip Architectures* (ACM, 2015), pp. 39–42
15. A.K. Mishra, N. Vijaykrishnan, C.R. Das, A case for heterogeneous on-chip interconnects for cmps, in *ACM SIGARCH Computer Architecture News*, vol. 39 (ACM, 2011), pp. 389–400
16. S. Dighe, S. Vangal, N. Borkar, S. Borkar, Lessons learned from the 80-core tera-scale research processor. Intel Technol. J. **13**(4) (2009)
17. SONICS NoCk-Lock Security, www.sonicsinc.com/wp-content/uploads/NoC-Lock.pdf
18. Alteris FlexNoC Resilience Package, www.arteris.com/flexnoc-resilience-package-functional-safety
19. A. Saeed, A. Ahmadinia, M. Just, C. Bobda, An id and address protection unit for noc based communication architectures, in *Proceedings of the 7th International Conference on Security of Information and Networks* (ACM, 2014), p. 288
20. R. Tamhankar, S. Murali, S. Stergiou, A. Pullini, F. Angiolini, L. Benini, G. De Micheli, Timing-error-tolerant network-on-chip design methodology. IEEE Trans. Comput.-Aided Design Integr. Circuits Syst. **26**(7), 1297–1310 (2007)

21. S. Murali, *Designing Reliable and Efficient Networks on Chips* (Springer Science & Business Media, Berlin, 2009)
22. C. Feng, L. Zhonghai, A. Jantsch, M. Zhang, Z. Xing, Addressing transient and permanent faults in noc with efficient fault-tolerant deflection router. IEEE Trans. Very Large Scale Integr. (VLSI) Syst. **21**(6), 1053–1066 (2013)
23. J.K.A.P. Su, K.-J. Lee, J. Huang, G.-A. Jian, C.-A. Chien, J.-I. Guo, C.-H. Chen, Multi-core software/hardware co-debug platform with arm coresight, on-chip test architecture and axi/ahb bus monitor, in *2011 International Symposium on VLSI Design, Automation and Test (VLSI-DAT)* (IEEE, 2011), pp. 1–6
24. ARM Coresight, www.arm.com/products/solutions/CoreSight.html
25. M.B. Herve, E. Cota, F.L. Kastensmidt, M. Lubaszewski, Noc interconnection functional testing: using boundary-scan to reduce the overall testing time, in *10th Latin American Test Workshop. LATW'09* (IEEE, 2009), pp. 1–6
26. H. Bleeker, P. van Den Eijnden, F. de Jong, *Boundary-Scan Test: A Practical Approach* (Springer Science & Business Media, Berlin, 2011)
27. C. Ciordas, T. Basten, A. Rădulescu, K. Goossens, J. Van Meerbergen, An event-based monitoring service for networks on chip. ACM Trans. Des. Autom. Electr. Syst. (TODAES) **10**(4), 702–723 (2005)
28. C. Ciordas, K. Goossens, A. Radulescu, T. Basten, Noc monitoring: impact on the design flow, in *Proceedings of the IEEE International Symposium on Circuits and Systems. ISCAS* (IEEE, 2006), p. 4
29. K. Goossens, J. Van Meerbergen, A. Peeters, R. Wielage, Networks on silicon: combining best-effort and guaranteed services, in *Proceedings of the Design, Automation and Test in Europe Conference and Exhibition* (IEEE, 2002), pp. 423–425
30. K. Goossens, J. Dielissen, O.P. Gangwal, S.G. Pestana, A. Radulescu, E. Rijpkema, A design flow for application-specific networks on chip with guaranteed performance to accelerate soc design and verification, in *Proceedings of the conference on Design, Automation and Test in Europe*, vol. 2 (IEEE Computer Society, 2005), pp. 1182–1187
31. B. Vermeulen, K. Goossens, A network-on-chip monitoring infrastructure for communication-centric debug of embedded multi-processor socs, in *International Symposium on VLSI Design, Automation and Test. VLSI-DAT'09* (IEEE, 2009), pp. 183–186
32. C. Ciordas, K. Goossens, T. Basten, A. Radulescu, A. Boon, Transaction monitoring in networks on chip: the on-chip run-time perspective, in *International Symposium on Industrial Embedded Systems. IES'06* (IEEE, 2006), pp. 1–10
33. R. Abdel-Khalek, V. Bertacco, Functional post-silicon diagnosis and debug for networks-on-chip, in *Proceedings of the International Conference on Computer-Aided Design* (ACM, 2012), pp. 557–563
34. H. Yi, S. Park, S. Kundu, On-chip support for noc-based soc debugging. IEEE Trans. Circuits Syst. I: Regul. Pap. **57**(7), 1608–1617 (2010)
35. J. Gao, J. Wang, Y. Han, L. Zhang, X. Li, A clustering-based scheme for concurrent trace in debugging noc-based multicore systems, in *Proceedings of the Conference on Design, Automation and Test in Europe* (EDA Consortium, 2012), pp. 27–32
36. J.-S. Yang, N.A. Touba, Expanding trace buffer observation window for in-system silicon debug through selective capture, in *26th IEEE VLSI Test Symposium. VTS* (IEEE, 2008), pp. 345–351
37. E. Anis, N. Nicolici, Interactive presentation: low cost debug architecture using lossy compression for silicon debug, in *Proceedings of the Conference on Design, Automation and Test in Europe* (EDA Consortium, 2007), pp. 225–230
38. R. Abdel-Khalek, V. Bertacco, Diamond: distributed alteration of messages for on-chip network debug, in *2014 Eighth IEEE/ACM International Symposium on Networks-on-Chip (NoCS)* (IEEE, 2014), pp. 127–134
39. A.M. Gharehbaghi, M. Fujita, Transaction-based post-silicon debug of many-core system-on-chips, in *2012 13th International Symposium on Quality Electronic Design (ISQED)* (IEEE, 2012), pp. 702–708

40. A. Helmy, L. Pierre, A. Jantsch, Theorem proving techniques for the formal verification of noc communications with non-minimal adaptive routing, in *2010 IEEE 13th International Symposium on Design and Diagnostics of Electronic Circuits and Systems (DDECS)* (IEEE, 2010), pp. 221–224
41. A. Zaman, O. Hasan, Formal verification of circuit-switched network on chip (noc) architectures using spin, in *2014 International Symposium on System-on-Chip (SoC)* (IEEE, 2014), pp. 1–8
42. Julien Schmaltz, Dominique Borrione, A functional formalization of on chip communications. Form. Asp. Comput. **20**(3), 241–258 (2008)

Chapter 17
Post-Silicon Validation of the IBM POWER8 Processor

Tom Kolan, Hillel Mendelson, Amir Nahir and Vitali Sokhin

17.1 Introduction

The post-silicon validation phase in a processor's design life cycle is our last opportunity to find functional and electrical bugs in the design before shipping it to customers. Despite innovation in design verification methodology, the gap between design and verification continues to be sizable, and the role of post-silicon validation continues to grow.

High-end processors, such as the IBM POWER8 processor described in this chapter, are planned to go through more than one tape-out. We know in advance that we cannot find all bugs in the pre-silicon stage and plan a post-silicon phase in the project in which we target finding the remaining bugs, while leveraging fabricated hardware.

Post-silicon validation differs from pre-silicon verification in many aspects. With post-silicon validation being conducted on real hardware, the system's scale and speed are far greater than what is available in pre-silicon. That said, the silicon platform offers little to no ability to observe or control the state of the design under

[1] In reality, post-silicon validation and bring-up are two different things. The former term refers to the effort of finding the remaining functional bugs in the design. The latter may either refer to the activity of gaining system stability or is an encompassing name for all post-silicon activities. In this chapter, we used the terms interchangeably.

T. Kolan (✉)
IBM Research, Schindler 5, 3499610 Haifa, Israel
e-mail: tomk@il.ibm.com

H. Mendelson
IBM Research, 3087500 Kibbutz Hahotrim, Israel
e-mail: hillelm@il.ibm.com

A. Nahir
IBM Research, 3087500 POB 174, Megadim, Israel
e-mail: amir.nahir@gmail.com

V. Sokhin
IBM Research, Got Levin 26, 3292226 Haifa, Israel
e-mail: vitali@il.ibm.com

P. Mishra and F. Farahmandi (eds.), *Post-Silicon Validation and Debug*,
https://doi.org/10.1007/978-3-319-98116-1_17

test (DUT). These differences dictate a need for different approaches for stimuli generation, checking, and debugging. Every hardware problem found in the bring-up lab[1] may have dire consequences on the ability to ship the processor to customers on time. Accordingly, streamlining the lab work methodology and properly preparing all the tools used in the lab are of the highest priority when it comes to ensuring that lab work is effective.

In this chapter, we provide a high-level overview of the post-silicon methodology and technologies utilized as part of the POWER8 bring-up. Our focus is the *functional validation* of POWER8. We describe our results and list the primary factors that contributed to this highly successful bring-up.

Bare-metal exercisers served as our primary vehicle for test-case generation in the POWER8 post-silicon validation phase. An exerciser is a software application that is loaded on the system and then continuously generates test cases, executes them, and checks their results. Some exercisers are controlled via parameter files, while others are driven by a test template [1]. The combination of an exerciser and a specific control file is referred to as an *exerciser shift*.

We leveraged our in-house accelerated simulation platform to ensure that the exercisers were well prepared for the bring-up. Using synthesizable coverage monitors, we measured and analyzed our exerciser coverage, and selected the set of exerciser shifts that we wanted to use in the lab. This exercisers-on-accelerators (EoA) effort is conducted at the pre-silicon stage, so it also helps find bugs before the first tape-out. Roughly, 1% of all functional bugs in POWER8 were found in EoA. We also used accelerators to prepare debug aids for the lab and to train the lab team in advance.

The lack of observability into the state of the design renders pre-silicon checking techniques impossible in the post-silicon environment. We relied on several checking techniques to detect bugs during the POWER8 bring-up phase. Several hardware-synthesized checkers have proven highly successful in catching problems and were easier to debug. Most of our exercisers employed the multi-pass consistency end-of-test checking technique, while one exerciser used an on-platform reference model to predict results. We also had several shifts that included manually written assertions in the test code (self-checking tests).

Debugging a post-silicon test-case failure is a notoriously hard problem. The debugging process is a repetitive one, in which a failing test case is executed multiple times under different conditions to gain insight into the primary contributing factors that trigger a specific bug. Using the embedded debug logic in the POWER8 processor, as well as a dedicated debug mode and our in-house accelerators, we were able to determine the root cause of over 60% of all high severity bugs in less than a week.

The POWER8 bring-up is considered a very successful one [2]. The team was able to keep a bug discovery and resolution rate equal to or better than the previous POWER processor bring-up efforts, with significantly fewer resources. Half of all POWER8 post-silicon bugs were found in the first 3 months of the bring-up.

We attribute the success of the POWER8 post-silicon validation to the following factors:

1. The well-established pre-silicon verification methodology, leveraging state-of-the-art formal methods, and dynamic techniques to ensure that the chip is taped-out at a high functional level;
2. The on-going engagement, starting early in the project's life cycle, between the exerciser team, the design team, and the verification team;
3. The systematic preparation of the exercisers and their tests based on an Instruction Set Simulator (ISS), and leveraging synthesizable coverage monitors and accelerators to enable a coverage-driven approach to test development and selection;
4. The use of bare-metal tools as the key driver for test generation on silicon;
5. The cycle reproducible environment and the use of accelerators to recreate lab failures;
6. Last, but not least, the expertise of the lab team members and their dedication to the validation task.
   In POWER9, which was recently announced, we used a very similar bring-up process. We consider the POWER9 bring-up to be successful as well, which shows that the methodology described in this chapter is effective in getting highly complex chips to market quickly and effectively.

17.2 POWER8

POWER8 is designed for high-end enterprise-class servers, and it is one of the most complex processors ever created. The POWER8 chip is fabricated with IBM's 22 nm silicon-on-insulator technology using copper interconnects and 15 layers of metal. The chip is 650 mm^2 and contains over 5 billion transistors.

Figure 17.1 depicts a high-level block diagram of the POWER8 processor chip. Each processor chip has 12 cores. Each core is capable of 8-way simultaneous-multithreading (SMT) operation and can issue up to 10 instructions every cycle.

The POWER8 memory hierarchy includes a per-core L1 cache, with 32 KB for instructions and 64KB for data, a per-core SRAM-based 512 KB L2 cache, and a 96MB eDRAM-based shared L3 cache. In addition, a 16 MB off-chip eDRAM L4 cache per memory buffer chip is supported. There are 2 memory controllers on the chip supporting a sustained bandwidth of up to 230 GB per second.

The chip also holds a set of hardware accelerators, including a cryptographic accelerator and a memory compress/de-compress mechanism.

Finally, the chip includes a PCIe adapter and bridge to a bus supporting a coherent connection to an FPGA. The FPGA is directly accessible to user applications through a hashed table translation.

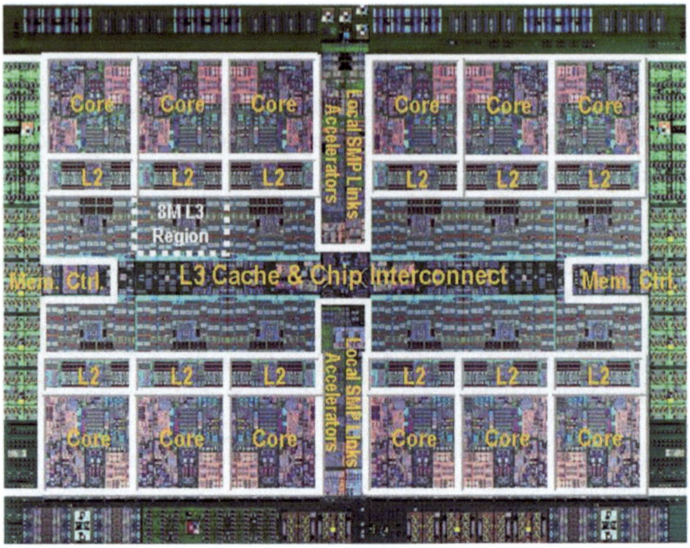

Fig. 17.1 POWER8 processor chip

17.3 Unified Methodology

In recent years, we have seen more evidence that pre- and post-silicon verification cannot achieve their goals on their own; pre-silicon, in terms of finding all the bugs before tape-out, and post-silicon, in terms of finding the bugs that escaped pre-silicon. This creates an increasing need to bridge the gap between these two domains by sharing methodologies and technologies and building a bridge allowing easier integration between the domains.

At IBM, we used a unified methodology for pre- and post-silicon verifications [3]. The methodology extends the well-established *coverage-driven verification* (CDV) methodology [4] to the post-silicon verification domain. CDV is based on three main components:

1. A verification plan comprises a large set of features in the *Design Under Verification* (DUV) that need to be verified;
2. Random stimuli generators that are directed toward the verification goals using test templates [5] (i.e., general specifications of the desired test structure and properties);
3. Coverage analysis tools [6] that detect the occurrence of events in the verification plan and provide feedback regarding the state and progress of the verification process.

The unified methodology calls for pre- and post-silicon to share the same verification plan, use similar languages to define test templates for both domains, and use the same coverage models to measure the status and progress of the verification

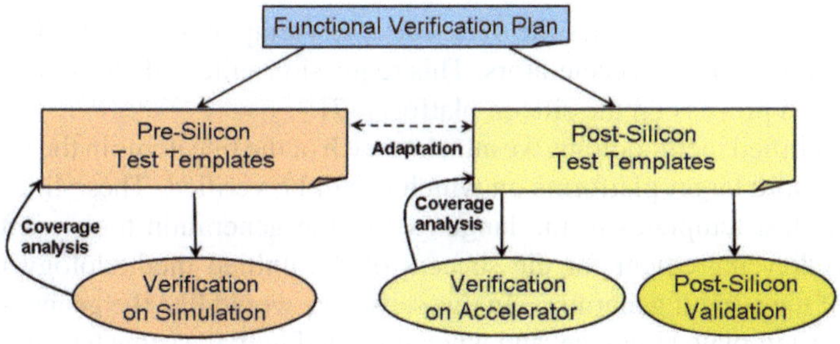

Fig. 17.2 A unified verification methodology

process. The methodology provides many advantages to its users. It provides commonality and allows sharing of effort in the creation of the verification plan. It can also facilitate mutual feedback between the two domains, for example, in the activity of bug hunting.

To better integrate post-silicon validation to the overall verification process and improve its synergy with pre-silicon verification, we need a unified verification methodology that is fed from the same verification plan source. A key ingredient for the success of such methodology is providing common languages for the pre- and post-silicon aspects of it in terms of test specification, progress measure, etc. Figure 17.2 depicts such a methodology. This verification methodology leverages three different platforms: simulation, acceleration, and silicon. The methodology requires three major components: a verification plan, directable stimuli generators suited to each platform, and functional coverage models. Note that important aspects in any verification methodology, such as checking, are omitted from the figure to maintain focus on the main aspect of our contribution, namely, stimuli generation.

The verification plan includes a long list of line items, each targeting a feature in the DUV that needs to be verified. Each such feature is associated with coverage events that the verification team expects to observe during the verification process and the methods to be used to verify the feature. The verification plan is implemented using random stimuli generators that produce a large number of test cases, and coverage tools that look for the occurrence of events in the verification plan. The random stimuli generators are directed toward the verification goals using *test templates*. The test templates allow the generators to focus on areas in the DUV ranging from large generic areas, like the floating-point unit, to very specific areas, like a bypass between stages of the pipeline. Coverage analysis identifies gaps in the implementation of the plan. Its feedback is used to modify test templates that do not fulfill their goals and create new ones.

Extending this methodology to post-silicon validation is difficult because the limited observability of the silicon does not allow to measure coverage on the silicon. To overcome this problem, we leverage the acceleration platform to measure coverage of post-silicon tools. To take advantage of the coverage information collected by the accelerators and use it in the post-silicon, shortly before first silicon samples come back from the fab, a regression suite of exerciser test templates is created based on the coverage achieved on the accelerators. This regression suite is then used to continue the verification process on the silicon platform [7].

With the unified methodology, we attach to each of the line items in the verification plan one or more target platforms on which it will be verified. These line items are converted to test templates in the languages of the generation tools used by each platform. A key ingredient for the success of the unified methodology is similar operation of the stimuli generators. In this sense, we would like the generators to use the same test-template language, and when provided with the same test template, we would like the tools to produce similar (though not identical) test cases. Of course, the different platforms provide different opportunities and put different constraints and requirements on the generation tools, but whenever possible, there are advantages to having similar tools. First, the pre- and post-silicon teams can share the task of understanding the line items in the verification plan and planning ways to test them. In addition, the common language allows for easier adaptation of test templates from one platform to another. For example, when a bug is detected on the silicon platform, narrowing down the test template and hitting it on the simulation platform eases the root cause analysis effort.

It is important to note that the differences between platforms also dictate differences in the way test templates are written for pre- and post-silicon tools. A test template could be very specific and describe a small set of targeted tests or it could be more general leaving more room for randomization. The validation engineer writing test templates for a post-silicon exerciser must bear in mind the fact that the test template is used to generate a huge number of test cases and get many processor cycles. To effectively use these test cycles, the test template must allow for enough interesting variation. A test template that is too specific will quickly "run out of steam" on silicon and start repeating similar tests. A pre-silicon test template on the other hand would typically be more directed to ensure that the targeted scenarios are reached within the fewer cycles that are available on simulation.

17.4 Exercisers

Bare-metal exercisers were the primary vehicle for our test-case generation in the POWER8 post-silicon validation. An exerciser is a software application that is loaded to the system and then continuously generates test cases, executes them, and checks their results, without a need to reset or reinitialize the hardware between test cases.

The characteristics of the post-silicon platforms create challenges and impose tradeoffs that affect the design of these exercisers. While post-silicon platforms offer a huge number of execution cycles, their low availability, short lab time, and high-cost calls for maximizing the time spent executing test cases and minimizing overhead. Accordingly, the exerciser team developed a set of exercisers, each capturing a different tradeoff between generation complexity and checking capabilities. Some of the exercisers are controlled via parameter files, while others are driven by a test template [8]. The combination of an exerciser and a specific control file is referred to as an *exerciser shift*.

In pre-silicon verification, tests run on a test bench that incorporates checkers to detect most of the failures. However, most checkers are not available when running on the post-silicon platform. Specifically, a reference model is not available due to its high complexity. Exercisers employ a technique called *multi-pass* consistency checking. Each generated test case is executed several times (passes). We validate that system resources, including registers and allocated memory, hold the same values at the end of each pass. This way, the first pass essentially serves as a reference model to the next ones. It is important to note that this method would not detect "1 + 1=3" kind of bugs, but we believe that these sort of bugs should be detected much earlier by pre-silicon environments. Exercisers are geared toward finding delicate timing bugs, for which the multi-pass method is highly suitable. Another possible caveat of end-of-test checking is bug masking. This could be avoided by keeping the tests relatively short, or by storing some resources midway through longer ones [9].

The majority of the exercisers in use are bare metal. This means that the exerciser runs on hardware, or on the DUT model in acceleration, with no operating system (OS). Bare-metal exercisers possess a list of attributes that make them highly suitable for post-silicon validation. They are inherently "self-contained," meaning that once the exerciser shift is loaded onto the platform, be it an accelerator or a silicon one, it can run "forever" with no interaction with the environment. This significantly reduces the overheads related to initializing the platform, loading the exerciser, and so forth.

From a software perspective, bare-metal exercisers are very simple. The exercisers are written in either assembly language or in a combination of assembly and C. They usually implement simple interrupt handlers, system calls, and even printouts/reporting, typically provided by the OS. This simplicity is required in the early stage of bring-up, when the silicon platform is still highly unstable. For example, in the beginning of the bring-up process, before access to memory is enabled, the bring-up team uses exercisers that have a small footprint in terms of code size; these exercisers fit inside the POWER8 L2 cache. This simplicity greatly assists in debugging, since there are no abstraction layers between the exerciser code and the hardware. Furthermore, this direct interaction with the hardware allows for better control over the state of the hardware. Unlike OS-based test suites, exercisers enable creating targeted tests that stress specific areas of the micro-architecture like pipeline flushing or atomicity, rather than stress more general features, e.g., compression, video playback, etc.

One way of keeping exercisers simple involves offloading many of the more difficult calculations to an offline pre-run stage. In this stage, we can solve any number of computationally difficult problems without wasting precious machine time. Some examples are memory allocations (to improve cache line sharing), translations, and floating-point input generation.

Another important aspect of exercisers is high utilization. In order to increase scalability, test generation should be a concurrent program. That is, every thread generates its own portion of the multi-threaded test case, with a minimum amount of barriers. For this, techniques like *joint pseudo-random seed* are used.

The attributes listed above enable bare-metal exercisers to effectively utilize accelerators, both in the pre-silicon phase (as explained in Sect. 17.5), and when we recreate lab failures (as explained in Sect. 17.10).

Overall, during the POWER8 pre- and post-silicon development process, as in the case of other previous processors, the bare-metal exercisers approach proved highly successful. This approach is key in our post-silicon validation strategy. In fact, we generally view bugs that were found with OS-based tests as escapes from our validation strategy.

17.5 Preparing for the Lab

We began our work preparing for the post-silicon validation phase well before silicon samples were available. A high-level description of the preparation process is shown in Fig. 17.3. The exerciser team, which was responsible for preparing the tools and tests for the lab, began its work as soon as the High-Level Design (HLD) exit milestone was reached. This was the stage at which the key features of POWER8 were determined.

The interlock between the exerciser team and the design and verification teams continued throughout the processor's life cycle. The goals of the interlock changed depending on which phase of the design was happening.

The first discussion, shortly after HLD-exit, between the exerciser and design teams, was required to make sure the exerciser team gained a better understanding of the new features to be supported by POWER8, so they could determine how to support them in the exercisers. It was also important to ensure that the design team understood the validation requirements and incorporated them in the design.

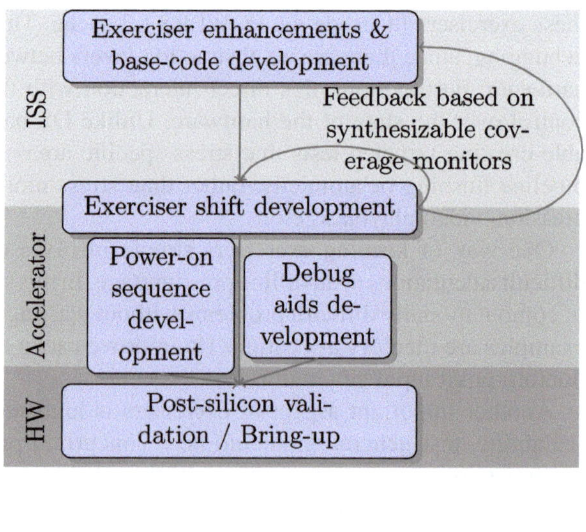

Fig. 17.3 Preparing for the lab: primary steps and technologies

Our exerciser development relied heavily on the use of an Instruction Set Simulator (ISS) [10]. We used the ISS to verify the correctness of the exerciser software code, and for test developers to observe that the generated test cases captured the scenario's verification intent. It was also used to collect architectural coverage, to verify that the tool generator covers interesting aspects of the test space [11].

When sufficient functional stability was achieved, the pre-silicon verification team used our in-house Awan simulation acceleration platform [12] to run the exerciser shifts. This phase is termed Exercisers-on-Accelerators (EoA) and has several goals. First, running exerciser shifts on the DUT model helped finding software bugs in the exerciser code (most software bugs are found on ISS, but some aspects of the design are not covered by it). In addition, accelerators enable running many-core models which are too big to simulate. Since this effort was conducted during the pre-silicon stage, the exercisers had the potential of hitting logic defects that could be fixed at low cost. During the POWER8 pre-silicon verification phase, we found roughly 1% of all functional bugs by running exercisers-on-accelerators. We note that the accelerator's speed depends on the size of the model the accelerator is running. To ensure high cycle volume, much of the EoA ran partial chip models, such as a core-only model, or a model that includes a single core and the memory hierarchy components.

In the next stage, we leveraged a set of coverage monitors synthesized into the DUT model. We added these coverage monitors into the logic running on the accelerator; they were not synthesized into the silicon itself, as they would consume significant area and power, and introduce timing issues.

Leveraging acceleration-synthesized coverage monitors enabled us to use a pre-silicon-like coverage-driven methodology for post-silicon test development [7]. Additional details about this approach, and how we used it to select shifts for the post-silicon validation lab, are described in Sect. 17.8.

The exerciser team used these coverage monitors to ensure that the overall set of shifts provided good coverage on the entire DUT. Many of the coverage goals, implemented through the acceleration-synthesized coverage monitors, were also targeted by the pre-silicon verification team. Thus, this methodology also helped the pre-silicon verification team guarantee high DUT coverage.

We also used accelerators to prepare the different procedures and tools for the lab team. This included the validation of the different steps required to boot the system (Virtual Power-On, or VPO), track the execution of an exerciser as it runs, and dump and format debug data from the embedded trace arrays. In addition to preparing the different tools, we used a complete lab-like environment, where accelerators stand in for real hardware, to train the lab team in processing and debugging lab failures in conditions similar to the ones they would encounter during the post-silicon validation phase.

All of the above steps were taken to ensure that once the first silicon samples were in the lab and ready to be used, all the supporting means—including the exercisers, the debug aids, and the team—would be ready to start the post-silicon validation of the system.

17.6 Triggering Bugs, Stimuli Generation

In order to trigger hard-to-hit timing bugs, exercisers use several techniques. One technique is called *thread irritation* [13]. This consists of a main scenario, to be run on part of the threads, and a shorter stressing scenario, to be performed in a loop, by the rest of the threads (e.g., a single cyclic load, store, or cache invalidation). The irritator works in a tight infinite loop, while the main thread completes its scenario. When the main thread is done, it releases the irritator from its loop. This achieves two important goals. First, it ensures full utilization—one thread is not idly waiting for the other to finish. Second, it ensures concurrent execution of the two scenarios, without adding any barriers. This level of concurrency is very difficult to reach by standard OS-based testing.

Another important technique is the wide use of macros. These are handcrafted code snippets which are designed to trigger interesting micro-architecture events (flush, load-hit-store, etc.). Each exerciser contains dozens to hundreds of macros, collected over the years, which could be randomly injected within any given scenario. This increases the chances of creating new dependencies and timing sequences, unforeseen by the test writer.

In addition to bare-metal exercisers, we leveraged a set of hardware irritators embedded in the design. A hardware irritator is a piece of logic that can trigger micro-architectural events at random. The irritators are initialized during the processor's power-on sequence, and randomly inject events as the processor executes instructions. For example, an irritator can be used to flush the pipeline at random, without having the executed instruction stream to create the conditions required for this event. Furthermore, irritators can mimic large system behavior. For example, in a single chip system, an irritator can inject a random Translation Lookaside Buffer (TLB) invalidate-entry event as if it were coming from a different chip.

Different components in the POWER8 chip support nonfunctional running modes, introduced for the sake of validation. These modes further assisted the bring-up team in stressing the design. For example, the POWER8 L2 cache supports a mode in which almost every access would trigger a cast out. By setting the processor to this state, we could aggravate the stress on the L3 cache.

Since hardware irritators and nonfunctional modes are embedded in the design, and, accordingly, have an overhead in terms of area and power, they must be carefully thought of and designed as part of the processor.

17.7 Observing Bugs, Checking

The goal of checking is to detect failures as close as possible to the point of origin and to report details that will help debug and discover the failure's root cause.

We initially observed erroneous behavior either through a checker embedded in the hardware or when it was flagged by a checking mechanism in the exerciser. Since we made limited use of OS-based testing, we rarely needed to debug software crashes (also known as core dumps), where the operating system provides little to no data regarding the nature of the fail.

Hardware-based checkers were designed and embedded into the POWER8 processor. These checkers cover some generic erroneous behavior such as access out of memory and a timer-based checker for instruction completion. The former checker validates that all memory accesses performed by the processor are within the bounds of the existing physical memory attached to the DUT. The instruction completion timer is a simple per-hardware-thread counter that is reset to some predefined value upon the completion of any instruction of the associated hardware thread. For any cycle where the thread does not complete, the counter decrements by 1. If the counter reaches the value 0, this means that some predefined number of cycles have passed with no instructions completing on that thread. Upon this event, the checker fires.

During the POWER8 post-silicon validation, it was easier to debug failures triggered by hardware-based checkers. This was because when a failure was triggered by such a checker, the DUT was stopped fairly close to the origin of the bug. Accordingly, the debug logic, once configured to trace the relevant component in the design, typically contained valuable hints as to the origin of the bug. Furthermore, some checkers provide an initial indication when they fire. For example, when the access out of memory checker fires, the transaction that triggered this is captured in the bus' debug logic and points to the hardware thread and address that triggered the fail.

Bare-metal exercisers employ a checking technique called *multi-pass consistency checking* [1, 14]. In this technique, every test case is executed several times (passes). The first pass is referred to as a *reference pass*, and the values of certain resources, such as architected registers and some memory, are saved at the end of the execution of this pass. After following passes, the exerciser compares the end-of-pass values with those of the reference pass. In case an inconsistency is detected, the error is reported and the execution is stopped. The multi-pass consistency checking technique imposes restrictions on test generation. For example, write–write collisions between different threads to the same memory location, where the order of thread accesses cannot be determined may yield inconsistent results and are therefore not supported. In case we want to test such behaviors, we can run the test in a single pass, without checking its results. This will allow us to find bugs using hardware-based checkers or exercisers self-checkers.

Multi-pass consistency failures were very hard to debug. This occurred primarily because the checker flags the error millions, and sometimes billions, of cycles after the error first occurred.

Despite the restrictions and the difficulty in debugging such failures, multi-pass consistency checking has proven useful in finding some of the toughest functional bugs in POWER8.

In addition to multi-pass consistency checking, exerciser developers introduce self-checking statements into some of the scenarios. Such checks were only applicable in directed scenarios and required manual labor, but were useful in faster localization of the fail, preventing errors from being masked, and extending the attributes checked for such scenarios.

Finally, to ensure the memory model and certain computational aspects of the design, one of the exercisers [15] employed an on-platform golden model (similar to an ISS) to determine the expected end of test values.

17.8 Coverage Closure

One of the key activities in preparing for the post-silicon lab is the selection of the content (in our case, in the form of exerciser shifts) that will be used to validate the chip. Since there is always a shortage of working silicon samples in the lab, it was important for us to optimize the set of exerciser shifts we use in the lab.

The basic idea is to create a regression suite with coverage properties in pre-silicon verification, where acceleration-synthesized coverage monitors enable the collection of coverage information. We rely on pseudo-random exercisers as the main vehicle of silicon validation, and so, accordingly, the regression suite we construct is a probabilistic one [16].

In standard pre-silicon verification environments, coverage is detected and collected by the test bench, i.e., by the environment itself. This requires intensive interactions between the DUV and the environment, which in turn significantly slow down acceleration or emulation. An alternative solution is to embed the coverage monitors, which are responsible for detecting and collecting coverage events, in the DUV itself and execute them on the accelerator or emulator with the rest of the DUV. This solution has some disadvantages. First, embedding the coverage monitors in the DUV affects the raw execution speed, but this slowdown is typically smaller than the case when the coverage is implemented in the verification environment. The second disadvantage is the additional complexity of implementing the monitors in the synthesizable code of the DUV instead of the environment. Finally, embedding the coverage monitors in the DUV may lead to capacity issues in the accelerator or emulator.

The flow of the method used during the verification and validation of the IBM POWER8 Processor is shown in Fig. 17.4. In general, the process is divided into three steps: reaching coverage closure with the exercisers-on-acceleration (or emulation) platform; harvesting test definitions that can reproduce the coverage on silicon; and running the exercisers with the harvested test definitions on silicon.

The first step is similar to the general flow for achieving coverage closure in pre-silicon functional verification [17]. We start by activating the exercisers on the DUV and running test cases created by the exercisers using an existing set of test definitions. Next, we collect coverage data from these runs and perform coverage analysis. If we achieved coverage closure, that is, all the important coverage events are hit, then this step is completed and we move on to the next step. Otherwise, the causes for coverage holes are investigated, and corrective measures aiming at closing the coverage holes are taken. These measures can include addition of new

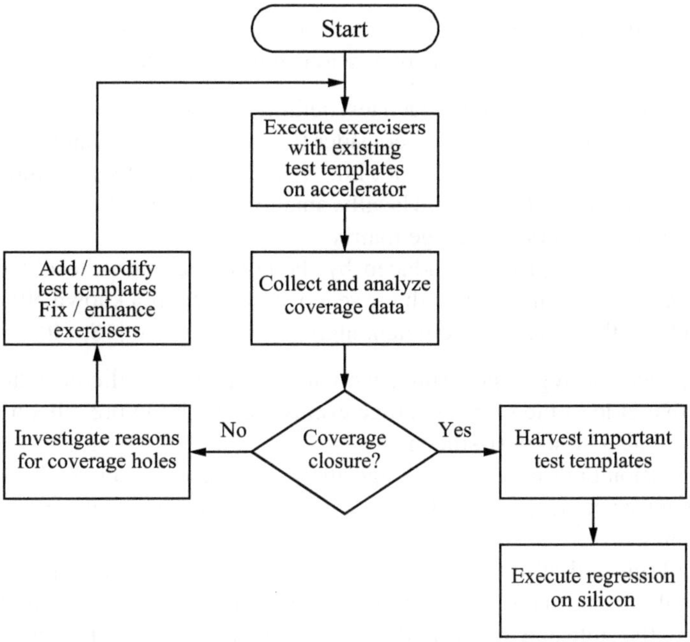

Fig. 17.4 Flow of post-silicon coverage guarantee

test definitions that target areas are not covered or lightly covered, modification to existing test definitions that do not reach their coverage targets, and corrections and enhancements to the exercisers themselves when needed.

After reaching coverage closure, the next step is harvesting a small set of test definitions that can guarantee coverage closure in post-silicon validation to create the requested regression suite. Reference [16] describes several algorithms for this purpose that are based on the probability of hitting each event in a single run by each test definition. These algorithms do not fit our needs. In our case, the exact probability of hitting a coverage event by a given test definition in a single pre-silicon run is not really important because of the big difference between the execution speed in pre- and post-silicon. For example, if the execution speed ratio is 10^4 (e.g., the accelerator speed is 300 K cycles/second and the silicon runs at 3G cycles/second) and we estimate the probability of hitting the target coverage event in a 10 min run of a given test definition at 1% (e.g., we hit the event once in 100 runs of 10 min each), then the probability of not hitting the event in 10 min on silicon with the same test definition is $(1 - 1/100)^{10,000} = 2 \cdot 10^{-44}$, which is practically 0. On the other hand, since we do not have a feedback from the silicon indicating whether we actually hit an event or not, we would like to minimize missing events. Going back to our previous example, it is hard to say if the event that was hit once in 100 runs indicates 1% probability of hitting the event or that this event was hit by pure luck. Therefore, we would like to reduce the reliance on events that are not hit in enough pre-silicon runs of a specific test definition. Given these two factors, the method for harvesting the test definitions and building the regression suite is the following:

- For each coverage event and for each test definition, if the event is hit by more than N runs of the test definition, then consider the event as hit by that test definition. Otherwise, consider the event as not hit. N is determined by how much hitting events by luck we want to avoid. Usually, small values of 1–3 for N are sufficient. This step results in a 0/1 coverage matrix.
- Solve the *set cover* problem induced by the coverage matrix to obtain the requested regression suite. Although the set cover problem is a known NP-complete problem [18], there are many efficient algorithms to solve it [18, 19].

Once we create the regression suite, we can run it on the silicon with (almost) guarantee that at least the same coverage events that are hit in pre-silicon using the exercisers are also hit in post-silicon.

The regression built based on the algorithm above is augmented with shifts that have hit many, or unique, bugs in the EoA stage, as well as shifts that are chosen by expert reviews.

The list of selected shifts is periodically revisited throughout the preparation stage, with the goal of optimizing the regression suite. In many cases, the team decides to add a shift to the suite due to new findings, or remove one in case there is confidence that similar coverage is guaranteed by other shifts. Even after the post-silicon stage start, changes can be introduced.

The process described above is used to construct the regression suite for the first post-silicon validation of the chip. Since there are more than one planned tape-out, the team needs to construct regression suites for the following tape-outs as well. The process for this is essentially the same, where the original suite may be changed following the learning process from preceding bring-ups. One noteworthy difference is that shifts that have hit bugs in one bring-up will be allocated with an extended run time during all following post-silicon validation efforts, to ensure that the bugs have been fixed.

17.9 Triage

During bring-up, exerciser shifts are being run by an automatic dispatcher. This dispatcher monitors the bring-up machines and dispatches a random shift whenever a machine goes idle. Each shift runs either for a certain amount of time or till a fail was detected. Upon detecting a fail, the dispatcher saves the log files and reports the fail to a central database. The report includes the shift name, fail type, runtime till the fail, run logs, and more.

The lab team uses this database for triaging the fails. A typical triage process consists of reviewing fail logs and clustering them by similar fail signatures. In contrary to the pre-silicon stage, in post-silicon it is utterly important to mark a fail in the correct cluster. This is true, since in numerous cases, a bug will only be hit one or two times during the entire bring-up lab. Missing these fails, by erroneously grouping them with other fails, might allow bugs to escape to the field.

Each exerciser team is responsible for triaging its own fails. The reason is that knowledge of the tool itself may be needed in order to identify the fail cluster, especially if it is a unique fail. Hence, each exerciser team maintained their fails in control—by daily monitoring the fails and clustering them.

Correct triaging improves the efficiency of the debug process, in several aspects. First, we have several data points on the same bug (which are different fails in the same cluster), so we can zoom in on the required machine state for hitting the bug. Second, the team could come up with more ideas for experiments, trying to hit the bug faster or differently. Third, we did not waste precious time on debugging the same fail over and over again.

17.10 Debugging

Across academia and industry, debugging post-silicon failures is acknowledged as a major challenge, and has, accordingly, received significant attention. In this section, we describe the best practices used in the IBM post-silicon lab during the validation of the POWER8 processor.

When a test case failed in the lab, the team analyzed it to determine the cause of the failure. The origin of the failure could be one of the numerous reasons, including a manufacturing issue, erroneous machine setup, an electrical bug, a functional bug, or a software bug in the exerciser.

One key feature facilitating effective post-silicon debug in POWER8 is the existence of the *cycle reproducible* environment. This environment is a special hardware mode in which executing the same exerciser shift would reproduce the exact same results. The cycle reproducible environment is limited to the domain including the processor core, its private L2 cache, and an 8MB region of the L3 cache (annotated in Fig. 17.1). Accordingly, many design problems could not be debugged this way. These include, for example, bugs outside this part of the design (for example, memory controller bugs), or bugs that require interaction across more than one core. However, the majority of the functional bugs could be found and debugged using this method, which proven its effectiveness.

The team first re-ran the same failing exerciser test in order to filter out problems originating from manufacturing or setup issues. If the original fail was, for example, due to a manufacturing problem in one of the cores, the same exerciser test would not fail on a good core. Another example may be two (or more) different shifts, failing on the same processor, for a similar reason. This could also be suspected as a manufacturing problem, in this specific processor. We ran the failing shift on a different machine, in the cycle reproducible environment, to see if it reproduces, which could also identify some manufacturing problems.

Then, the exerciser team tried to make sure that the fail is not a software bug in the exerciser. They did initial debugging of the fail, trying to localize the bug and come up with some conjectures of its origin. This process enabled them to identify several software bugs, by finding erroneous code sequences, or assert with high probability that this is indeed a hardware bug.

In the remainder of this section, we focus on failures originating from functional bugs in the design.

If the fail occurred in the cycle reproducible environment, the team could re-run the same exerciser shift repeatedly with different trace array configurations. In different runs, the trace arrays were configured to either collect data on different parts of the design, or to terminate the run at a different time, based on, for example, a cycle counter [20]. The latter option was used to aggregate data from multiple runs, enabling a BackSpace-like [21] approach to post-silicon debug, and has proven instrumental to the validation of the POWER8 processor. It is important to note that the debugging process is made much easier, if the fail detection occurs as close to the fail reason as possible. Therefore, in some cases, we had to adapt the shift to fail differently (e.g., turn a software livelock into a machine stop), or at least to point to a small range of cycles in which the bug manifested. For example, we could add printouts of the cycle counter in runtime, to get a hold of a fairly small range in which we had to look for the bug.

The hardest to debug cases are when the fail does not occur in the cycle reproducible environment. In these cases, additional experiments need to be conducted to determine key aspects in the hardware configuration that were required to hit the fail. Such experiments can be done by changing the hardware setup, e.g., disabling some of the processor cores, limiting cache sizes, or modifying the number of active hardware threads in the core. Some experiments call for making changes to the exerciser shift, for example, adding more instructions of a certain type, or creating specific conditions, like instruction sequences or dependencies, required to hit the bug. Following every experiment, the team reviewed the results. This included whether the execution of the exerciser shift passed or failed, and the data gathered from failed runs. Based on this data, the team determined which additional experiments were required. This process was done manually, meaning that the experiments were suggested (and sometimes hand-coded on-the-fly) by the team, according to the information we had so far; and that the analysis of the results was also done by the team, in order to raise conjectures about the bug root cause.

Another approach to debugging failures was to reproduce them on the acceleration platform or on the cycle reproducible platform. This approach relied heavily on the use of tools, specifically, bare-metal exercisers, that can effectively run on both the hardware and an accelerator. Because the hardware platform is over four orders of magnitude faster than the acceleration platform, a failing shift could not be migrated from silicon to acceleration as is. Typically, we required a set of experiments to fine-tune the exerciser shift in order to make it hit the bug on silicon fast enough to enable recreation on acceleration. With the enhanced observability capabilities of the acceleration platform, recreating a bug there was sufficient in order to provide the logic designer all the data required to determine the root cause of the bug. If reproducing in the acceleration platform was unsuccessful, we tried reproducing the fail in the cycle reproducible environment. The drawback to this method was that it was only possible if the bug was able to be hit in this environment, in contrary to the acceleration platform, where we emulate the full system.

Finally, if none of the approaches described above succeeded, the lab team had to drive the debug process based only on the data available from the on-chip trace arrays. The POWER8 debug logic, which is similar to the one described by Riley et al. [20, 22], has three key attributes that enable effective debug. First, the trace array can be configured, as part of the hardware's initialization, to track different parts of the design. For example, if the observation under debug was suspected to originate from a bug in the Load-Store Unit (LSU), the debug logic could be configured to trace more data from the LSU at the expense of data from other units. Second, the debug logic can be configured to trace compound events. Finally, the events on which the trace arrays store their input can also be configured. Thus, instead of saving inputs every clock cycle, the trace arrays can be configured to latch data only when some event is detected. This enabled noncontinuous tracing of events, which has proven useful in some hard to debug cases.

The methods described above made it possible for us to root cause all of the bugs we found in the Power8 lab. Each bug required a different amount of manual effort, and varied in elapsed time to root cause. The lab team used the methodology and techniques that were used and developed in debugging each bug, to learn and improve the methodology of debugging of the following ones.

17.11 Results

The POWER8 bring-up is considered as a very successful one. The team was able to keep a bug discovery and resolution rate equal to or better than previous POWER processor bring-up efforts, with significantly less resources.

The results of the POWER8 bring-up are partly depicted in Fig. 17.5 and Table 17.1. Every point in the figure accounts for one bug. The table contains information on all of the bring-up bugs, not only the functional ones. Table 17.1 summarizes the data presented in Fig. 17.5. The table shows, per bug severity, the average and 90%-ile of debug time. For example, medium severity bugs required, on average, 10.07 days to root cause, and 90% of all medium severity bugs required 20, or less, days to root cause. Overall, about 1% of all POWER8 bugs were found in the post-silicon validation phase.

The bugs are divided into three classes, based on their severity. The severity of the bug was determined based on a combination of the impact of the bug on the functional behavior of the system, the performance cost of working around the bug (if that was possible), and the complexity of the fix. The severity of a bug could only be determined after its root cause was found and a fix was suggested.

The lab team was constantly under high workload. Since not all bugs could be processed at the same time, the team worked based on priorities. The debug time in Fig. 17.5 accounts for the *gross time* required to find the root cause of the bug, that is, the number of days that elapsed from the first time the bug was hit to its root cause

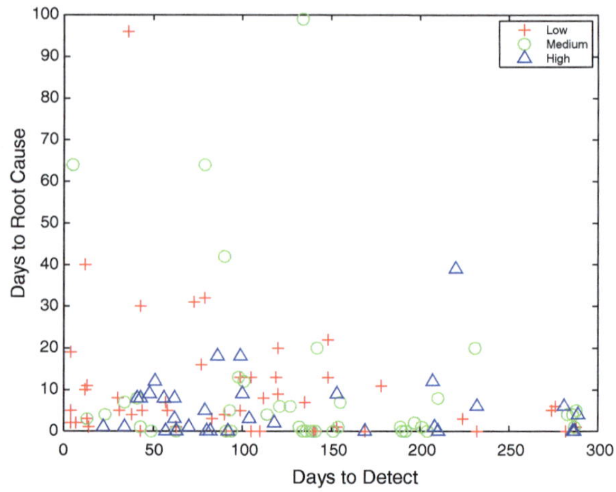

Fig. 17.5 POWER8 Bring-up Results: *The location of the point on the x-axis, termed detect time, relates to the number of days from the beginning of bring-up to the day the bug was first hit. A point's location on the y-axis, termed debug time, relates to the number of days from the time the bug was first hit to the time its root cause was determined*

Table 17.1 Statistics of days to root cause

	Average	90%
Low	9.92	22
Medium	10.07	20
High	5.96	12

day. In some cases, after a bug was initially detected, priorities dictated that other issues be handled before debugging could continue. The exact time and effort that was spent debugging every problem was not tracked. Interestingly, Fig. 17.5 shows that the debug time for high severity bugs is significantly lower, on average, than that of less important ones. Over 60% of all high severity bugs were root caused in under a week. This indicates that the lab team effectively speculated the severity of each bug at a very early stage. This is attributed to the expertise of the lab team members; based on the limited data available, they were able to infer the real nature of the bug and its expected severity soon after the bug was first detected.

In theory, all of the bugs could be found on the first day of the bring-up. In practice, several factors cause bugs to be identified at later stages of the process. These factors include triage and manual debug time of the lab team (which is finite), overcoming setup and manufacturing problems, changes in the lab configuration during bring-up, changes applied to the hardware or software to workaround bugs found, rareness of the bug, and additional tape-outs of the chip. Therefore, a bring-up lab is considered a "marathon" rather than a "sprint," and requires a constant effort from the lab team.

Figure 17.5 also shows how the rigorous preparations for the bring-up paid off. Half of all POWER8 post-silicon bugs were found in the first 3 months of the bring-up. This is considered a very good result, since during the first 2 months of the bring-up the team had to dedicate a lot of time to overcoming hardware stability issues and to "screen" the manufactured chips for good functioning ones.

Overall, the success of the POWER8 post-silicon validation is attributed to the following factors: *(1)* Well-established pre-silicon verification methodology, leveraging state-of-the-art formal methods, and dynamic techniques to ensure that the chip is taped-out at a high functional level; *(2)* On-going engagement, starting shortly after HLD-exit, between the exerciser team, the design team, and the verification team; *(3)* Systematic preparation of the exercisers and their tests based on an ISS, and the leveraging of synthesizable coverage monitors and accelerators to enable a coverage-driven approach to test development and selection; *(4)* Usage of bare-metal tools as the key driver for test generation on silicon; *(5)* Cycle reproducible environment and the use of accelerators to recreate lab failures; *(6)* Expertise of the lab team members and their dedication to the validation task.

17.12 Future Challenges

Many topics in post-silicon validation merit further research. In this section, we list several important ones.

While using synthesizable coverage monitors in acceleration provided us with good means of selecting shifts for the lab and ensure processor coverage, in-silicon coverage monitoring is needed to provide real feedback on the coverage of the fabricated chip. Since any in-silicon coverage monitor has area, power, and timing implications, the number of such monitors has to be kept low. Approaches such as configurable monitors [23] may be the right direction. However, the selection of which events should be monitored remains an important research question.

As explained above, the nature of the lab debug work is highly repetitive. Currently, the aggregation of data from these experiments, both passing and failing ones, is a manual process conducted by experts. Adopting techniques such as statistical debugging [24] to automate the debug process could reduce both debug time and the workload on the lab team.

Every bug found during the POWER8 bring-up was hit more than once, and often by several exerciser shifts. Since every failure must be processed, the team spent a great deal of time performing basic analysis on failures to ensure they map to known problems. Automation of this process could save precious time. Specifically, we aim to develop a system where lab team members can feed in complete or partial *signatures* of known bugs, and have the system classify failures based on these signatures. In addition to classifying a failure, such a system would have to report the confidence level of its classification. Beyond assisting in the triage process, such a system could greatly help in gaining a complete understanding of the nature of existing bugs.

With the creation of the OpenPOWER Consortium [25], the verification and validation of the system IBM is building with its partners is expected to be even more challenging.

17.13 Conclusion

The growing importance of the post-silicon validation phase to ensure the processor's quality prior to its shipment has increased the need to streamline the bring-up lab's work methodology and establish technologies for best practices. We presented the approach taken by the IBM team during the bring-up of the POWER8 processors. The meticulous preparation process, coupled with state-of-the-art technologies for stimuli generation, checking, and debugging, made the POWER8 bring-up a success. POWER9 used a similar approach, with some adjustments and enhancements, and was successful as well. We were able to achieve significant technological improvements to POWER9, within a very challenging schedule, to cope with an ever-growing competition and business models changes.

References

1. A. Adir, M. Golubev, S. Landa, A. Nahir, G. Shurek, V. Sokhin, A. Ziv, Threadmill: a post-silicon exerciser for multi-threaded processors, in *DAC* (2011), pp. 860–865
2. A. Nahir et al., Post-silicon validation of the IBM POWER8 processor, in *DAC* (2014), pp. 1–6
3. A. Adir et al., A unified methodology for pre-silicon verification and post-silicon validation, in *DATE* (2011), pp. 1590–1595
4. H.B. Carter, S.G. Hemmady, *Metric Driven Design Verification: An Engineer's and Executive's Guide to First Pass Success* (Springer, Berlin, 2007)
5. M.L. Behm, J.M. Ludden, Y. Lichtenstein, M. Rimon, M. Vinov, Industrial experience with test generation languages for processor verification, in *DAC* (2004), pp. 36–40
6. A. Piziali, *Functional Verification Coverage Measurement and Analysis* (Springer, Berlin, 2004)
7. A. Adir, A. Nahir, A. Ziv, C. Meissner, J.Schumann, Reaching coverage closure in post-silicon validation, in *Haifa Verification Conference* (2010)
8. A. Adir, A. Nahir, A. Ziv, Concurrent generation of concurrent programs for post-silicon validation. IEEE Trans. CAD Integr. Circuits Syst. **31**(8), 1297–1302 (2012)
9. D. Lee et al., Probabilistic bug-masking analysis for post-silicon tests in microprocessor verification, in *DAC* (2016), pp. 24–29
10. P. Bohrer et al., Mambo: a full system simulator for the powerpc architecture. SIGMETRICS Perform. Eval. Rev. **31**(4), 8–12 (2004)
11. O. Lachish et al., Hole analysis for functional coverage data, in *DAC* (2002), pp. 807–812
12. J.A. Darringer, E.E. Davidson, D.J. Hathaway, B. Koenemann, M.A. Lavin, J.K. Morrell, K. Rahmat, W. Roesner, E.C. Schanzenbach, G.E. Téllez, L. Trevillyan, EDA in IBM: past, present, and future. IEEE Trans. CAD Integr. Circuits Syst. **19**(12), 1476–1497 (2000)
13. A. Adir et al, Advances in simultaneous multithreading testcase generation methods, in *HVC* (2010), pp. 146–150
14. J. Storm, Random test generators for microprocessor design validation, http://www.oracle.com/technetwork/systems/opensparc/53-rand-test-gen-validation-1530392.pdf (2006). Accessed 01 Sept 2013
15. A. Nahir, A. Ziv, S. Panda, Optimizing test-generation to the execution platform, in *ASP-DAC* (2012), pp. 304–309
16. S. Fine, S. Ur, A. Ziv, A probabilistic regression suite for functional verification, in *Proceedings of the 41st Design Automation Conference* (2004), pp. 49–54
17. B. Wile, J.C. Goss, W. Roesner, *Comprehensive Functional Verification - The Complete Industry Cycle* (Elsevier, London, 2005)

18. M. Garey, D. Johnson, *Computers and Intractability: A Guide to the Theory of NP-Completeness* (W.H. Freeman, San Francisco, 1979)
19. E. Buchnik, S. Ur, Compacting regression-suites on-the-fly, in *Proceedings of the 4th Asia Pacific Software Engineering Conference* (1997)
20. M.W. Riley, N. Chelstrom, M. Genden, S. Sawamura, Debug of the CELL processor: moving the lab into silicon, in *ITC* (2006), pp. 1–9
21. F.M. de Paula, A.J. Hu, A. Nahir, nuTAB-BackSpace: Rewriting to normalize non-determinism in post-silicon debug traces, in *CAV* (2012)
22. F.M. de Paula, A. Nahir, Z. Nevo, A. Orni, A.J. Hu, Tab-backspace: unlimited-length trace buffers with zero additional on-chip overhead, in *DAC* (2011), pp. 411–416
23. M. Abramovici, P. Bradley, K.N. Dwarakanath, P. Levin, G. Memmi, D. Miller, A reconfigurable design-for-debug infrastructure for SoCs, in *DAC* (2006), pp. 7–12
24. J.A. Jones, M.J. Harrold, J.T. Stasko, Visualization of test information to assist fault localization, in *ICSE* (2002), pp. 467–477
25. OpenPOWER announcement, http://www-03.ibm.com/press/us/en/pressrelease/41684.wss. Accessed: 18 Nov 2013

Part VI
Conclusion and Future Directions

Chapter 18
SoC Security Versus Post-Silicon Debug Conflict

Yangdi Lyu, Yuanwen Huang and Prabhat Mishra

18.1 Introduction

Post-silicon validation and debug is essential to detect and diagnose potential bugs in the design. To facilitate the testing/debugging process of fabricated integrated circuit, testability features are added to control or observe internal nodes. Trace buffer is a commonly used Design-for-Debug (DfD) structure. Similarly, scan chain is widely used as a Design-for-Test (DfT) structure. Scan chains are able to control and observe the values of internal registers during "test mode" by connecting all the registers in a chain. Similarly, trace buffers embedded in a System-on-Chip (SoC) trace a small set of internal signals during execution whose values are used during post-silicon (offline) debug. Observability provided by these DfD and DfT structures are crucial during post-silicon debug to detect and fix errors. Debug engineers are trying to put more and more such structures to maximize observability without violating various design constraints such as area and energy budget.

However, there is an inherent conflict between security and observability. While debug engineers would like to have better observability, the security experts would like to enforce limited or no visibility with respect to the security modules in an SoC design. It is accepted in the research community that there is a strong link between observability/testability and security [12]. The structures inserted for the benefit of debugging can be a source of information leakage. As scan chains have both observability and controllability, the scan-based attacks have been extensively

Y. Lyu (✉) · P. Mishra
Department of Computer Science and Engineering, University of Florida, Gainesville, USA
e-mail: lvyangdi@ufl.edu

Y. Huang
VMware Inc., 3401 Hillview Ave, Palo Alto, CA 94304, USA
e-mail: yuanwenhuang@ufl.edu

P. Mishra and F. Farahmandi (eds.), *Post-Silicon Validation and Debug*,
https://doi.org/10.1007/978-3-319-98116-1_18

explored especially on cryptographic primitives, including Data Encryption Standard (DES) [28] and Advanced Encryption Standard (AES) [29], stream ciphers [18], and RSA [8, 19]. Trace buffer attack as a new area has been shown to be successful in leaking secret information from AES by Huang et al. [13, 15].

In this chapter, we discuss how the debugging structures can be a threat to security. In particular, this chapter describes both scan-based attacks and trace buffer attack. There are a wide variety of security threats and potential countermeasures [9, 10, 14, 17] that this chapter will not cover. The rest of this chapter is organized as follows. Background on AES and signal selection is covered in Sect. 18.2. Section 18.3 presents recent scan-based attacks. Section 18.4 describes trace buffer attack on AES. Section 18.5 presents the experimental results, and Sect. 18.6 concludes this chapter.

18.2 Background

18.2.1 AES Specification

AES works on a block size of 128 bits and a key size of 128, 192, or 256 bits, which are referred to as AES-128, AES-192, and AES-256, respectively.[1] We briefly review AES-128 here, for further details, readers can refer to [11].

The encryption flow of AES is shown in Fig. 18.1. AES accepts a 128-bit plaintext and a 128-bit user key, and generates a 128-bit ciphertext. The encryption proceeds through an initial round and subsequent 10-round repetition of four steps. These steps are *SubBytes*, *ShiftRows*, *MixColumns*, and *AddRoundKey*. In the final round, *MixColumns* step is skipped. For each of these rounds, separate 128-bit round subkeys are needed. The initial round takes the primary key, and each of the following 10 rounds uses a different round key. The round key generation follows the Rijndael's key expansion algorithm to generate the next four-word round key $[RK_{i+1,1}, RK_{i+1,2}, RK_{i+1,3}, RK_{i+1,4}]$ based on the current four-word round key $[RK_{i,1}, RK_{i,2}, RK_{i,3}, RK_{i,4}]$ using Eq. 18.1. The *lcs* represents a one-byte left circular shift operation, and the *sbox* function is byte-to-byte substitution according to a 16×16 lookup table.

$$\begin{aligned}
RK_{(i+1,1)} &= RK_{(i,1)} \oplus sbox(lcs(RK_{(i,4)})) \oplus RC_i \\
RK_{(i+1,2)} &= RK_{(i,2)} \oplus RK_{(i+1,1)} \\
RK_{(i+1,3)} &= RK_{(i,3)} \oplus RK_{(i+1,2)} \\
RK_{(i+1,4)} &= RK_{(i,4)} \oplus RK_{(i+1,3)}
\end{aligned} \tag{18.1}$$

The plaintext is organized as a 4×4 column-major order matrix, which is operated through the AES rounds. The *SubBytes* step uses a nonlinear transformation on every

[1] For the rest of the chapter, unless explicitly specified, we will use AES-128 and AES interchangeably.

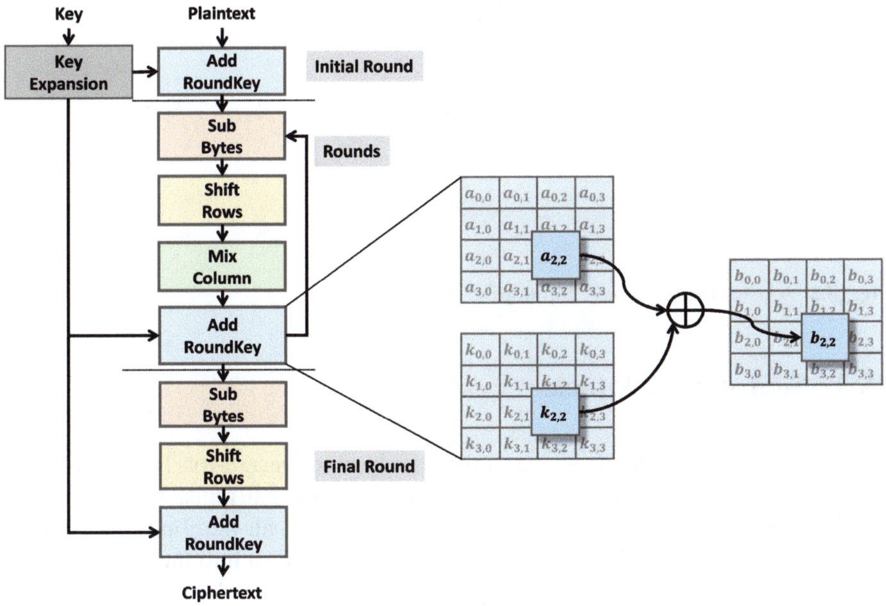

Fig. 18.1 AES encryption flow [15]

element of the matrix. The nonlinear transformation is defined by an 8-bit substitution box, also known as Rijndael S-box. The *ShiftRows* step cyclically shifts the bytes in each row by a certain offset. In the *MixColumns* step, each column is multiplied by a fixed matrix. In the *AddRoundKey* step, each byte of the matrix is exclusive-ORed with each byte of the current round subkey. This is shown graphically in Fig. 18.1.

18.2.2 Signal Selection

The goal of trace signal selection is to obtain a set of signals, which can restore the maximum number of internal states in the chip. There has been many signal selection techniques over the years [2–7, 16, 20–27]. For example, Basu et al. [3] proposed a metric-based algorithm that employs total restorability for selecting the most profitable signals. Chatterjee et al. [7] proposed a simulation-based algorithm which is shown to be more promising than metric-based approaches. Li and Davoodi [16] proposed a hybrid approach which combines the advantages of metric- and simulation-based approaches. Recently, Rahmani et al. [21, 25] demonstrated that machine learning can lead to fast and scalable signal selection without sacrificing restorability.

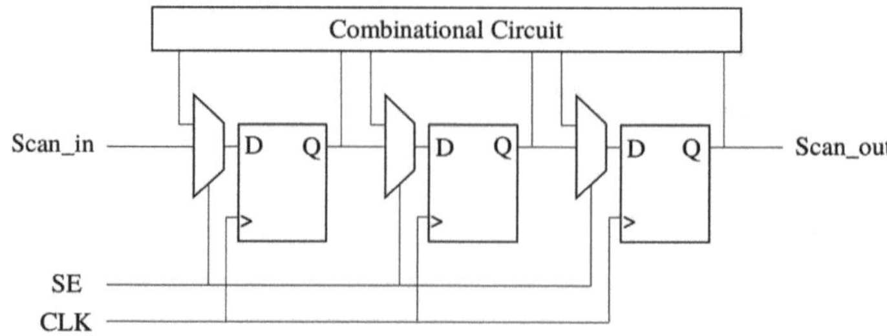

Fig. 18.2 A scan chain with three flip-flops

18.3 Scan-Based Attacks

The insertion of scan chains is one of the widely used Design-for-Test (DfT) techniques that can control and observe the values of internal registers as shown in Fig. 18.2. When the scan enable (SE) signal is asserted, all the flip-flops are connected via a shift chain. The input signal (scan_in) is driven into the chain, and the output signal (scan_out) is observed by the outside at each cycle.

As scan chains provide a way to observe the internal registers, sensitive information stored in the internal registers can be shifted out during test mode. Existing attacks include DES [28], AES [29], stream ciphers [18], and RSA [8, 19]. We briefly outline the attack in [29] to show how scan chains can be used to reveal secret information.

Yang et al. [29] proposed an attack on a hardware implementation of AES, which utilizes the difference between two test output vectors to recover the secret key. The first step of the attack is to figure out where the intermediate ciphertexts ($b_{i,j}$ in Fig. 18.1) are stored. First, run the chip in normal mode for one clock cycle with a specific plaintext. The plaintext is processed by the initial round and round 1. Then, the signals in all registers are extracted in test mode, forming the scan vector s_1. Next, a different plaintext, which is different from the first plaintext in only one byte, is chosen to repeat the process and retrieve the second scan vector s_2. The difference of s_1 and s_2 is localized. Let plaintext p be arranged as a 4×4 matrix as shown in Fig. 18.1. Assuming two plaintexts are different in $p_{1,1}$, the intermediate ciphertexts will be different in $(b_{0,0}, b_{1,0}, b_{2,0}, b_{3,0})$. By trying different plaintext pairs and inspecting the 1's in $s_1 \oplus s_2$, we are able to determine which 32 bits in the scan vector correspond to $(b_{0,0}, b_{1,0}, b_{2,0}, b_{3,0})$. Experimental results show that it requires six plaintexts to determine each group of 32 bits on average. Note that one-to-one mapping from the bits in intermediate ciphertext and the bits in scan vector is not constructed.

The second step is to recover the secret key from the scan output. We continue with the example in the first step. We use the superscript to distinguish the two different encryption processes. From Fig. 18.1, we can get the following equations:

$$b^1_{i,0} = a^1_{i,0} \oplus k_{i,0}, \quad b^2_{i,0} = a^2_{i,0} \oplus k_{i,0} \tag{18.2}$$

Table 18.1 Number of 1's in $b^1 \oplus b^2$ and the $(c^1_{1,1}, c^2_{1,1})$ pairs that can be uniquely determined by [29]

Number of 1's in $b^1 \oplus b^2$	9	12	23	24
$(c^1_{1,1}, c^2_{1,1})$ pairs	(226, 227)	(242, 243)	(122, 123)	(130, 131)

$$\begin{aligned} b^1 \oplus b^2 &= b^1_{i,0} \oplus b^2_{i,0} = (a^1_{i,0} \oplus k_{i,0}) \oplus (a^2_{i,0} \oplus k_{i,0}) \\ &= (a^1_{i,0} \oplus a^2_{i,0}) \oplus (k_{i,0} \oplus k_{i,0}) \\ &= a^1_{i,0} \oplus a^2_{i,0} \end{aligned} \tag{18.3}$$

From Eq. 18.3, the number of 1's in $b^1 \oplus b^2$ is not related to the subkey of round 1. Let c denote the output of initial round, which is the plaintext XORed with the primary key. As pointed out in [29], the number of 1's in $b^1 \oplus b^2$ is dependent only on $c^1_{1,1}$ and $c^2_{1,1}$. If the number of 1's in $b^1 \oplus b^2$ happens to be 9, 12, 23, or 24, the pairs $(c^1_{1,1}, c^2_{1,1})$ can be uniquely determined as shown in Table 18.1.

The authors proposed to use two plaintexts that are different in only one bit, i.e., $(2m, 2m + 1)$ as $(p^1_{1,1}, p^2_{1,1})$. They keep trying different pairs until the number of 1's in $b^1 \oplus b^2$ is in Table 18.1. Then, one byte of the primary key can be retrieved as $c_{1,1} \oplus p_{1,1}$. Experimental results [29] show that 32 plaintexts are used to retrieve one byte on average. By inspecting different parts of b, all bytes are retrieved using 512 plaintexts.

18.4 Trace Buffer Attacks

Trace buffer is able to improve the observability of circuit and thus assists post-silicon debug and analysis [10, 22, 24, 25]. It is a buffer that traces (records) some of the internal signals in a silicon chip during runtime. If an error is encountered, the content of trace buffer would be dumped out through JTAG interface for offline debug and error analysis. Due to design overhead constraints, the number of trace signals is only a small fraction of all internal signals in the design. The size of the trace buffer directly affects the observability that we can get from the trace buffer.

Figure 18.3 illustrates how the trace buffer is used during post-silicon validation and debug. Signal selection is done during the design time (pre-silicon phase). Let us assume that $S_1, S_2, \ldots, S_n$ are the selected trace signals. Figure 18.3 shows a trace buffer with a total size of $n \times m$ bits, which traces n signals (buffer width) for m cycles (buffer depth). For example, the ARM ETB [1] trace buffer provides buffer sizes ranging from 16Kb to 4Mb. In this case, a 16Kb buffer can trace 32 signals for 512 cycles (i.e., $n = 32$ and $m = 512$). Once the trace signals are selected, they need to be routed to the trace buffer. A trigger unit is also needed that decides when to start and stop recording the trace signals based on specific (error) events. The trace buffer

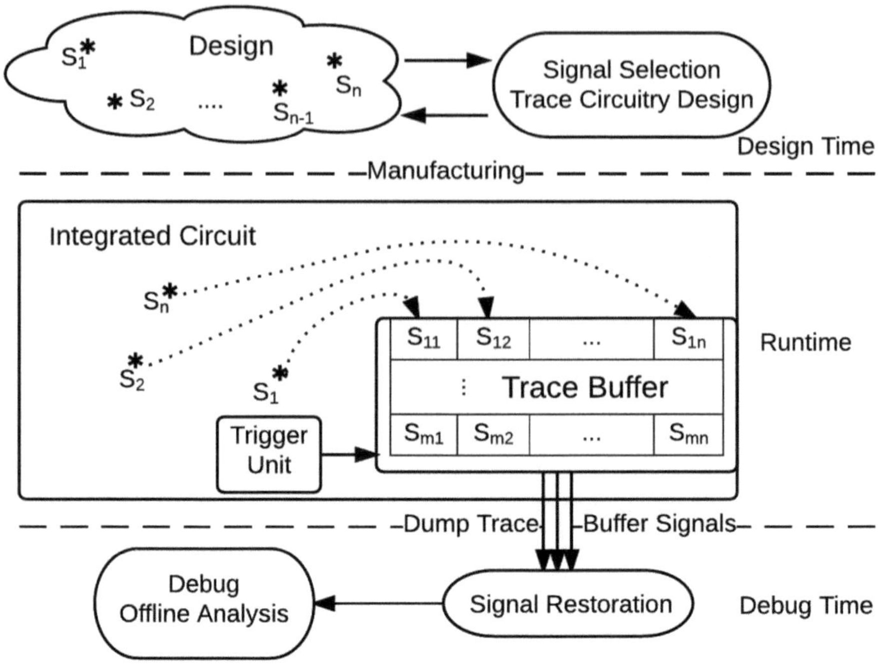

Fig. 18.3 Overview of trace buffer in system validation and debug. S_1, S_2, ..., S_n are the selected trace signals. The trace buffer traces n signals (buffer width) for m cycles (buffer depth) with a total size of $n \times m$ bits [15]

records the states of the traced signals during runtime. During debug time, the states of traced signals will be dumped out through the standard JTAG interface. Signal restoration is performed to restore as many states as possible, which is to maximize the observability of the internal signals in the chip. The offline debug and analysis would be based on the traced signals and the restored signals.

Huang et al. [15] proposed trace buffer attack on AES with/without the knowledge of RTL implementation. The details of the attack are introduced in the following sections. Huang et al. [15] utilized one of the state-of-the-art signal selection techniques [22], rather than manually selecting the signals that are easier for the attack.

18.4.1 Trace Buffer Attack Utilizing the RTL Implementation

Assuming that the Register-Transfer Level (RTL) implementation is available, the proposed attack proceeds in two phases. In the first phase, the correspondence between the signal values in trace buffer and variables in the AES design is established. In the second phase, the signal values are fed to the restoration algorithm to restore internal signals and eventually recover bits in the user-specified primary key.

18.4.1.1 Step 1: Determine Trace Buffer Signals

The challenge for trace buffer attack is that the attacker does not know what signals are recorded in the trace buffer. As the attacker has access to a few test chips and the RTL description of the AES design, the one-to-one mapping between the traced signals and the registers in RTL description can be established by running some test chips and matching with RTL simulation using Algorithm 1.

Algorithm 1: MAP SIGNALS TO REGISTERS IN RTL [15]

```
Input: AES RTL implementation, AES test chip
Output: Identified signals in trace buffer
while true do
    Select a random plaintext T_itr, a random key K_itr
    Run RTL simulation with T_itr and K_itr for c cycles
    Run the test chip with T_itr and K_itr for c cycles
    for Each traced signal S_i in trace buffer do
        Represent S_i as a vector of c values
        for Each register R_j in RTL do
            Represent R_j as a vector of c values
            if the vectors of S_i and R_j are the same then
                (S_i, R_j) is a possible match
            end
        end
        if S_i has a unique match R_j then
            (S_i, R_j) is a verified match
        end
    end
    if Every signal in S has a unique match then
        Break
    end
end
return Identified signals in trace buffer
```

For each iteration, a random input plaintext T_{itr} and a random key K_{itr} are selected to run in both the test chip and the RTL simulation for c cycles. Each traced signal will have a vector of c values stored in the trace buffer. For each traced signal, its vector is compared with that of all the registers from RTL simulation. If a unique match is found, this traced signal is identified in the RTL description. The process is repeated until all the traced signals are uniquely identified.

18.4.1.2 Step 2: Signal Restoration

After the signals in the trace buffer are identified, the next step is to run the chip in the working mode with the secret primary key and take advantage of the trace buffer to amount the attack. The attacker dumps out the signal states recorded in the

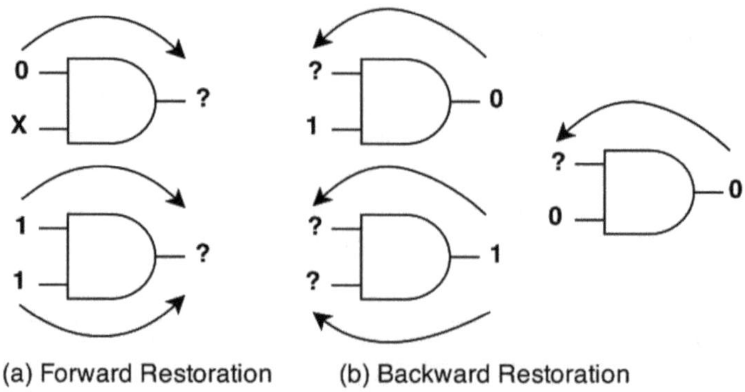

Fig. 18.4 Illustration of signal restoration for an AND gate [15]

buffer during online encryption and restores as many other signals as possible, and eventually obtain the primary key.

The signals can be reconstructed from the traced signals in two directions:

1. Forward restoration: pushes the restoration of signals from input to output, as shown in Fig. 18.4a.
2. Backward restoration: infers input values if some outputs are known, as shown in Fig. 18.4b.

The rightmost restoration in Fig. 18.4b shows an unsuccessful example. The restoration for registers (flip-flops) is that the state at current cycle is related to the state at previous cycle as specified by their truth tables.

Algorithm 2 outlines the major steps in a typical restoration algorithm. Starting from the trace buffer content, forward and backward restorations are applied repeatedly to construct value assignments for un-traced nodes until no change happens in one iteration. Although this algorithm has exponential complexity, in reality, it completes the process very fast since the number of new values created decreases significantly after each iteration.

18.4.2 Trace Buffer Attack Without RTL Implementation

Without the knowledge of its RTL implementation, the one-to-one mapping between the traced signals and the registers cannot be determined. Instead, the traced signals are mapped to the variable bits in the AES algorithm and the primary key can be retrieved utilizing Rijndael's key scheduling.

To illustrate the step of trace buffer attack on AES without RTL implementation, the iterative AES-128 chip is used as an example which has a trace buffer (32×512) of width 32 and depth 512 and takes 13 cycles to complete one encryption.

Algorithm 2: SIGNAL RESTORATION ALGORITHM [15]

```
Input: Trace buffer content, AES netlist
Output: Restored signal (node) values
Read in the AES circuit and form a hypergraph
Put all traced nodes into the UnderProcess queue
Update the traced nodes with their known values (0/1)
Update all other nodes with unknown values (x)
while UnderProcess is not empty do
    Take a node N from the UnderProcess queue
    for each node in N's BackwardNeighbors do
        Backward Restoration for this neighbor node
        if value at any cycle is restored then
            Add this neighbor node to UnderProcess
        end
    end
    for each node in N's ForwardNeighbors do
        Forward Restoration for this neighor node
        if value at any cycle is restored then
            Add this neighbor node to UnderProcess
        end
    end
end
return
```

18.4.2.1 Mapping Signals to Algorithm Variables

As the intermediate encrypted text and the round keys are most beneficial for signal restoration in recovering the primary key bits, we would like to find out whether any signals (bits) in the trace buffer are from them. C implementation of AES is used to simulate variable values from the 10 rounds after the initial round in Fig. 18.1. The values of each bit form a 10-bit vector. The length of each traced signal in the chip is 13, which is the number of cycles to finish one encryption operation. Therefore, if one signal is from the intermediate encrypted text or the round key, the resulting 10-bit string would be a substring of the same bit/signal in the trace buffer.

Algorithm 3 shows the details about how the signals from trace buffer bits are mapped to algorithm variable bits. The algorithm tries to identify as many signals as possible, and it will terminate when matched signals are uniquely identified and no more unique match can be found.

18.4.2.2 Signal Restoration Utilizing Rijndael's Key Expansion

From Eq. 18.1, two important rules are extracted for signal values restoration between cycles (rounds) as shown in Eq. 18.4.

Algorithm 3: MAP SIGNALS TO BITS IN AES VARIABLES [15]

Input: AES C implementaion, AES test chip
Output: Identified signals in trace buffer
while *true* **do**
    Select a random plaintext T_{itr}, a random key K_{itr}
    Run the C program of AES-128 with T_{itr} and K_{itr}
    Run the test chip with T_{itr} and K_{itr}
    for *Each traced signal S_i in trace buffer* **do**
        Represent S_i as a 512-bit binary string
        for *Each variable V_j in AES algorithm* **do**
            Extract V_j across the 10 encryption rounds
            for *Each bit $V_{(j,k)}$ in V_j* **do**
                Represent $V_{(j,k)}$ as 10-bit binary string
                if *$V_{(j,k)}$ is a repeating pattern in S_i* **then**
                    (S_i, $V_{(j,k)}$) is a possible match
                end
            end
        end
        if *S_i has a unique match $V_{(j,k)}$* **then**
            (S_i, $V_{(j,k)}$) is a verified match
        end
    end
    if *Every signal in S have either unique or no match* **then**
        Break
    end
end
return *Identified signals in trace buffer*

$$
\begin{aligned}
\text{Rule 1:} \quad & sbox(lcs(RK_{(i,4)})) = RK_{(i,1)} \oplus RK_{(i+1,1)} \oplus RC_i \\
\text{Rule 2:} \quad & RK_{(i+1,j-1)} = RK_{(i,j)} \oplus RK_{(i+1,j)}, \ j = 2, 3, 4
\end{aligned}
\tag{18.4}
$$

In Rijndael's round key expansion, the fourth word of current round key is the seed word for generating the next round key. The *lcs* and *sbox* operations on the fourth word are the sources to introduce unpredictable randomness to round keys. With the knowledge of all bits of the fourth word, the primary key can be obtained through the process of Table 18.2.

However, as the experimental results in Sect. 18.5.1.2 show, the trace buffer contains only partial information (25 bits) of the fourth word. The one-to-one bijection in *sbox* lookup table is critical in restoring the primary key. For example, if we have $sbox([1001x0x1]) = [100x0001]$, four combinations of $[1001x0x1]$ are fed into the *sbox* function, and the consistent result with $[100x0001]$ leaks the unknown bits. Fortunately, there is only one unique solution $sbox([10010001]) = [10000001]$ in the lookup table. With more unknown bits, more candidate solutions may be found. In that case, we have to evaluate all possible mappings. Experimental results suggest that the chances of multiple candidates are very rare. Algorithm 4 shows the steps to restore the primary key from the available trace buffer content. Compared to the

Table 18.2 The process of key recovery assuming full knowledge of the fourth word of all round keys. Round keys are represented as hexadecimal digits and "X" means "unknown" [15]

(A) Assume the fourth word of all rounds known				
RK_1	XXXXXXXX	XXXXXXXX	XXXXXXXX	62636363
RK_2	XXXXXXXX	XXXXXXXX	XXXXXXXX	F9FBFBAA
RK_3	XXXXXXXX	XXXXXXXX	XXXXXXXX	0B0FAC99
RK_4	XXXXXXXX	XXXXXXXX	XXXXXXXX	7E91EE2B
RK_5	XXXXXXXX	XXXXXXXX	XXXXXXXX	F34B9290
RK_6	XXXXXXXX	XXXXXXXX	XXXXXXXX	6AB49BA7
RK_7	XXXXXXXX	XXXXXXXX	XXXXXXXX	C61BF09B
RK_8	XXXXXXXX	XXXXXXXX	XXXXXXXX	511DFA9F
RK_9	XXXXXXXX	XXXXXXXX	XXXXXXXX	4C664941
RK_{10}	XXXXXXXX	XXXXXXXX	XXXXXXXX	6F8F188E
(B) Apply Rule 2 to recover RK_4				
RK_1	XXXXXXXX	XXXXXXXX	XXXXXXXX	62636363
RK_2	XXXXXXXX	XXXXXXXX	9B9898C9	F9FBFBAA
RK_3	XXXXXXXX	696CCFFA	F2F45733	0B0FAC99
RK_4	EE06DA7B	876A1581	759E42B2	7E91EE2B
RK_5	7F2E2B88	F8443E09	8DDA7CBB	F34B9290
RK_6	EC614B85	1425758C	99FF0937	6AB49BA7
RK_7	21751787	3550620B	ACAF6B3C	C61BF09B
RK_8	0EF90333	3BA96138	97060A04	511DFA9F
RK_9	B1D4D8E2	8A7DB9DA	1D7BB3DE	4C664941
RK_{10}	B4EF5BCB	3E92E211	23E951CF	6F8F188E
(C) Use Eq. 18.1 to get $RK_3 \sim RK_1$, and RK_0 (the primary key)				
RK_0	00000000	00000000	00000000	00000000
RK_1	62636363	62636363	62636363	62636363
RK_2	9B9898C9	F9FBFBAA	9B9898C9	F9FBFBAA
RK_3	90973450	696CCFFA	F2F45733	0B0FAC99
RK_4	EE06DA7B	876A1581	759E42B2	7E91EE2B

process in Table 18.2, additional Step 2 utilizes the unique mapping property of *sbox* to recover missing bits of the fourth word.

18.5 Experimental Results

This section presents the experimental results on both iterative AES-128 and pipelined AES ciphers.

Algorithm 4: RESTORE MISSING BITS IN ROUND KEYS [15]

```
Input: Identifid signals in round keys from trace buffer
Output: Restored round key bits
Update identified bits with values (1/0) from trace buffer
Update all other bits with unknown values (x)
/* Step 1: Apply Rule 2                                          */
for j ← 4 to 2 do
    for i ← 1 to 9 do
        RK(i+1,j−1) = RK(i,j) ⊕ RK(i+1,j)
    end
end
/* Step 2: Apply Rule 1                                          */
for i ← 4 to 9 do
    Use the bijection property of sbox to recover missing bits in RK(i,4)
end
/* Step 3: Apply Rule 2 one more time                            */
for j ← 4 to 2 do
    for i ← 1 to 9 do
        RK(i+1,j−1) = RK(i,j) ⊕ RK(i+1,j)
    end
end
/* Step 4: Use Eq. 1 to get the primary key                      */
for i ← 9 to 1 do
    RK(i−1,2) = RK(i,2) ⊕ RK(i,1)
    RK(i−1,3) = RK(i,3) ⊕ RK(i,2)
    RK(i−1,4) = RK(i,4) ⊕ RK(i,3)
    RK(i−1,1) = RK(i,1) ⊕ sbox(lcs(RK(i−1,4))) ⊕ RC(i−1)
end
return PrimaryKey = RK0
```

18.5.1 Case Study 1: Iterative AES-128

The iterative AES-128 design has 530 flip-flops and about 25,000 basic logic gates. The 530 flip-flops (registers) include two 128-bit registers to hold plaintext and ciphertext, four 32-bit registers to hold round keys, sixteen 8-bit registers to hold intermediate states, and other control and temporary signals.

18.5.1.1 Attack with RTL Implementation

Trace buffer attack is powerful with the knowledge of RTL implementation. Table 18.3 shows the results on the iterative AES-128 cipher, with different trace buffer sizes. The trace buffers with a buffer width of 32 and a buffer depth no less than 128 are able to recover the full primary key in a few minutes.

Figure 18.5a shows the number of bits in the user key leaked with different buffer sizes. Figure 18.5b shows the total number of internal states restored (debug observability) during restoration. The number of restored primary key bits increases with

Table 18.3 Iterative AES-128: Number of bits in the key recovered and memory/time requirements for signal restoration [15]

BufferDepth / BufferWidth		64	128	256	512
8	**Leaked key** (bits)	6	6	6	6
	Memory (MB)	116.4	161.4	252.0	432.0
	Time (mm:ss)	0:27.75	0:56.07	1:50.35	3:43.26
16	**Leaked key** (bits)	18	25	28	28
	Memory (MB)	116.4	161.4	252.0	432.0
	Time (mm:ss)	0:27.82	0:55.94	1:51.00	3:44.10
32	**leaked key** (bits)	98	128	128	128
	Memory (MB)	116.4	161.4	252.0	432.0
	Time (mm:ss)	0:28.01	0:55.98	1:52.81	3:51.38

bigger buffer width. For the same buffer width, the number of restored key bits increases slightly as the trace cycles increase, and it will be saturated after buffer depth is big enough (256 cycles or more). The fact that the 32×512 trace buffer can restore all 128-bit primary key is not surprising. The success of recovering the full primary key is due to the observability provided by the trace buffer. The iterative AES-128 design[2] has relatively short pathways with only 530 flip-flops in total.

18.5.1.2 Attack Without RTL Implementation

After applying Algorithm 3, 30 out of 32 signals are identified in the trace buffer, including 2 bits from the intermediate register and 28 bits from the round key register. The 28 bits from the 128-bit round key register include 1 bit from the first word, 2 bits from the third word, and 25 bits from the fourth word. The primary key is successfully extracted by Algorithm 4 from the selected signals. The results show that although the signal selection in [22] only greedily choose signals that are best for observability, the selected signals contribute greatly to information leakage. Security-aware signal selection techniques are an area for future research to increase brute force effort in recovering the key.

The attack without RTL implementation is harder compared to when RTL implementation is available in two ways. First, the attack with RTL can identify all signals traced in the buffer. It means the attack with RTL has more information to start with. Second, the restoration in Algorithm 2 (with RTL implementation) can deterministi-

[2]For iterative implementation, the restoration is clearly able to recover the key and we expect the same trend to follow for AES-192 and AES-256.

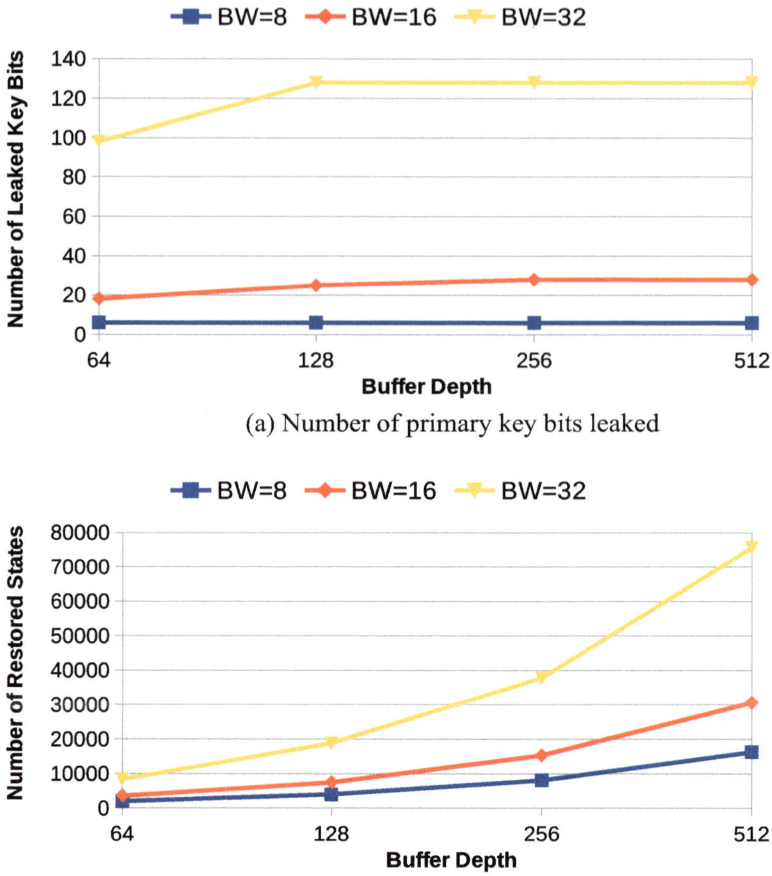

(a) Number of primary key bits leaked

(b) Number of flip-flop states restored

Fig. 18.5 Iterative AES-128: security and observability trade-off using Buffer Widths (BW) of 8, 16, and 32, and buffer depths of 64, 128, 256, and 512. The 32 × 128, 32 × 256, and 32 × 512 trace buffers are able to recover the full primary key [15]

cally propagate values forward and backward in the AES circuit, while the restoration in Algorithm 4 (without RTL) would need a lot more brute force effort to test and verify all possible mappings if the *sbox* lookup table cannot find a unique mapping.

18.5.2 Case Study 2: Pipelined AES Ciphers

The main difference from the iterative version is that the pipelined implementation unrolls all the encryption rounds to be independent hardware units, which makes the pipelined version about 10–15 times as large as the iterative. For example, the

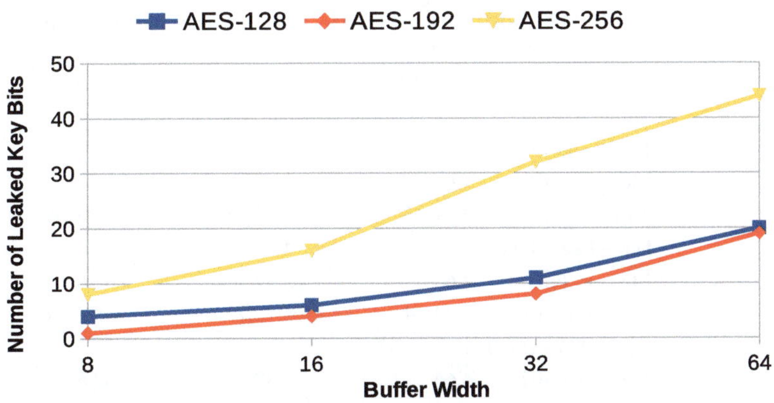

(a) Number of primary key bits leaked

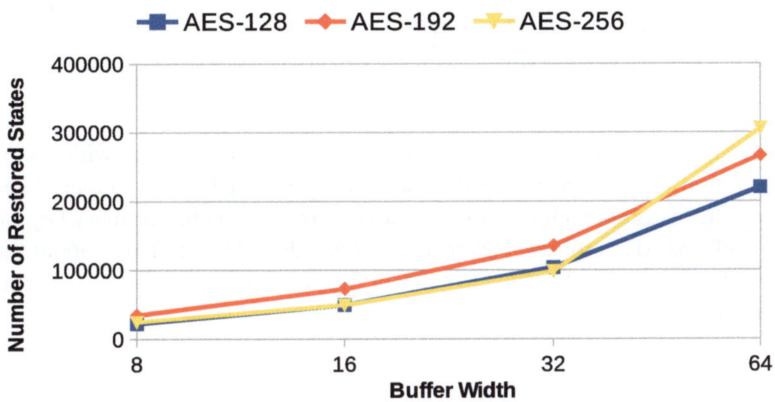

(b) Number of flip-flop states restored

Fig. 18.6 Pipelined AES-128, AES-192, and AES-256 ciphers: security and observability trade-off [15]

pipelined AES-128 cipher has 6720 flip-flops and about 290,000 logic gates, which is roughly 10 times (10 encryption rounds) as large as the iterative AES-128. This poses a greater challenge for the restoration process, because many signal values are not inferable due to the long pathways from the known signals. Only signals that are very close to the input can be propagated backward and possibly restore the primary key bits.

Table 18.4 and Fig. 18.6 show the experimental results on the pipelined implementation of AES-128, AES-192, and AES-256 ciphers using the attack method with RTL implementation. The trace buffer sizes vary with different buffer widths but fix the buffer depth to be 512 which should be suitable for the pipelined AES ciphers. We are able to restore 20, 19, and 44 bits of the primary key for AES-128,

Table 18.4 Pipelined AES-128, AES-192, and AES-256: Number of bits in the key recovered and memory/time requirements for signal restoration. [15]

BufferWidth / AESciphers		AES-128	AES-192	AES-256
8	**Leaked key** (bits)	4	1	8
	Memory (GB)	4.66	5.37	6.56
	Time (h:mm:ss)	3:51:45	4:29:05	6:38:06
16	**leaked key** (bits)	6	4	16
	Memory (GB)	4.66	5.37	6.56
	Time (h:mm:ss)	3:44:14	4:12:22	6:22:59
32	**Leaked key** (bits)	11	8	32
	Memory (GB)	4.66	5.37	6.56
	Time (h:mm:ss)	3:19:12	4:10:25	6:31:08
64	**Leaked key** (bits)	20	19	44
	Memory (GB)	4.66	5.37	6.56
	Time (h:mm:ss)	3:42:02	4:08:43	6:03:15

AES-192, and AES-256, respectively, in a few hours using a buffer width of 64. As the trace buffer width increases, both observability and the leaked number of key bits increase. The restoration algorithm is not able to restore the full primary key for any of the pipelined AES ciphers. Nevertheless, considerable knowledge about the key is gained, which does not suffice to recover the secret though, can aid other modes of cryptanalysis.

18.6 Conclusion

This chapter illustrates how the structures that are designed for testing/debugging can assist security attacks. We briefly outlined a scan-based attack. We also described a recently proposed trace buffer attack [15] in detail. With a trace buffer size of 32×128, the full key of the iterative AES-128 can be restored in a few minutes with signal restoration. For pipelined AES, partial key can be restored in a few hours. These experimental results demonstrate that the observability provided by the testing/debugging structures makes the AES vulnerable. This work illustrates the need for security-aware trace signal selection and highlights the need for further research in understanding the trade-off between security and debug observability.

References

1. Arm embedded trace buffer, http://infocenter.arm.com/help/index.jsp?topic=/com.arm.doc.dai0168b/ar01s03s03.html. [Online]
2. K. Basu, P. Mishra, Efficient trace signal selection for post silicon validation and debug, in *2011 24th Internatioal Conference on VLSI Design* (2011), pp. 352–357
3. K. Basu, P. Mishra, Rats: restoration-aware trace signal selection for post-silicon validation. IEEE Trans. Very Larg. Scale Integr. (VLSI) Syst. **21**(4), 605–613 (2013)
4. K. Basu, P. Mishra, P. Patra, Efficient combination of trace and scan signals for post silicon validation and debug, in *2011 IEEE International Test Conference* (2011), pp. 1–8
5. K. Basu, P. Mishra, P. Patra, Constrained signal selection for post-silicon validation, in *2012 IEEE International High Level Design Validation and Test Workshop (HLDVT)* (2012), pp. 71–75
6. K. Basu, P. Mishra, P. Patra, A. Nahir, A. Adir, Dynamic selection of trace signals for post-silicon debug, in *2013 14th International Workshop on Microprocessor Test and Verification* (2013), pp. 62–67
7. D. Chatterjee, C. McCarter, V. Bertacco, Simulation-based signal selection for state restoration in silicon debug, in *2011 IEEE/ACM International Conference on Computer-Aided Design (ICCAD)* (2011), pp. 595–601
8. J. Da Rolt, A. Das, G. Di Natale, M.-L. Flottes, B. Rouzeyre, I. Verbauwhede. A new scan attack on RSA in presence of industrial countermeasures (Springer, Berlin, 2012), pp. 89–104
9. F. Farahmandi, Y. Huang, P. Mishra, Trojan localization using symbolic algebra, in *2017 22nd Asia and South Pacific Design Automation Conference (ASP-DAC)* (2017), pp. 591–597
10. F. Farahmandi, R. Morad, A. Ziv, Z. Nevo, P. Mishra, Cost-effective analysis of post-silicon functional coverage events, in *Design Automation and Test in Europe (DATE)* (2017)
11. FIPS 197, Advanced Encryption Standard, http://csrc.nist.gov/publications/fips/fips197/fips-197.pdf (2001). [Online]
12. D. Hely, M. L. Flottes, F. Bancel, B. Rouzeyre, N. Berard, M. Renovell, Scan design and secure chip [secure ic testing], in *Proceedings of the 10th IEEE International On-Line Testing Symposium* (2004), pp. 219–224
13. Y. Huang, S. Bhunia, P. Mishra, Mers: statistical test generation for side-channel analysis based trojan detection, in *Proceedings of the 2016 ACM SIGSAC Conference on Computer and Communications Security*, CCS '16 (ACM, New York, 2016), pp. 130–141
14. Y. Huang, A. Chattopadhyay, P. Mishra, Trace buffer attack: security versus observability study in post-silicon debug, in *2015 IFIP/IEEE International Conference on Very Large Scale Integration (VLSI-SoC)* (2015), pp. 355–360
15. Y. Huang, P. Mishra, Trace buffer attack on the AES Cipher. J. Hardw. Syst. Secur. **1**(1), 68–84 (2017)
16. M. Li, A. Davoodi, A hybrid approach for fast and accurate trace signal selection for post-silicon debug. IEEE Trans. Comput.-Aided Des. Integr. Circuits Syst. **33**(7), 1081–1094 (2014)
17. Y. Lyu, P. Mishra, A survey of side-channel attacks on caches and countermeasures. J. Hardw. Syst. Secur. **2**(2), 33–50 (2017)
18. D. Mukhopadhyay, S. Banerjee, D. RoyChowdhury, B.B. Bhattacharya, Cryptoscan: a secured scan chain architecture, in *14th Asian Test Symposium (ATS'05)* (2005), pp. 348–353
19. R. Nara, K. Satoh, M. Yanagisawa, T. Ohtsuki, N. Togawa, Scan-based side-channel attack against RSA cryptosystems using scan signatures. IEICE Trans. Fundam. Electron. Commun. Comput. Sci. **E93.A**(12), 2481–2489 (2010)
20. K. Rahmani, P. Mishra, Efficient signal selection using fine-grained combination of scan and trace buffers, in *2013 26th International Conference on VLSI Design and 2013 12th International Conference on Embedded Systems* (2013), pp. 308–313
21. K. Rahmani, P. Mishra, Feature-based signal selection for post-silicon debug using machine learning. IEEE Trans. Emerg. Top. Comput. (99), 1 (2017)
22. K. Rahmani, P. Mishra, S. Ray, Scalable trace signal selection using machine learning, in *2013 IEEE 31st International Conference on Computer Design (ICCD)* (2013), pp. 384–389

23. K. Rahmani, P. Mishra, S. Ray, Efficient trace signal selection using augmentation and ILP techniques, in *2014 Fifteenth International Symposium on Quality Electronic Design*, pp. 148–155
24. K. Rahmani, S. Proch, P. Mishra, Efficient selection of trace and scan signals for post-silicon debug. IEEE Trans. Very Larg. Scale Integr. (VLSI) Syst. **24**(1), 313–323 (2016)
25. K. Rahmani, S. Ray, P. Mishra, Postsilicon trace signal selection using machine learning techniques. IEEE Trans. Very Larg. Scale Integr. (VLSI) Syst. **25**(2), 570–580 (2017)
26. P. Thakyal, P. Mishra, Layout-aware selection of trace signals for post-silicon debug, in *2014 IEEE Computer Society Annual Symposium on VLSI* (2014), pp. 326–331
27. P. Thakyal, P. Mishra, Layout-aware signal selection in reconfigurable architectures, in *18th International Symposium on VLSI Design and Test* (2014), pp. 1–6
28. B. Yang, K. Wu, R. Karri, Scan based side channel attack on dedicated hardware implementations of data encryption standard, in *2004 International Conference on Test* (2004), pp. 339–344
29. B. Yang, K. Wu, R. Karri, Secure scan: a design-for-test architecture for crypto chips. IEEE Trans. Comput.-Aided Des. Integr. Circuits Syst. **25**(10), 2287–2293 (2006)

Chapter 19
The Future of Post-Silicon Debug

Farimah Farahmandi and Prabhat Mishra

19.1 Summary

Given the widespread acceptance of System-on-Chip (SoC) designs in the electronic industry, it is critical to ensure their correctness from both functional and nonfunctional perspectives. This book provides a comprehensive reference for SoC designers, validation engineers as well as researchers interested in post-silicon validation and debug of heterogeneous SoCs. This book contains contributions from post-silicon validation and debug experts. Different chapters cover a wide variety of state-of-the-art SoC debug infrastructure, post-silicon validation methods, and post-fabrication error detection and correction techniques. The topics covered in this book can be broadly divided into the following categories.

19.1.1 Post-silicon SoC Validation Challenges

Chapter 1 highlighted the importance of the validation of SoC designs from both functional and nonfunctional perspectives in the modern era of SoC design. Validation activities of SoC-based designs can be classified into three categories: (i) pre-silicon validation, (ii) post-silicon validation and debug, and (iii) survivability and on-field debug. Post-silicon validation is widely acknowledged as a significant bottleneck in SoC design methodology—many recent studies suggest that it consumes more than 50% of SoCs overall design effort (total cost) [12]. We need to

---

F. Farahmandi (✉) · P. Mishra
Department of Computer and Information Science and Engineering,
University of Florida, Gainesville, USA
e-mail: ffarahmandi@ufl.edu

P. Mishra and F. Farahmandi (eds.), *Post-Silicon Validation and Debug*,
https://doi.org/10.1007/978-3-319-98116-1_19

consider five important activities during post-silicon validation, including powering the device, logic validation, hardware/software co-validation, electrical validation, and speed-path validation. There are several challenges associated with each of these activities including the limited observability and controllability of the silicon. Different debug infrastructures are utilized on silicon to improve observability and reduce post-silicon debug effort.

19.1.2 Debug Infrastructure

The next five chapters described efficient techniques to design debug architecture. The goal of these approaches is to decrease the complexity of post-silicon validation and debug efforts.

- *On-chip Instrumentation*: Chapter 2 surveyed activities for planning for post-silicon readiness. The chapter reviewed different Design-for-Debug (DfD) architectures including scan chains, trace buffers, coverage monitors, and performance monitors to increase the observability and controllability of the post-silicon design.
- *Metric-based Signal Selection*: Chapter 3 discussed trace signal selection algorithms that utilize the structure of the design to increase the observability of the overall design.
- *Simulation-based Signal Selection*: Chapter 4 presented approaches that measure the capability of selected signals using the information gathered by simulation.
- *Hybrid Signal Selection*: Chapter 5 reviewed hybrid signal selection approaches that combine the merit of both metric-based and simulation-based techniques.
- *Signal Selection using Machine Learning*: Chapter 6 demonstrated that machine learning techniques can be deployed to improve the performance of hybrid approaches.

19.1.3 Generation of Tests and Assertions

The next four chapters presented effective techniques to generate tests and assertions to validate the silicon.

- *Observability-aware Post-silicon Test Generation*: Chapter 7 presented a directed test generation methodology which considers transaction-level models as well as design-for-debug infrastructures to generate penetrating tests. The proposed approach is beneficial in activating a particular behavior as well as propagating its effect to the observable points.
- *On-chip Constrained-Random Stimuli Generation*: Chapter 8 proposed an on-chip constraint-random test generation approach that is programmable to exploit the different functionalities of the design during runtime. The technique is based on

translating user-specified constraints to various masks for random test sequences.

- *Test Generation for Memory Consistency Validation*: Chapter 9 outlined a post-silicon validation method for shared-memory chip multiprocessor. This method records and analyzes shared-memory interactions and cache accesses to detect any possible memory consistency violations.
- *Selection of Post-silicon Hardware Assertions*: Chapter 10 described metrics to rank pre-silicon assertions regarding their capabilities to detect bit flips during silicon execution. The selected assertions can be synthesized in hardware to assist real-time bit-flip error detection.

19.1.4 Post-silicon Debug

The next five chapters presented efficient approaches for post-silicon debug for localizing, detecting, and fixing post-silicon errors.

- *Debug Data Reduction Techniques*: Chapter 11 reviewed techniques for compressing data stored in trace buffers and other debug infrastructure. Data reduction techniques enable efficient utilization of on-chip debug architectures to enhance overall design observability and improve debugging time.
- *High-level Debug of Post-silicon Failures*: Chapter 12 presented a debugging framework that utilizes high-level models to localize the source of error. Once an error is detected, the related faulty sequences are mapped to high-level models for faster bug localization.
- *Post-silicon Fault Localization with Satisfiability Solvers*: Chapter 13 reviewed usage of satisfiability solvers for post-silicon fault localization. Using values of trace buffers, a symbolic representation of the runtime erroneous behavior is constructed for various sliding windows. The temporal location of the bug is located if the resulting formula of a specific window is unsatisfiable. The spatial location of bug can be detected using the maximal subset of the circuit that matches with the erroneous behavior of the chip in the defined time window.
- *Coverage Analysis of Post-silicon Tests with Virtual Prototypes*: Chapter 14 outlined metrics to evaluate the coverage of post-silicon tests using virtual prototypes. The method is based on collecting execution data from the virtual prototype within a virtual platform for a specific test. Next, it computes the estimated test coverage by replaying the captured data on the virtual prototype as well as using various software and hardware specific coverage metrics.
- *Utilization of Debug Infrastructure for Post-silicon Coverage Analysis*: Chapter 15 utilized on-chip debug infrastructure to perform cost-effective post-silicon functional coverage analysis.

19.1.5 Case Studies

The next two chapters provide a thorough discussion on post-silicon validation efforts for two case studies, a network-on-chip (NoC) and IBM Power8 processors.

- *Network-on-Chip Validation and Debug*: Chapter 16 outlined unique challenges in post-silicon validation of NoC architectures. The challenges arise from inherent parallel communication nature, security concerns as well as reliable data transfer requirements.
- *Post-silicon Validation of the IBM Power8 Processor*: Chapter 17 provided an overview of post-silicon validation phases in a Power8 processor's design life cycle to find both electrical and functional bugs.

Finally, Chap. 18 discussed the inherent conflict between design-for-debug and security vulnerabilities. It demonstrated that unprotected debug infrastructure can lead to asset leakage and threaten the security and trust of an SoC.

19.2 Future Directions

Although there have been significant efforts to improve the post-silicon validation and debug activities, there are still several challenges ahead to be addressed in this field. For example, there are insufficient prior efforts in post-silicon security validation. Moreover, the effective utilization of design-for-debug and design-for-test infrastructures such as trace buffer, scan chains, performance monitors, and event triggers should be considered. Furthermore, machine learning can also reduce the complexity of post-silicon validation significantly. We briefly outline some of the challenges and possible research directions in post-silicon validation and debug.

19.2.1 Effective Utilization of Debug Infrastructure

It is important to effectively utilize the available debug infrastructure to reduce the overall post-silicon validation effort. There are many on-chip solutions to improve observability including scan chains [6, 15], performance monitors, and trace buffers [1, 4, 5, 14]. Existing trace compression techniques target both depth and width of trace buffer. Depth compression [18] approaches deal with selecting the cycles where the data are erroneous and store the data for only those cycles. On the other hand, width compression [2] compresses the signal values that are stored per cycle. Some recent approaches focus on designing programmable on-chip debug hardware to enable faster error localization [11, 13]. Existing work on observability improvement instruments the silicon with circuitry for recording, triggering, and aggregating execution traces. Trace frameworks such as Intel TraceHub [7] and ARM CoreSight [16] facilitate configuration, transport, and visualization of software, firmware, and hardware traces. Current approaches primarily use the content of the trace buffer for post-silicon debug and ignore a wide variety of debug structures that are available in silicon or use them in an ad hoc fashion. For example, *freeze and dump* feature was introduced in early 1990 for dumping memory arrays. Similarly, scan chains are widely used for manufacturing testing. Clearly, there is a need to effectively utilize existing DfD infrastructure to improve observability and significantly reduce the overall post-silicon debug effort.

19.2.2 Security versus Design-for-Debug

While effective utilization of DfD infrastructure improves the post-silicon observability, it does not address the inherent conflict between security and observability. A debug engineer would like to maximize observability through DfD infrastructure to reduce debug time, whereas a security engineer would prefer to have zero visibility of internal signals to enhance security and privacy. Existing approaches ignore many crucial design constraints such as security and observable policy that restricts observation of certain signals at a certain time by specific individuals based on their privilege levels. Recent works have shown that unless "security" is treated as a first-class objective while designing DfD infrastructure, it can lead to various security attacks such as trace buffer attack [3, 8, 9] and scan chain attack [17]. For example, it is necessary to consider information flow tracking as well as on-chip debug structures during signal selection to maximize observability without violating security constraints. Moreover, exploring scrambling and encryption methods to protect the data stored in the trace buffers to prevent information leakage is also promising.

19.2.3 Machine Learning for Post-silicon Validation

Once SoC design has successfully passed the initial functional testing phase, it is possible and likely that the design will go through modifications over time. Specific features may be added, removed, or enhanced. Making such changes may have unintended consequences about the design functionality. The process of identifying undesirable behavior after specific modifications is called regression testing. Verification engineers run regression tests on their simulation environments during pre-silicon verification. When regression testing fails, the source of error should be root caused. Using the existing data in verification environment traces would be beneficial to reduce debugging efforts [10]. Moreover, existing data in verification environment traces can be clustered into several buckets such that each bucket contains traces that have failed as a result of the same cause. Therefore, debug of known failures can be avoided. Machine learning can be effectively utilized for trace data analytics as well as classifying a new failure into already known buckets. If mapping is successful, the known solutions (fix) can be utilized. A machine learning framework can be tuned to handle both known and unknown (e.g., minor variation of known ones) SoC failures.

References

1. M. Abramovici, P. Bradley, K. Dwarakanath, P. Levin, G. Memmi, D. Miller, A reconfigurable design-for-debug infrastructure for socs, in *Proceedings of the 43rd Annual Design Automation Conference* (ACM, New York, 2006), pp. 7–12
2. E. Anis, N. Nicolici, On using lossless compression of debug data in embedded logic analysis, in *2007 IEEE International Test Conference (ITC)* (IEEE, Santa Clara, 2007), pp. 1–10
3. J. Backer, D. Hely, R. Karri, Secure and flexible trace-based debugging of systems-on-chip. ACM Trans. Des. Autom. Electron. Syst. (TODAES) **22**(2), 31 (2017)
4. K. Basu, P. Mishra, Efficient trace signal selection for post silicon validation and debug, in *2011 24th International Conference on VLSI Design* (IEEE, 2011), pp. 352–357
5. F. Farahmandi, R. Morad, A. Ziv, Z. Nevo, P. Mishra, Cost-effective analysis of post-silicon functional coverage events, in *2017 Design, Automation and Test in Europe Conference and Exhibition (DATE)* (IEEE, 2017), pp. 392–397
6. A.B. Hopkins, K.D. McDonald-Maier, Debug support for complex systems on-chip: a review. IEE Proc.-Comput. Digit. Tech. **153**(4), 197–207 (2006)
7. https://software.intel.com/en-us/intel-platform-analysis-library. Intel Platform Analysis Library
8. Y. Huang, A. Chattopadhyay, P. Mishra, Trace buffer attack: security versus observability study in post-silicon debug, in *2015 IFIP/IEEE International Conference on Very Large Scale Integration (VLSI-SoC)* (IEEE, 2015), pp. 355–360
9. Y. Huang, P. Mishra, Trace buffer attack on the aes cipher. J. Hardw. Syst. Secur. **1**(1), 68–84 (2017)
10. A. Jindal, B. Kumar, K. Basu, M. Fujita, Elura: a methodology for post-silicon gate-level error localization using regression analysis, in *2018 31st International Conference on VLSI Design and 2018 17th International Conference on Embedded Systems (VLSID)* (IEEE, 2018), pp. 410–415
11. Y. Lee, T. Matsumoto, M. Fujita, On-chip dynamic signal sequence slicing for efficient post-silicon debugging, in *2011 16th Asia and South Pacific Design Automation Conference (ASP-DAC)* (IEEE, Yokohama, 2011), pp. 719–724
12. P. Mishra, R. Morad, A. Ziv, S. Ray, Post-silicon validation in the soc era: a tutorial introduction. IEEE Des. Test **34**(3), 68–92 (2017)
13. S.-B. Park, T. Hong, S. Mitra, Post-silicon bug localization in processors using instruction footprint recording and analysis (ifra). IEEE Trans. Comput.-Aided Des. Integr. Circuits Syst. **28**(10), 1545–1558 (2009)
14. F. Refan, B. Alizadeh, Z. Navabi, Bridging presilicon and postsilicon debugging by instruction-based trace signal selection in modern processors. IEEE Trans. Very Larg. Scale Integr. (VLSI) Syst. **25**(7), 2059–2070 (2017)
15. B. Vermeulen, T. Waayers, S.K. Goel, Core-based scan architecture for silicon debug, in *Proceedings International Test Conference* (IEEE, 2002), pp. 638–647
16. www.arm.com. CoreSight On-Chip Trace and Debug Architecture
17. B. Yang, K. Wu, R. Karri, Secure scan: a design-for-test architecture for crypto chips. IEEE Trans. Comput.-Aided Des. Integr. Circuits Syst. **25**(10), 2287–2293 (2006)
18. J.-S. Yang, N.A. Touba, Expanding trace buffer observation window for in-system silicon debug through selective capture, in *26th IEEE VLSI Test Symposium (VTS 2008)* (IEEE, San Diego, 2008), pp. 345–351

Index

P. Mishra and F. Farahmandi (eds.), *Post-Silicon Validation and Debug*,
https://doi.org/10.1007/978-3-319-98116-1

MIX
Papier aus verantwortungsvollen Quellen
Paper from responsible sources
FSC® C105338

If you have any concerns about our products,
you can contact us on
ProductSafety@springernature.com

In case Publisher is established outside the EU,
the EU authorized representative is:
Springer Nature Customer Service Center GmbH
Europaplatz 3, 69115 Heidelberg, Germany

Printed by Libri Plureos GmbH
in Hamburg, Germany